"There is a societal need to diversify our thinking about the future, moving beyond the troubled metaphors of blank slates, open roads and sleek machinery. This collection seeks to broaden and deepen our philosophical understanding of anticipation, helpfully questioning how creativity and imagination can serve to reframe the present."

Cynthia Selin, *Associate Professor, School for the Future of Innovation in Society/School of Sustainability/ Consortium for Science, Policy and Outcomes/Arizona State University; Associate Fellow, Said Business School/University of Oxford*

A Creative Philosophy of Anticipation

This edited collection highlights the valuable ontological and creative insights gathered from anticipation studies, which orients itself to the future in order to recreate the present.

The gathered essays engage with many writers from speculative metaphysics to poetic philosophy, ancient writing systems to the fringes of pataphysics. The book situates itself as a creative intervention in and with various thinkers, designers, artists, scientists and poets to offer insight into ways of anticipating. It brings together philosophical practices for which creativity is both a fundamental area of consideration and a mode of working, a characterization of recent Continental Philosophy which takes a departure from traditional futures studies thinking.

This book will be of interest to scholars and research in futures studies, anticipation, philosophy, creative practice and theories about creative practice, as well as the intersections between philosophy, creativity and business.

Jamie Brassett is Reader in Philosophy, Design and Innovation at University of the Arts London, UK. He has worked at Central Saint Martins since 1995, across all disciplines. Jamie started MA Innovation Management in 2008 and ran it for 11 years. He DJs better now than in the 1990s.

John O'Reilly is Senior Lecturer in Practice as Research, Teaching and Learning Exchange, University of the Arts London, UK. Philosopher, editor and journalist, John has specialisms in illustration, popular culture and recent Continental Philosophy.

Routledge Research in Anticipation and Futures

https://www.routledge.com/Routledge-Research-in-Anticipation-and-Futures/book-series/RRAF

Series editors: Johan Siebers and Keri Facer

The Promise of Nostalgia
Reminiscence, Longing and Hope in Contemporary American Culture
Nicola Sayers

Indigenous Futures and Learnings Taking Place
Edited by Ligia (Licho) López López and Gioconda Coello

Sustainable and Democratic Education
Opening Spaces for Complexity, Subjectivity and the Future
Sarah Chave

A Creative Philosophy of Anticipation
Futures in the Gaps of the Present
Edited by Jamie Brassett and John O'Reilly

Anticipation, Sustainability, Futures and Human Extinction
Ensuring Humanity's Journey into the Distant Future
Bruce E. Tonn

A Creative Philosophy of Anticipation

Futures in the Gaps of the Present

Edited by Jamie Brassett and John O'Reilly

Routledge
Taylor & Francis Group
LONDON AND NEW YORK

First published 2021
by Routledge
2 Park Square, Milton Park, Abingdon, Oxon OX14 4RN

and by Routledge
52 Vanderbilt Avenue, New York, NY 10017

Routledge is an imprint of the Taylor & Francis Group, an informa business

British Library Cataloguing-in-Publication Data
A catalogue record for this book is available from the British Library

Library of Congress Cataloging-in-Publication Data
Names: Brassett, Jamie, author. | O'Reilly, John, 1963- author.
Title: A creative philosophy of anticipation: futures in the gaps of the present/edited by Jamie Brassett and John O'Reilly.
Description: Abingdon, Oxon; New York, NY: Routledge, 2021. | Series: Routledge research in anticipation and futures | Includes bibliographical references and index.
Identifiers: LCCN 2020049083 (print) | LCCN 2020049084 (ebook)
Subjects: LCSH: Expectation (Philosophy)
Classification: LCC B105.E87 C74 2021 (print) | LCC B105.E87 (ebook) | DDC 155.2/4--dc23
LC record available at https://lccn.loc.gov/2020049083
LC ebook record available at https://lccn.loc.gov/2020049084

ISBN: 978-0-367-23456-0 (hbk)
ISBN: 978-0-367-75735-9 (pbk)
ISBN: 978-0-367-23459-1 (ebk)

Typeset in Bembo
by Deanta Global Publishing Servies, Chennai, India

Contents

Figures

Tables

Contributors

Nathaniel J.P. Barron holds a PhD in Philosophy from the University of Central Lancashire, UK, and is currently Teaching Associate in Sociology at the University of Birmingham, UK, where he teaches classical and contemporary social theory. His doctoral thesis develops a materialist theory of language from the concepts that make up Ernst Bloch's utopian philosophy. He is currently editing this work for publication with Brill, which, all things being equal, is due to be completed by the end of 2020, and which has a working title of *Language in Ernst Bloch's Speculative Materialism: Situating Bloch in the Marxist Tradition*. A condensed version of his thesis was published in 2017 by the *Bloch Almanach*, while he was Visiting Researcher at the Ernst Bloch Centre, Ludwigshafen. He has been Visiting Researcher at the Hegel Archive, Bochum, and the Walter Benjamin Archive, Berlin. He has recently completed a Visiting Fellowship at The Institute of Modern Languages Research, University of London, UK, where together with writing the present chapter, he compiled research notes for a forthcoming piece which probes the place of England in a triadic concept of European revolution, as developed by the German poet Heinrich Heine.

Jamie Brassett is Reader in Philosophy, Design and Innovation at the University of the Arts London (UAL), UK, where he has worked since 1995. He is also a member of the research team at Central Saint Martins (CSM). With his wife, Joanna, Jamie shares a Visiting Professorship in integrated design at Anhalt University of the Applied Sciences, Dessau, Germany. He is also a member of the external advisory board for UNIDCOM Research Centre at IADE, Lisbon, Portugal. Jamie started the MA in innovation management at CSM in 2008 and was its course leader for its first 11 years. Jamie holds a PhD in philosophy (University of Warwick, UK, 1993) with a thesis titled *Cartographies of Subjectification*. Jamie co-edited with Betti Marenko (2015) *Deleuze and Design* for Edinburgh University Press. Jamie has written extensively on design from his philosophical perspective: especially using the works of Deleuze and Guattari, Souriau and Whitehead, Serres and Lucretius, Simondon and Spinoza; and oftentimes with his dear friend and colleague John O'Reilly. Alongside this book, Jamie has been working on another for Routledge, co-editing with Richard Reynolds a collection on superheroes and philosophies of excess.

Mark Donoghue is an independent scholar researching the connections between pictures and perception, particularly the relationship between spatial projection and cognitive models of space. He completed his PhD at Chelsea College of Art, the University of the Arts London, UK, in 2017 on the ontology of multi-projection space in nineteenth-century Japanese woodcut prints. This doctoral research explored the varieties of spatial projection utilized in the Japanese artist Katsushika Hokusai's (1760–1849) print compositions. Because these prints were at an intersection of Eastern Asian and European picture making practices, they became an ideal subject to explore the concept of picture space. Mark examined this intersection of spatial practices through the process philosophies of Gilles Deleuze (1925–1995) and William James (1842–1910). Continuing from his PhD Mark's current research interests focus on the philosophical and psychological issues around pictorial space with particular focus upon the theoretical implications of Bayesian theories of cognition for picture perception.

Derek Hales is a design philosopher working in the field of speculative design, his PhD in Cultural Studies from the Royal College of Art, UK, assimilated the philosopher Gilles Deleuze's concept of the technological object as 'objectile' to a pataphysical clinamen. He is a chartered architect, a former chair of the Research and Innovation Group of the Royal Institute of British Architects and is Programme Director of the practice-based doctorate in the Built Environment at the University of Salford, UK. His research is transdisciplinary and is directed towards a methodological interest in how fictions, fictional worlds and fictive materialities, widely conceived, become sites of artistic research. Derek has been engaged in practice-based and artistic research on the technical object in abstract culture, since 1995, developing artists' R&D commissions, research residencies, festivals and workshops in sound art, noise art, device art, fictive materialities and other projects funded by the Arts Council of England, UK research councils and through various innovation awards. There are two principal vectors to his recent research interest: fictional spaces, fictional objects and imaginary media in histories and theories of art and architecture; the fictional worlds, fictive data and possible futures of speculative design and abstract culture.

Lucy Kimbell is Professor of Contemporary Design Practices and director of the Social Design Institute at the University of the Arts London, UK. She has previously worked at the University of Brighton, UK, and University of Oxford, UK. Her research explores the capacity of design practices to change public policy issues and social and organizational action including through design thinking, service design and design for policy. Her artworks, which involve playing with data include, *Audit* (Bookworks, 2002), *The LIX Index* (Film and Video Umbrella/Channel 4, 2002–3), *Pindices* (ZKM Karlsruhe, 2005) a collaboration with sociologist Andrew Barry and

Air Pollution Toile (Modern Art Oxford, 2018). She has published widely across a number of different disciplinary areas, including Design Futures and Policy Studies. Lucy's book, *The Service Innovation Handbook: Action-Oriented Creative Thinking Toolkit for Service Organizations* came out in 2015 (BIS Publishers). New projects include using design practices to enable cross-disciplinary, co-produced research to address organizational challenges and public issues such as AI. Lucy originally studied engineering design and appropriate technology, then computing in art and design, and then a PhD in design studies. Lucy held an Arts and Humanities Research Council Research Fellowship (2014–15) in the Policy Lab, a special design-intensive policy innovation unit in the Cabinet Office of the UK Government.

Anne Marchais-Roubelat is Associate Professor at the *Conservatoire National des Arts et Métiers*, Paris, France, which she joined in 1993. Her research areas include strategic management of action processes in public and private organizations, scenario planning and transformation processes. She is a member of the research team in strategy, foresight, innovation and development of Lirsa Lab (*Laboratoire interdisciplinaire de recherches en sciences de l'action*). She is a board member of *Flux*, international journal of research in networks and territories and of *Prospective et Stratégie*, a French-language future-oriented journal published on Cairn.info academic platform. She co-developed an action-based scenario foresight methodology and has published widely in the field of scenario planning in *Futures, Technological Forecasting and Social Change, Foresight* and *Society and Business Review* journals. She has published two books on decision and action processes: *De la décision à l'action. Essai de stratégie et de tactique* (Economica, 2001) and *La Décision. Figures, symboles et mythes* (Apors, 2012); and co-authored (with Olivier Crette), *Analyse critique de l'expertise et des normes: théorie et pratique* (Editions L'Harmattan, 2015) and (with Nicolas Adas, Bettina Bouchayer, Thierry Devaux and Sylvain Mondon) *L'organisation. Démarches et outils* (Apors, 2018).

John O'Reilly is Senior Lecturer in Practice as Research at Teaching and Learning Exchange, the University of the Arts London (UAL), UK. He was previously Course Tutor on the MA Innovation Management at Central Saint Martins, UK, running the 'Exploration and Experimentation' unit. His teaching, research and client-focussed work focuses on helping others develop novel forms of knowledge production and insight through cultivating an understanding of their practice. After a PhD in Philosophy at the University of Warwick, UK, John has taught in a number of universities (as well as UAL) including the University of Greenwich, UK, the Royal College of Art, UK, and the China Academy of Arts, China, on courses ranging from psychoanalysis to the philosophy of science fiction to creative uncertainty. He has worked in journalism, trend analysis and branding for nearly 20 years being a regular contributor to *The Guardian* and *The Independent* on Art, Media and Music, working with Oliviero Toscani on fashion label Benetton's *Colors Magazine*. He has written essays and catalogues for artists such as Martin Creed, image makers such as Noma Bar and exhibitions such as *Communicate: Independent British Graphic Design since the Sixties* at the Barbican. His books include: *Recharge Your Design Batteries* (2009) with Tony Linkson; *No Brief: Graphic Designers' Personal Projects* (2002); and *Illusive 4* (2015), an overview of directions in the made image.

Fabrice Roubelat is Associate Professor at the IAE School of Management, the University of Poitiers, France, which he joined in 2003. His research, teaching and PhD supervision areas focus on strategic foresight, anticipation and creativity practices, and innovation marketing. He is a member of the international strategic intelligence research team of Cerege Lab. He is a member of the editorial boards of *Futures, the Journal of Policy, Planning and Futures Studies*, of *Stratégique*, the leading French military strategy journal, and of Wiley journal *Futures and Foresight Science*. He is the editor of *Prospective et Stratégie*, a French-language future-oriented journal published on Cairn.info, the French academic platform. After his role as scenario planner at EDF foresight unit (1992–1997), he led the development of a foresight methodology to design action-based scenarios. He recently served as an expert member of a business and economic climate change impact focus group for the French national climate change observatory (Onerc). He has published in the field of scenario planning in *Futures*, *Long Range Planning*, *Foresight*, *Technological Forecasting and Social Change*.

Acknowledgements

We would like to thank the organizers, reviewers and participants at the international Anticipation conferences in Trento (2015), London (2017) and Oslo (2019) for their thoughtful comments in review and helpful questioning in person during our various presentations and curated sessions. We would also like to thank the series editors, Keri Facer and Johan Siebers, for their support and critical feedback, as well as the many peer reviewers – both official and personal – for critical insights on the whole and the parts, which have proven invaluable to the development of this book. Special thanks also to Danielle Tran for guidance. Many thanks go to the members of our editorial team at Routledge – Rebecca Brennon, Oindrila Bose, Felisa Salvago Keyes and Julia Pollacco – for their guidance and patience, especially close to the end of the process when we were all rocked by COVID-19. Thank you to our families, whose love and support have been indispensable.

1 Introduction to a creative philosophy of anticipation

Jamie Brassett and John O'Reilly

Opening gaps

There were many anticipations creating this book. We start with the Second International Conference on Anticipation, held at Senate House in London 2017, where we (Jamie Brassett and John O'Reilly) organized and delivered a curated discussion panel – though this is by no means the earliest encounter germane to this story. Anticipation took the form of an atmosphere in this 1930s Art Deco building. Senate House was initiated by William Beveridge when he was elected to the post of Vice-Chancellor of the University of London in 1926. Beveridge's work anticipated and therefore created a new timeline of the future as the thinker of the welfare state that enabled heterogenous futures. Beveridge also had a different, more homogenous version of the future; like many of the pre-war left in Britain who were motivated by the idea of eugenics, the engineering out of undesired genetic qualities (Freedland, 2012; Renwick, 2019). For Beveridge, the anticipatory power of the welfare state gave way to the predictive power of eugenics.

The 19-floor building of Senate House was designed by vegetarian Quaker Charles Holden, a specialist in the atmosphere of the subterranean, having designed over 40 underground stations in London. The slightly dystopic atmosphere generated by the severe beauty of the building's Portland Stone materials haunted the future in the present in perfectly anticipatory fashion. George Orwell's Ministry of Truth in *Nineteen Eighty-Four* was modelled on Senate House. The strategy of organizational threat as a mode of political control by Orwell's Ministry of Truth, anticipates the operational logic behind the strategies of the US and UK governments nearly 20 years after the fictional year of the book – especially with the War on Terror. 'Fear is the anticipatory reality in the present of a threatening future,' writes Brian Massumi in his essay 'The future birth of the affective fact'. He continues: 'It is the felt reality of the nonexistent, loomingly present as the affective fact of the matter' (Massumi, 2010, p. 54). In everyday experience such prophecies can become self-fulfilling, but this disguises the affective, material power of threat as an atmosphere. The affective fact, Massumi argues, opens the way to generate and legitimate many different kinds of actions, actors and their networks to pre-empt the event they actualize as a threat. While *Nineteen Eighty-Four* is remembered as a future fiction of its written present experiences of totalitarian government, of systemic lying as a strategy, of war as a mode of being, it is a story about love and its anticipation; a story taken up by Jamie in his chapter.

There were other futures anticipated by Senate House; the most globally affective one is that of Gotham City. *Batman Begins* (Nolan, 2005) and *The Dark Knight Rises* (Nolan, 2012) make use of Senate House in their vision of Gotham. A city whose urban planning mostly consists of the creative act and atmosphere of chiaroscuro, from the bat signal to the shadows from which the hero emerges. Nolan's vision – about the breakdown of urban government, the anger of citizens finding expression in populist forms and autocracy – suggest perhaps that Senate House's architecture was in fact plagiarized from the future (as Derek Hales might say, in his pataphysical way, later in Chapter 9).

Our panel at Anticipation 2017 comprised four writers in this collection – Mark Donoghue, Anne Marchais-Roubelat, Fabrice Roubelat and John O'Reilly – with Jamie Brassett acting as chair. The panel also included Bettina Bouchayer who co-presented with Anne but who was unable to participate in this book. Its title, 'Inserting the Future in the Crannies of the Present', was a reference to a passage in philosopher, mathematician and physicist Alfred North Whitehead's *Adventures of Ideas* (1967b). In Chapter 12, titled 'Past, Present, Future', he writes: 'Immediate experience requires the insertion of the future in the crannies of the present' (Whitehead, 1967b, p. 191). What was most striking for us about this is the tone, its style. Its brash confidence disguises profound implications through a spatialization of time. Through this we become capable of acting on the demand of time; a demand that anticipates time because without this demand, this anticipation outside of time, there would be no time for time. Anticipation as the creation of time. (Both Marchais-Roubelat and O'Reilly deal with this.) Whitehead's phrase struck us as an important precursor to more contemporary accounts of anticipation (to which we will refer later in this book), where the future is not only met but brought into the present as a creative act.

If anticipation is understood this way, the novelty it creates cannot be predicted simply by adding the given components. The panel inserted the future in the crannies of the present by creatively substituting expectation with anticipation. Moreover, the passage from Whitehead resonated with our understanding and approach to design-intensive innovation (Lambert and Flood, 2018) and its management. We have wondered: if design contributes to, 'immediate existence' (or even wholly constructs it) then, at least, mapping and locating the possibilities for such existence to manifest itself is an important act of creativity and futures thinking coming together (Brassett and O'Reilly, 2015, 2018a, b). Like this introduction, and the book itself, the mysterious landscape of the future described by Whitehead did not pre-fix the anticipations of the panellists. Whitehead's generous topography enabled different crevices to open in a specific present (that of a conference on anticipation), with creative potentials for our participants to fill. Mark brought the ontologies of the picture space into a mix with the ecological psychology of J.J. Gibson; Fabrice spoke through the divinations of Plutarch; Anne and Bettina presented Heidegger's conceptions of time in collision with specific forms of branding and marketing; and John the creative anticipation of Lucretius's *clinamen* in relation to the innovation of tennis players and *Napoleon Dynamite* (2004). Our then-present required us to find and fill some crannies in which to place various, unexhaustive and non-totalizing models of potential, possible, creative anticipation.

Here we are, some years later, having – at the very least – modelled various ways of working together: teaching, writing and practising. With that experience creating a space for diverse individuals and disciplines to collide and ricochet in many different directions. Furthermore, this experience has allowed us to develop a set of specific approaches to anticipation and the future which, while coalescing in a number of outcomes – the conference, this book, some journal articles and special issue editing – have refrained from homogenizing.

There is more. This book collects more than just those speakers from a specific panel at *Anticipation 2* (2017). Some of our relationships are new to this book, others more long-standing. So here we have, in addition to those of us presenting in the Whiteheadian curated session at that conference: philosopher of the future and Ernst Bloch scholar, Nathaniel Barron; architect, philosopher, digital design expert and innovation practitioner Derek Hales (whose chapter here has links with his presentation at the first Anticipation conference at the University of Trento, Italy, 2015); social design expert and pioneer of policy-focused design innovation, Lucy Kimbell (who ran another panel at *Anticipation 2*, with the stand-up comedian Trevor Locke). All our contributors, their thoughts, practices and other creative moments entangle at various points and spin off away from each other at others – we will note some of these when we discuss the chapters in a little more detail below.

If we think of anticipation as an activity partly fostered by pressure from the future (thinking of anticipation also as an ecology of time) we might want to think again of atmosphere, but this time also in meteorological terms. In 'Affective Atmospheres', Ben Anderson (2009, p. 77) cites Marx's address to a gathering celebrating the fourth anniversary of the *Peoples' Paper*: 'the atmosphere in which we live weighs upon everyone with a 20,000-pound force, but do you feel it? No more than European society before 1848 felt the revolutionary atmosphere enveloping and pressing it from all sides' (Marx, 1978, p. 577). Atmospheres – even, or especially, those emanating from the future – argues Anderson, are produced by bodies (human, non-human, material, immaterial) but exceed them. There is something creative about the urge from the future that an atmosphere, felt as revolutionary, must exert on us. The atmospheres of this book began with the bodies of Senate House, our panel, the anticipation conference, meetings yet to come, all haunting of the future of this book – one time was anticipated in the moment of you reading this now. Other times got lost, haunting this book with other futures that may now be impossible. While another time had an atmosphere created by the bodies of COVID-19.

But there is something that we need to note, to anticipate. For alongside the contextualization that we give of our completing the manuscript for this book as the time of the COVID-19 pandemic, another important event arose just before we submitted our manuscript: the killing of African American George Floyd by a white police officer. While the Black Lives Matter movement has been active for some years, and the issues it has coalesced around even longer, the protests and demonstrations galvanized throughout early 2020 by George Floyd's killing have forced many to reflect upon their privileges. And so, as we have arrived at the end of producing this book, we are deeply aware of the whiteness of the voices we have collected. This was never our goal and there are many reasons why this book has arrived in this condition, but no excuses. We offer the story here, therefore, not to excuse ourselves, rather as a cautionary tale.

From signed contract to required submission the process was particularly short, we felt we had no time to send out an open call for chapters, working with the group we had gathered around the Anticipation conferences. Beginning with an equal balance of gender positions the group was impacted by colleagues unable to participate, while others we approached found the time constraints too tight. We are sure that this book provides a set of creative perspectives on anticipation and futures studies that we feel is exciting, but we nevertheless recognize that we would have liked to have more scholars who identify as non-typical within our academic sectors, including people of colour. It should be noted that the scholars collected here exist at the intersection of a number of non-typical heritages, classes, socio-economic abilities and cultural positionalities: ours are not wholly and entirely privileged voices. Furthermore, we have – deliberately – engaged with scholarship and practices that are activist or non-typical. A few examples are: O'Reilly's examination of the activist, performative and theoretical work of Black Quantum Futurism and the Black Space Agency; Marchais-Roubelat's discussion of the poetry and religion of ancient indigenous people of northern Europe in relation to futurist approaches to organizational studies; Hales also encounters Afrofuturism alongside outsider theorists and practitioners; Kimbell's contextualization of her work in relation to age, gender, Alzheimer's disease and dementia. Notwithstanding this, we editors are cognizant of our product's partiality and that there are many other ways that our stories, concepts and practices could be told. We, Jamie and John, shall try to find the hope that is alive in the cracks and crannies of this work to create better futures; and therefore, to recombine new presents.

Inserting futures (with a creative philosophy)

Even in its recent development within futures studies, Anticipation has a recognizable characterization, which can be stated fairly simply: deceptively so. A system that anticipates is one that models a future, brings it into its present and bases decision-making upon this model. This modelling reorganizes the system and its present in such a way that one can say that the future is causing a present state. This is at the heart of key progenitor of these studies, Robert Rosen (1934–1998): a theoretical, relational biologist and mathematician. His most important work in this area is *Anticipatory Systems: Philosophical, Mathematical and Methodological Foundations* (2012; first published 1985) and, in it, he and colleagues attest to the difficulties wrought on his career as a scientist by propounding and sticking with something as so patently, apparently, ridiculous as the future causing the present. This is what you will see in any book on this topic, especially in its guise in futures studies – which is the context for this volume.

The other key figure in the development of anticipation studies is philosopher, futures scholar and thinker of complex systems, Roberto Poli. The scope of his work is astounding, which is manifest in his (2017) book, *Introduction to Anticipation Studies*. Poli is a figure with a significant presence in this volume, with explicit critical evaluations of his work in some chapters (Barron, Brassett, Brassett and O'Reilly, for example) and more lightly felt in others. Poli paves the way for a philosophical approach to anticipation, which we follow in this book. His own philosophical interests and expertise, however, highlight some gaps in his mapping that we are pleased to explore given our various but aligning perspectives. Chief among these is the work of philosopher Gilles Deleuze, on his own and with his sometime co-writer, the psychoanalyst and political activist Félix Guattari.[1] While Poli (2017, pp. 87–88) does make mention of Deleuze's philosophy – which comes as one of seven 'philosophical samples' (Poli, 2017, pp. 78–88) in Chapter 5 'Anticipation in Philosophy' – he is open about his own lack of acceptance of Deleuze's positions. It is interesting – stylistically and, maybe, psychoanalytically – that Poli's own voice is most personal and clear when writing of Deleuze. For example, he writes: 'According to Deleuze, this virtual form of causation is categorically different from ordinary mechanical causation. *I have to admit I fail to see why*' (Poli, 2017, p. 88; emphasis added). Now, of course, there is nothing at all wrong with such a statement; but, in this volume, there are many of us who seek to investigate aspects of Deleuze's – and Deleuze and Guattari's – philosophies that we find *do* offer insights into futures in general and anticipation studies in particular. This is not to say that Poli is therefore wrong in his analyses. Our approach is more open than that (and Brassett may say, in the terms put into play by his chapter in this volume, 'sympathetic'), an approach itself influenced by an insight from Deleuze (1995). In a 1986 interview about Foucault's work – just two years after Foucault's death and at the time that Deleuze (1988; originally 1986) published his book on his friend, titled simply, *Foucault* – titled 'Breaking Things Open, Breaking Words Open' (pp. 83–93), Deleuze (1995, p. 87) says this: 'Never interpret; experience, experiment'. The urge not to interpret allows different interpretations to live together, for when the judgements it requires are not the most important act to deliver, then multiple interpretations can abound. That no one should be getting philosophically extremist about canon, is a valuable take away we have from the philosophies (and other attitudes: literature, psychoanalysis, politics, for example) that Deleuze and Guattari create.

This leaves us with, 'experience, experiment' – on which the translator's note says: 'The single French verb *expérimenter* means at once to experience, and to experiment' (Deleuze, 1995, p. 197, n. 13). This provides a different attitude to philosophy than the interpretative. Ours is a philosophy of making, doing, creating, living and is found joyfully expressed from Deleuze's early work on Spinoza (1990b; originally published 1968) and Nietzsche (1983; originally published 1962) to the collaborations with Guattari (1984, 1988, 1994). Philosophy makes, does things with the concepts it encounters and experiences. It experiments; it *creates*. This is carried even further in Deleuze and Guattari's last book together, *What is Philosophy?* (1994; originally published 1991) where they argue that philosophy can be characterized by its powers of concept creation. Thus, an important aspect of anticipation – that a model of the future is used to recreate the present – is also an aspect of philosophy as we will be using it. This allows us to follow a tributary from the style of philosophy that Poli produces to another, one different to his, which – it is our hope – will offer something else in addition. Important here is the philosopher of science Isabelle Stengers, for whom Deleuze and Guattari's works are influential. In an essay on *What is Philosophy?* Stengers (2005, p. 152) writes the following:

> It is his [Deleuze's] one book that addresses its reader as if he were perhaps a friend, but at that twilight hour 'when one distrusts even the friend,' even the one who had most enthusiastically followed the great opening of possibilities that Deleuzian themes have inspired. Here, the crucial problem may well be that 'we lack resistance to the present,' [Deleuze and Guattari, 1994, p. 108] and to resist here does not mean to criticize or to denounce but to construct.

This is a passage that gestures towards Nietzsche in a way that we will develop in the final chapter of the book, with further concepts taken up more fully in Brassett's chapter. Still, for now, our own 'resistance' to Poli's review of Deleuze's work is a thoroughly constructive one. The passage from which Stengers quotes about resistance to the present is worth citing in full. Deleuze and Guattari (1994, p. 108) write:

> we do not lack communication. On the contrary, we have too much of it. We lack creation. *We lack resistance to the present.* The creation of concepts in itself calls for a future form, for a new earth and a people yet to come. (Original emphasis.)

When resistance is creation, 'resisting with concepts' can be another way of saying 'philosophy'; a philosophy that calls for a 'people yet to come' to come. A future can only be made to come here, to the present, if we open ourselves to the possibilities that we, here, now, our 'immediate existence', may be recreated by the arrival of the 'people yet to come'. Stengers is surely one who has been a sympathetic friend of the ontological reconstruction on offer in Deleuze and Guattari's work; she may even be the friend 'who had most enthusiastically followed the great opening of possibilities that Deleuzian themes have inspired' now distrusted or, at least, resisted. Indeed, their phrase 'a free and wild creation of concepts' – characteristic for them of 'English Philosophy' (Deleuze and Guattari, 1994, p. 105) – ends up as a subtitle of her *Thinking with Whitehead. A Free and Wild Creation of Concepts* (2011; originally published 2002).

Stengers's philosophy is an adventure, an exploratory and experiential voyage into the future in a way that brings the future back to hand. She writes, in an article on Whitehead's 1927–28 work *Process and Philosophy* :

> The adventure of the creation of a conceptual agency cannot be disentangled from the experiential adventure of the philosopher experimenting disclosure. We can speak here of 'experimenting' because the disclosure is part of a process that can be described as conceptually 'lured'. Each concept has to be designed and redesigned, as the point is not of adequacy to some kind of pre-existent matter of fact but, rather, that of two questions which are always at work together: is the conceptual agency succeeding in doing what the philosopher wants it to do, and are those aims an adequate expression of the challenge she has decided to confront?
>
> (Stengers, 2008, p. 97)

We can say that, in anticipation, a future is 'lured' to the present; with 'lure' in Stengers's sense to work utterly in registers of adventure, speculation and experimentation. She writes: 'For the value of concepts is to lure new feelings, to induce "sheer disclosure" as a new way for experience to come to matter' (Stengers, 2008, p. 100). A lure is modelled on a future state where a thing to be caught will find it attractive and bites. If a concept is to 'lure' it is to be launched into all our presents as a future possibility that, once allowed in, can recreate experience, or even modify existence. It will be an adventure and an experiment; it will speculate and allure; it will resist and create. We return to Whitehead's (1967b, p. 191), '[i]mmediate existence requires the insertion of the future in the crannies of the present'. This points in another, ontological, direction that is of importance to us in this book.[2] Before we come to that, there is a little more to say about types of creativity.

Up till now we have directed our attention to two important pillars of this volume: philosophy and creativity; both of which have valuable links to anticipation, as well as with each other. Many of us in this volume have some sort of relationship to the creative industries and education (which often now coalesce as both creativity and education – along with government – become business as a function of neoliberalism. From a design perspective see: Julier, 2017). And at one point 'design' was going to be another pillar of this book. Indeed, some of us here refer to particular design practices (for example: Kimbell on textiles, printmaking and design fictions; Hales on architecture and speculative design; O'Reilly on design-intensive innovation education;

and Barron on design philosophy). The trajectories that each chapter takes, however, tend to accentuate the philosophical and creative, rather than a commonly discussed and recognized creative practice: like design. (Though, even to describe design as a 'recognized creative practice' is to stretch facts a little too far, as its ever-changing multiplicity is seen as problematic by some (Verganti, 2009) or a benefit (Brassett, 2015).) However, we do not provide a sustained interrogation of 'design' for it to warrant an important place as a focus for this book. Along with some exemplifications of our own conceptual experimenting of anticipation through design (as mentioned just above), we do refer to other forms of creative practice: textile design and painting (Kimbell, Donoghue); organizational design and design thinking as a business practice (Marchais-Roubelat, O'Reilly, Roubelat); literature (Marchais-Roubelat, Hales, Brassett); and, even, science (Brassett, Donoghue). So, the role of various creative practices is important here, too. Before positioning the chapters in more detail – but with greater emphasis on where some ideas put forward might lead – we will say a little more about a final, crucial aspect of this book.

As we parse a number of creative practices from a focus upon creativity, so do we highlight a particular aspect of philosophy: ontology – the philosophical questioning of being, often thought of as the asking and attempt at answering the question, 'why is there something rather than nothing?' Once again, we are guided here by Poli, who has specific articles on this (Poli, 2010a, 2011; Poli and Seibt, 2010) and who notes in Chapter 7, 'Ontological Sketches' of his *Introduction to Anticipation Studies* (2017, p. 103) that 'ontology is the theory of entities and their structures' that often involves, formally, 'dealing with categories like *thing*, *process*, *matter*, *form*, *whole*, *part*, and *number*' (Poli, 2017, p. 102; original emphases). We will not be arguing about ontology from first principles, however, neither will we attempt to establish categories as those Poli mentions just above. We will be working with (experimenting and experiencing) these ontological categories in the company of some philosophers who have developed key approaches to these (and more). You will find us discussing 'thing' with Latour (2008) or 'process' with Whitehead (1978), 'wholes' and 'parts' with Stengers (1997) or time with Heidegger (2007), many of which fold back and through the works of Deleuze and Guattari. Oftentimes, Poli examines these ontologies and categories through the works of Edmund Husserl (1859–1938), Nicolai Hartmann (1882–1950) and Ernst Bloch (1885–1977). Such, respectively, phenomenologist, critical realist and Marxist critical theorist approaches take different directions from those we present here. Except for one: Nathaniel Barron. A Bloch scholar, he provides a critique of Poli's approach to Bloch that allows him (Barron) to develop some important conclusions about the relationships between anticipation, ontology and creativity. For Barron, Poli is closer to Adorno (or even Kant) than Bloch, even while Bloch plays an important part in his (Poli's) philosophy of anticipation. Barron states it is important that 'if anticipation studies is concerned with the ontological reason for anticipation's existence, then it should also understand that that reason involves the question of goal' (personal communication, with permission).

But we are running ahead of ourselves a little. Here are the main points to take, then: that our interaction with anticipation will be largely philosophical and then with a great focus upon ontology; we will use these to focus upon the conceptualization of creativity and its role in anticipation. Now we move to the chapters a little more closely.

Anticipating chapters

The framing and situating remarks for this book are done with and we will next look over the chapters themselves. Traditionally, editors would be expected to provide some summary of the individual chapters. We will do this minimally, as we are seeking neither to interpret nor to overpower our readers' own interpretative flourishes. Here we devote more space to following 'lines of escape/flight' (Deleuze and Guattari, 1984, 1988) from these chapters in directions that are variously suggested to us, especially insofar as they approach issues of creative practice. To do this is to pick up some key concepts and run with them to see where they take us; to launch them into new areas and see what splashes they make; to insert them as cogs in different machines to see what is produced.

The discourse of the 'spoiler' in cinema is pertinent here: to spoil the plot, the narrative, the ending for those who have yet to see the film by those who have. Might this introduction be to the book what spoilers are to films: telling the reader what is to come, anticipating the readers' reading, predicting and thereby owning the 'best bits' for the reader? We will come back to this notion of 'best' later. Corporate futurism is often, literally, spoiling. The future as the site of corporate concerns does not necessarily align with that characterized by the concerns of people in the future to come. Although such scenarios make it so, make people subject to corporate concerns. The originary myth of Shell's scenario planning in the 1970s may have been wrapped up in the need to avoid another energy crisis, but they did not see the ecological catastrophe of the future-present they were creating.

We are not looking to spoil, though. Our recasting of this book's origins we present as a tale of chances rather than one of authenticity. And our mapping of the chapters we hope is more to do with seeing what, else, might be of interest. In a way, each of the paragraphs below that discuss the chapters could be read immediately after the chapters themselves; maybe you can try it and head straight in now? This might allow us past any accusation of spoiling. But now, the chapters.

The nine chapters that follow in this book share concerns with anticipation, creativity and philosophy, however each one anticipates the concept with a style of research that affirms its disciplinary concern with anticipation but also anticipates the novelty of anticipating with creativity and with the future of a discipline different, even while they align at moments.

Picturing

Mark Donoghue's chapter 'Anticipation, creativity and picture perception' has developed from research he has been carrying out on the ontology of picture spaces, with reference to the works of Deleuze, nineteenth-century Japanese artist Katsushika Hokusai (1760–1849) and ecological psychologist James Gibson (1904–1979). His chapter here focuses upon the logics of perception and the anticipatory ontology of picture spaces. In order to do this, Donoghue brings the philosophies of Henri Bergson (1859–1941) and William James (1842–1910) – especially their concepts of duration (*la durée*) and Pragmatism respectively – into collision with concepts from the Bayesian theory of probability. This, Donoghue readily recognizes, may seem like a paradoxical collection of concepts to bring together; that is, to imagine processes of prediction in relation to those of generation of novelty. He shows these paradoxes permeating both Bergson's and James's works and it is through their evaluation that Donoghue can begin to outline a future- and present-focused creativity. Thus, creativity is that which exceeds normal psychoneurological processes of probability calculation and predictive patterning, even while it finds a ground there. This provides Donoghue a position that not only develops from Bergson and James's philosophies, but also brings it into proximity to the work of Deleuze; particularly Deleuze's (1990a) *The Logic of Sense*, in the fifteenth chapter, 'Of Singularities' (pp. 117–125), where Deleuze navigates the different generating processes of persons, individuals, events and, of course, singularities. One of Deleuze's concerns here is in the energetic modifications and modulations of the ontological process of becoming and, at one moment, he quotes philosopher of technicity, Gilbert Simondon (1924–1989) thus: 'The living lives at the limit of itself, on its limit' (Simondon, 1964, p. 260; quoted in Deleuze, 1990a, p. 119). To live at the limit is to experience the limit and to experiment with exceeding it; it is to spread outwards from a ground to (and over) the edges of experience. An ambitious piece of work, this chapter manages to engage our shared subject with a transdisciplinary and creative flair. We wonder how the arguments Donoghue gives here might fare in relation to the speculative and process philosophy of Alfred North Whitehead (who was well disposed to both James and Bergson), as well as the evolutionary biologist and philosopher C. Lloyd Morgan (whose work on *Emergent Evolution* (1927) dealt positively with James, Bergson and Whitehead). To recognize creativity as that which outpaces any simplistic analysis of parts in a reductive manner, aligns with the various positions we shall adopt throughout this book: particularly those on anticipation.

Runes

Anne Marchais-Roubelat's chapter, 'Flowing or frozen anticipation? Runes and the creativity of time', brings to anticipation a critique of time in futures studies from the perspective of Heidegger's concepts of frozen and flowing time. Rich in reference to research in organization studies, Marchais-Roubelat also works her argument in relation to Old Norse poetry and Anglo-Saxon runes. If, she argues, the creativity alive in anticipation accesses time thought as fluid, complex and nontotalizing – as another option to time regarded as frozen, instrumental and measured – then an anticipatory organization should be able to loosen its rigid ontology and so find an attitude to uncertainty that is less terrifying and more fruitful. This aligns, we think, with the creative ontologies of innovation management and organization theory found, especially, in the work of Haridimos Tsoukas and Robert Chia (Tsoukas, 1998; Chia, 1999; Tsoukas and Chia, 2002) in particular and in Process Organization more generally (for example, Hernes, 2014). Yet Marchais-Roubelat's chapter heads in the most creative directions by engaging with runes, discussing their multiple nature of divination, magic and narrative production; and, even further, in regarding runes as event. This requires us to encounter the temporalities of magic, divination (Marenko, 2014, 2015) and a narrative production that weaves them all simultaneously, heterogenously, in their full complexity and creativity. This is an ambitious argument and provides avenues for future research that are rich in their possibilities. We would be particularly interested to see where anticipation as event might lead us, following Marchais-Roubelat's insight into the divinatory power of runes as events. This would manifest also an encounter with Whitehead, especially through his conception of the 'event'; and, indeed, Deleuze's (1993, pp. 76–82) relating of this to Leibniz. Events have 'actions, effects and influence', do not have a sense of their own 'function' and become problematized by their relationship both to singularities (see: Deleuze, 1990a) and to endurance (see: Stengers, 2011, pp. 185–200; also, Donoghue in this volume, in relation to Bergson's notion of 'duration'). Stengers is clear about Whitehead's atomism, especially as expressed in relation to his ontology of events. She quotes (Stengers, 2012, p. 188) from his *Science and the Modern World* (Whitehead, 1967a, p. 103) thus: 'We must start with the event as the ultimate unit of natural occurrence. An event has to do with all that there is, and in particular with all other events.' These are exciting times for an adventure into anticipation, as Stengers might say; and Marchais-Roubelat plays a part in taking us there.

Hope

Nathaniel Barron's chapter, 'Ernst Bloch's ontology of not-yet being: Intuiting the possibility of anticipation's fulfilment', is a closely argued interrogation of the ontology of hope that Bloch (2000) develops. Barron situates his examination in terms of developing a philosophy of design that, for him, has been lacking a thorough engagement with Bloch, one of the twentieth century's most accomplished philosophers of hope. Barron has a point. There is a phrase from the 1978 Nobel Laureate in Economics, Herbert Simon (1988, p. 67), which has become canon in design research: 'Everyone designs who devises courses of action aimed at changing existing situations into *preferred* ones' (emphasis added). Simon enhances the teleological attitude further: 'Design, on the other hand, is concerned with how things *ought* to be to be, with devising artifacts to attain goals' (Simon, 1988, p. 69; emphasis added). Simon's approach was influenced by his trust in methods and logics from the 'hard sciences' and the training he had in both the 'Chicago School' of economics and political science and as a consultant at the RAND Corporation (1951–76) (Huppatz, 2015; see also Bousbaci, 2008). Simon's centrality to the expression of design as a science is without question, but for those of us with perspectives on design infused with qualitative, nonlinear, emotional or artful energies, Simon's approach is limited (Celi and Morrison, 2017). It is notable that 'rational choice' in economics (Kay, 2013; Goodwin, 2014), 'mutually assured destruction' in military-political strategy (DeLanda, 1991) and a fact-oriented scientificity for design (Latour, 2008) have, since the 1990s, variously made way for more complex and qualitative approaches to their subjects, and more. The disruption that Barron offers, with Bloch's help, to the type of logical design Simon promotes, is one that highlights the philosophical, political and ethical power that are intimated by Simon's 'ought' and 'preferred'. Barron's is an ethics instantiated by the focus upon the action of self-developing self-critique: an ontological exposition of the ways and means that we may *be*. While Bloch's work is remembered more for its phenomenological and political power than its ethics, when Barron's Bloch comes towards creative practice such ethics (along with politics and ontologies of course) are foregrounded. For Barron, this is an approach that is activated by Bloch's insight (1985, p. 147) into the 'darkness of the just-lived moment' (Barron's translation). This Barron works through the concept of what he calls Bloch's ontology of 'not-yet being'. That there is a need for being to be future oriented is attested to by both its inability to find an existence that can be consummated fully in the present and an emotional charge around this inability that manifests as this 'darkness of the just-lived moment'. This is an extremely important issue for Barron, leading him to stress its ramifications as an imperative: anticipation must take seriously Bloch's insight that a not-yet being has potential for future actualization, for its (anticipation's) ontological promise to be fulfilled. Bloch and Barron provide another angle on the contemporary's perspective on present darkness that Brassett and O'Reilly develop in Chapter 10; and the hope, the light in the darkness, however minuscule, that lures anticipation on its creative journey. (Barron does this in relation to a critique of Poli's readings of Bloch, which we have already intimated. We will leave this for you to discover.) For us, this takes Bloch away from a recognizable – and maybe simplistic – dialectical motor where negation (a lack) is resolved through synthesis with what was negated. Both the untimely and the contemporary do not negate, theirs is not a nihilist moment but a creative one: the future alive in anticipation is that the not-yet is also a becoming. (This becoming, this ongoing transformation of existence into something more, something exceeding itself, is apparent not only here but in the next chapter too.) Barron positions Bloch's work within the anticipatory context of 'utopia', re-reading the history of philosophy and in particular concepts of 'creativity' and 'causality', in terms of anticipation.

The problem of anticipating a future creatively, argues Barron, is always one of an incapacity for 'beginnings'. Unfortunately, the art of revealing possibilities for renewal is all too often rendered insensible to us by powers which seek to perpetuate an idea of the eternality of the content of the present. It is capitalism's blind spot that Barron teases out: the invisible sensible. Indeed we might argue that it is this blind spot which disables neoliberal capital to see an 'outside' future, a *utopos*, that could predict anything other than 'neoliberalism 2.0' as the future of algorithms that talk to each other – smart cities, big data healthcare, driverless cars – because they already know who we are and what we will become. We wonder whether sensible presentation of Utopia is possible in forms of foresight delivered as scenarios and PowerPoint presentations; or whether the blind spot of corporate futurism lies in the digital imagination of Redmond and Cupertino.

Rules

In 'Are scenarios creative? Questioning movement and innovation in anticipation practices', Fabrice Roubelat uses his expertise in scenario planning – as a scholar and practitioner – to engage anticipation and creativity to imagine innovation differently. By developing an argument around the value of movement and action in futures thinking and practice, he cautions scenario designers not to think that their creative products are immutable and always relevant; similar to the famous line from Prussian field marshal Helmuth von Moltke (1800–1891) – that no plan survives the first encounter with the enemy. Roubelat's position aligns with a tendency in innovation management studies over the last 25–30 years or so that values a lithe and dynamic approach to organizational design, strategy and market orientation: notably the 'ambidextrous organization' (Tushman and O'Reilly, 1996); process ontologies (Tsoukas, 1998; Chia, 1999; Tsoukas and Chia, 2002; Hernes, 2014) and complexity (Brassett, 2015); but also, more recent trends in design thinking (two of the most influential texts are: Brown, 2009; Martin, 2009). But Roubelat is able to bring together thinking from Platonist philosopher Plutarch, philosopher of science Thomas Kuhn and scenario developer Herman Kahn (among others) in developing his perspective. Yet, this chapter is not as straightforward as it may seem, for Roubelat finds that his proposition of 'moving scenarios' shows that Kuhn's conception of scientific paradigms is not as robust as its reputation. The relation between a set of dynamic and complex scenarios and the rule-sets, practices and interests that manifest it provides a landscape that is neither as homogenous nor as recognizable as a paradigm. This is exacerbated even further, when we consider the push for a 'decolonized' account of the power of scientific relationality of the type advocated by, for example, philosopher of science Arun Bala (2006). The complex topology of interconnected, dialogic communities around the globe, now and throughout history, provides a space for thinking scenarios in a much less simplistic way. Even more, Roubelat's positioning of Anticipation is not so supportive as that of many of us in this volume: he is wary (rightly so in our opinion) of its teleological nature, but still finds useful its imaginative tendencies. So, while some of us have excitedly embraced the novelty of the practice of anticipation in relation to futures studies, Roubelat's steady location of it within a fairly well-recognized futures process as scenario design is a welcome critical contextualization.

Stitching

John O'Reilly's chapter is concerned with practice and its manifestations across research practice, creative practice, philosophical practice and pedagogical practice. These are situated between two separate events: NASA's development of the spacesuit for astronauts in the late 1950s and early 1960s and the instantiation of Black Quantum Futurism in 2015. Both involve obfuscated, obliviated or simply minoritarian discourses and practices in creative product development and creative activism. Encountering innovation processes, structures and deployments of power/knowledge and creative ontologies of anticipation, O'Reilly makes a case for the courage it takes to think and act otherwise than normal. To do this he highlights an observation by Argentinian Walter Miller, who has spent the last 30 years exploring ways to unpick the modern/colonial system and develop a thinking and politics of justice. In Miller's book (2011) *The Darker Side of Western Modernity: Global Futures, Decolonial Options,* he writes that: 'The illusion that Western Civilization could create the problem and solve it is facing its limits' (Miller, 2011, p. 282). It is a hopeful thought, though it will not be easy, because as Malcolm McLaren, innovation manager of Punk, realized ('quoting' the Situationists): 'capitalism sells our boredom back to us'. The capitalism engendered in the west (though currently with many characteristics) is particularly adept at folding its edgy – and possibly destructive – outsides back into the middle to energize its own creative approaches to surplus value production. A monomaniacal monopolization of capital will simply increase entropy unless the system is opened to the creative outsides (Serres, 1982). Deleuze and Guattari (1984, 1988) call this 'schizophrenization'. Just as the oil companies broke the planet notwithstanding their scenario and foresight work (however far they thought the unthinkable), there are corporations willing to capitalize our futures and sell them back to us through the micro targeting offered by abilities to manage large data sets. This 'corporate imagination' controls the intellectual, political and commercial currency of the future unless we can think and act otherwise, 'other-wise' as O'Reilly puts it. He writes:

> In the terms of this chapter, we might say that anticipation – as a form of creative researching – is also the capacity to anticipate beyond the given discourses of needs. Anticipation as research is always excessive. In exceeding and overflowing the given epistemologies the hyphen of the Otherwise is an artefact created by our research, and this hyphen which connects

and separates, begins to anticipate the kind of discipline anticipation might need to engage with if it is to stay open to the Otherness of the future.

As well as the 'others' in the present.

Love

Jamie Brassett uses the creative ontologies of anticipation, in the interleaving contexts of love and friendship, to drive a critique of the future's collusion with war. His chapter 'For a creative ontology of the future: an ode to love' interrogates Robert Rosen's speculative, relational biology in conjunction with Roman poet, scientist and philosopher Lucretius's creative atomist ontology. For Lucretius (2007), life is a materialization of love and for Rosen (1991, 2000, 2012) anticipation is one of the key characteristics of life. Why, Brassett contends, is much forward planning infused with strategy, a term of war? Lucretius, on the other hand, looks to the creative, generative power of love. And philosopher Michel Serres (1977), in his critical evaluation of Lucretius's work as a forerunner of more contemporary nondeterministic dynamics and negentropic complexity, argues the same. The concern then becomes: how might future-oriented work characterize itself in relation to love not war? Serres and Lucretius provide a way out of the *fœdus fati*, the atomic fall to death in which strategy is implicated, through the *fœdera naturae*, the natural treaties of a swerved, local, relational, creative universe. For Brassett, this is delivered – at least partially – through anticipation and the relational ontologies it articulates, both at its origin and in its contemporary possibilities. The sense of life, creative and loving, that we find in Lucretius encounters Rosen's relational biology of life: one that is always overflowing and exceeding the basic elements to which much contemporary physics seeks to reduce it. Serres is particularly vehement in locating identity and homogeneity with death and entropic stagnation; we might also require similar in ensuring anticipation (as an act and a developing disciplinary studies) a creative, loving future.

Emission

Lucy Kimbell manages to embody many different kinds of expertise: designer, artist, Science and Technology Studies scholar, social design expert and one of today's key developers of designing for policy. Kimbell brings both a highly affective perspective and rigorously argued examination of the ways in which an anticipatory art and design can be thought and practised. Her chapter, 'Inventive devices and public issues: The *Air Pollution Toile*', centres on her speculatively designed wallpaper toile that is simultaneously haunting and provocative. With patterns that change in relation to the various particulates of our polluted, metropolitan living, Kimbell's wallpaper locates itself within those design and design studies traditions that deliver both critical commentary on contemporary concerns and speculative accounts of possible action (for example: Dunne and Raby, 2013; Yelavich and Adams, 2014). Kimbell weaves the manifold pragmatics and critiques of her positioning into a text shot through with capacities for affecting and being affected that are crucial components of any relational ontology. While Kimbell's focus is turned elsewhere, with this mention of 'relational ontology' we (Jamie and John) see the relationality of the theoretical biology of Robert Rosen (who, as noted already, is a key figure in anticipation studies), as well as a brief mention of his tutor and mentor, Nicolas Rashevsky (1899–1972) who coined the term 'relational biology' and from whom Rosen took up the antireductionist cause (Nadin, 2010, 2011). It is worth noting here, briefly, that a relational ontology also suffuses the metaphysics and ethics of Baruch Spinoza (1996). We have investigated the ways in which Spinoza's ontology might be brought to futures and design (Brassett and O'Reilly, 2015; Brassett, 2016), ethics and social design (Brassett, 2018). We wonder, here, how these modes of thought may extend from and into Kimbell's *toile*. An intriguing aspect of Spinoza's ethics, especially for those of us involved in design and the arts is the dissolution of a body from a bounded entity to one that is predicated on its relations; those articulated as affecting and being affected. It is clear, for us, that Kimbell's *toile* is expressed both as a set of relations – pollution and its impacts; patterns and their mirroring in bodies and environments; and as a complex association of affects – between mothers and daughters, various business interests (e.g. fossil fuels, cigarettes) and the domestic interior, political discourses and direct action. It is also important, that Kimbell's work (the *toile* and its positioning here, in this book) is simultaneously, but not unambiguously – and therefore neither exhausted in any of these tendencies nor totalizing of all of them – a pragmatic, speculative and critical act of design.

Pataphysics

A tendency to ambiguity is important for anticipation scholar Mihai Nadin (2010, p. 26), who – citing Stephen Kercel (2007) – writes that 'ambiguity is an observable feature of complexity', with the consequence that anticipation is nothing if not complex and ambiguous. Our penultimate chapter works through complexity, ambiguity and anticipation through the work of philosophical prankster Alfred Jarry (1873–1907) and his intellectual contribution of 'Pataphysics. Author Derek Hales has been a leading thinker in this field for a number of years. In addition, Hales is also an important proponent of 'design fictions' (Hales, 2013) and their relation to 'speculative design'. Informed by all of these is Hales's innovation practice especially in relation to regional development and his disciplinary field of architecture. In this chapter, 'The anticipatory power of the *objectile*', Hales takes the reader through the paradoxical theory and practice of 'anticipatory plagiarism' as it is played out in the work of various architects, philosophers and mathematicians and in relation to 'Pataphysics. This is defined as the 'science of imaginary solutions, which symbolically attributes the properties of objects, described by their virtuality, to their lineaments' (Jarry, 1965, p. 145); furthermore, 'pataphysics stands in the same relation to metaphysics as metaphysics does to physics. Taught philosophy by Henri Bergson and with an ability to shock the bourgeoisie – which renowned 'Pataphysician Andrew Hugill (2015, p. 3) aligns with a possible etymology of the term: *épataphysique*, with *épater* being 'to shock' in French and the omitted 'e' leading to the preceding apostrophe – Jarry is probably most well-known for his play *Ubu Roi* (1968). There are a number of lines to take from this definition by Jarry. Thinking 'lineaments' as 'contours' we have some notions of Riemannian curved space – a mathematics that was important and popular as the nineteenth became the twentieth century. With 'virtuality', a gesture to Bergson; and, with a science characterized as 'imaginary', we confront many of the concerns of the other chapters in this book. Hales also re-examines the concept of the 'objectile' (see also: Hales, 2015). For Hales, the concept of the 'objectile' as developed by Gilles Deleuze (1993) regards an object not in any entirety but as a voyaging, processing of matter, a matter-flow (Hales, 2015). It therefore serves as a powerful pataphysical act: a throwing-out or throwing-up of thingness. This is, of course, also an act of drawing down the future into the present, as if it were the cup of a huge catapult machine only to be released to hurtle off again, goodness knows where. Hales brings to bear on such an anticipatory modelling two further concepts worth comment: anticipatory plagiarism and architectural fictions. These necessarily intertwine, but for brevity we can say that anticipatory plagiarism is the present's plagiarizing of future work. While couched in utterly pataphysical terms, this is also as good a definition of anticipation as one could have. Architectural fictions, strange, speculative acts that are creating a present with the future in, today, that will offer a future that is therefore already plagiarized. Hales manages this in a form of creative theory writing that is as awe inspiring and joyfully playful as any pataphysical text.

Mode

To close, there is the chapter by us, Jamie Brassett and John O'Reilly, first published in French, in the journal *Prospective et Stratégie* (Brassett and O'Reilly, 2018b). The chapter presented here is an amended and expanded English version of that article. We thank the General Editor of *Prospective et Stratégie* for permission to publish this version here. Now titled '2078/1978. Anticipation and the Contemporary', this chapter brings to anticipation a philosophical examination of the concept of 'the contemporary (person)' by Giorgio Agamben (2008) – itself informed by Friedrich Nietzsche's (1997) development of the 'untimely'. Agamben positions the contemporary person in a space situated both inside and outside the present and is regarded by both philosophers as a positioning that demands courage. In so doing, the contemporary stance allows the contemporary person to create a new present as a consequence of this position providing a view on the present's darkest moments. If this needs courage, we argue, so does anticipation. For it, too, demands that we adopt a position inside and outside the present in order to allow the future in and thus to make it anew. Such a positioning is mirrored in some futures work; notably, Riel Miller's (2011) urge to adopt a 'scenaric stance', which also requires a move away from an epistemological situation (searching for meanings) towards an ontological one (thinking and practising beings and becomings). As we have discussed earlier in this introduction, the work of Roberto Poli (2010b, 2011, 2017) has also become important in developing anticipation studies in recent years. For Poli, an anticipatory stance requires a certain 'irrelevance', a step outside of current norms to be able to get a perspective on the future. This is exactly the same word as Agamben uses to characterize the contemporary: 'irrelevance'. We develop an argument that positions untimely, contemporary and anticipatory in proximity to each other. In doing this, we overturn some thoughts and practices that we do not have time to cultivate, eager as we are to keep focus upon our main topics of enquiry. Chief amongst these is Whitehead's conception of 'speculative' philosophy. While this practice/concept has been prominent in recent years – Brassett (2016, p. 163) notes, among a variety of disciplines: 'speculative design' (also discussed by Hales and practised by Kimbell in this volume), 'speculative history', 'speculative realism' – a thorough investigation of the knot resulting from twining together Whitehead's speculative philosophy, anticipation studies and creative ontologies would be insightful future research.

Immediate existence (whenever that may be)

What each of these essays compose is a different way of styling the future (Brassett and O'Reilly, 2015) that has not already been predicted; with these styles generating atmosphere, as we noted above. Anticipation thought this way is a singular, creative and flourishing act; otherwise the future merely comes already prefabbed. Organization scholar Barbara Czarniawska writes in 'The Style and the Stylists of Organization Theory' (2005, p. 240):

> we speak of 'texts lacking style' that can be best described as compilations of words and phrases rather than as 'the work on the words'. Style is the writer's awareness of being engaged in writing, incorporated into the text itself (as opposed to lack of such awareness, but also to self-reflective or meta-reflexive texts).

Anticipation that truly engages in the unknown-unknown needs to create a different language with which to translate the future: to style the future in a way for the present, as we anticipated a few years ago (Brassett and O'Reilly, 2015). The different styles of each chapter are also – literally, cognitively, fashionably – modes of anticipation, modes of anticipation as ontological transformation: *Picturing; Runes; Hope; Rules; Stitching; Love; Emission; Pataphysics; Mode.* As much as each chapter is a situated engagement with the concept and practices of anticipation, we hope the reader will register theses different modes of anticipation. Anticipation is the capacity of futures studies for opening up to the multiple modes of future becoming, beyond the corporate 'unreal estate of the future', to making futures that are creative, ethical and lovely.

The timelines of this book as an anticipated work in all our futures are drawn back to the knots of our meetings in presents now past. Or they form as separate strands of practice, highlighting a manifold of anticipations that have communicated sometimes strongly, other times not, and every place in-between. This has been a fascinating process into and out of anticipation in thought and practice. George Orwell, in *Nineteen Eighty-Four*, tells us that '[t]he best books are the books that tell you what you know already'; these books, like algorithms, have already predicted us. But life cannot be reduced to or represented by an algorithm (Nadin, 2011). If this is what the best is, then we are hoping that this book on anticipation will become one of the worst.

Notes

1 We should note that futures scholar, Marcus Bussey (2013, 2017), engages with a philosophical focus very much in alignment with many of ours in this volume.
2 Futures and education scholar Keri Facer (2018) uses Whitehead's (1968, pp. 171–174) position on the future-focused role of the university to critique its neoliberalization (among other things) – thus bringing an important political aspect to play in issues around higher education and the future.

References

Agamben, G. (2008) 'What is the Contemporary?', in Agamben, G., *What is an Apparatus? and other Essays*. Trans. D. Kishik and S. Pedatella. Stanford, CA: Stanford University Press, pp. 1–24.
Anderson, B. (2009) 'Affective Atmospheres', *Emotion, Space and Society*, 2, pp. 77–81.
Bala, A. (2006) *The Dialogue of Civilizations in the Birth of Modern Science*. New York, NY: Palgrave Macmillan.
Batman Begins (2005) Directed by C. Nolan. [Feature Film]. Burbank, CA: Warner Bros. Pictures.
Bloch, E. (2000) *The Spirit of Utopia*. Trans. A.A. Nassar. Stanford, CA: Stanford University Press.
Bloch, E. (1985) *Philosophische Aufsätze zur objektiven Phantasie*. Frankfurt am Main: Suhrkamp.
Bousbaci, R. (2008) '"Models of Man" in Design Thinking: The "Bounded Rationality" Episode,' *Design Issues*, 24(4), (Autumn), pp. 38–52.
Brassett, J. (2018) 'Creating Affective Social Design. An Ethical and Ontological Discussion', *Cubic Journal*, 1, Special Issue: 'Design Social', pp. 146–159.
Brassett, J. (2016) 'Speculative Machines and Technical Mentalities: A Philosophical Approach to Designing the Future', *Digital Creativity*, 27(2), pp. 163–176.
Brassett, J. (2015) 'Poised and Complex. The Becoming Each Other of Philosophy, Design and Innovation', in Marenko, B. and Brassett, J. (eds.), *Deleuze and Design*. 'Deleuze Connections' Series. Edinburgh: Edinburgh University Press, pp. 31–57.
Brassett, J. and O'Reilly, J. (2018a) 'Collisions, Design and The Swerve', in Vermaas, P. and Vial, S. (eds.), *Advancements in Philosophy of Design*. 'Design Research Foundations' Series. Cham, Switzerland: Springer, pp. 71–98.

Brassett, J. and O'Reilly, J. (2018b) 'Retour à 2078: Réflexions sur l'anticipation et le contemporain', *Revue de Prospective et Stratégie*, 9, Édition Spéciale: 'Voir Loin', (December), pp. 11–22.

Brassett, J. and O'Reilly, J. (2015) 'Styling the Future. A Philosophical Approach to Scenarios and Design', *Futures*, 34, Special Issue: 'Scenarios and Design', pp. 37–48.

Brown, T. (2009) *Change by Design: How Design Thinking Transforms Organizations and Inspires Innovation*. New York, NY: HarperCollins Publishers.

Bussey, M. (2017) 'Anticipatory Aesthetics: The Senses and the Body in Anticipatory Theory and Practice', in Poli, R. (ed.), *Handbook of Anticipation*. Cham, Switzerland: Springer International Publishing AG. doi:10.1007/978-3-319-31737-3_84-1.

Bussey, M. (2013) 'Foresight Work as Bridge Building: Poetry, Presence and Beyond', *Journal of Futures Study*, 17(4), pp. 103–116.

Celi, M. and Morrison, A. (2017) 'Anticipation and Design Inquiry', in Poli, R. (ed.), *Handbook of Anticipation*. Cham, Switzerland: Springer International Publishing AG. doi:10.1007/978-3-319-31737-3_49-1.

Chia, R. (1999) 'A "Rhizomic" Model of Organizational Change and Transformation: Perspective from a Metaphysics of Change', *British Journal of Management*, 10, pp. 209–227.

Czarniawska, B. (2005) 'The Styles and Stylists of Organization Theory', in Knudsen, C. and Tsoukas, H. (eds.), *The Oxford Handbook of Organization Theory: Metatheoretical Perspectives*. Oxford: Oxford University Press, pp. 237–261.

DeLanda, M. (1991) *War in the Age of Intelligent Machines*. New York, NY: Zone Books.

Deleuze, G. [1986] (1995) 'Breaking Things Open, Breaking Words Open', in Deleuze, G., *Negotiations 1972–1990*. Trans. M. Joughin. New York, NY: Columbia University Press, pp. 83–93.

Deleuze, G. [1988] (1993) *The Fold. Leibniz and The Baroque*. Trans. T. Conley. London: The Athlone Press.

Deleuze, G. [1969] (1990a) *The Logic of Sense*. Trans. M. Lester with C. Stivale. Ed. C. Boundas. London: The Athlone Press.

Deleuze, G. [1968] (1990b) *Expressionism in Philosophy*: Spinoza. Trans. M. Joughin. New York, NY: Zone Books.

Deleuze, G. [1986] (1988) *Foucault*. Trans. S. Hand. London: The Athlone Press.

Deleuze, G. [1962] (1983) *Nietzsche and Philosophy*. Trans. H. Tomlinson. London: The Athlone Press.

Deleuze, G. and Guattari, F. [1991] (1994) *What is Philosophy?* Trans. G. Burchill and H. Tomlinson. London: Verso Books.

Deleuze, G. and Guattari, F. [1980] (1988) *A Thousand Plateaus: Capitalism and Schizophrenia 2*. Trans. B. Massumi. London: The Athlone Press.

Deleuze, G. and Guattari, F. [1972] (1984) *Anti-Œdipus. Capitalism and Schizophrenia 1*. Trans. R. Hurley, M. Seem, and H.R. Lane. London: The Athlone Press.

Dunne, A. and Raby, F. (2013) *Speculative Everything. Design, Fiction and Social Dreaming*. Cambridge, MA and London: The MIT Press.

Facer, K. (2018) 'The University as Engine for Anticipation: Stewardship, Modelling, Experimentation, and Critique in Public', in Poli, R. (ed.), *Handbook of Anticipation*. Cham, Switzerland: Springer International Publishing AG. doi:10.1007/978-3-319-31737-3_29-1.

Freedland, J. (2012) 'Eugenics: The Skeleton that Rattles Loudest in the Left's Closet', *The Guardian*, 17 February. Available at: https://www.theguardian.com/commentisfree/2012/feb/17/eugenics-skeleton-rattles-loudest-closet-left (Accessed: 17 June 2020).

Goodwin, N. (2014) 'The Human Element in the New Economics: A 60-Year Refresh for Economic Thinking and Teaching', *Real-world Economics Review*, 68, (August), pp. 98–118.

Hales, D. (2015) 'Redesigning the Objectile', in Marenko, B. and Brassett, J. (eds.), *Deleuze and Design*. 'Deleuze Connections' Series. Edinburgh: Edinburgh University Press, pp. 139–172.

Heidegger, M. [1915] (2007) 'The Concept of Time in the Science of History', in Kisiel, T. and Sheehan, T. (eds.), *Becoming Heidegger: On the Trail of His Early Occasional Writings, 1910–1927*. Evanston, IL: Northwestern University Press, pp. 61–72.

Hernes, T. (2014) *A Process Theory of Organization*. Oxford: Oxford University Press.

Hugill, A. (2015) *Pataphysics: A Useless Guide*. Cambridge, MA and London: MIT Press.

Huppatz, D.J. (2015) 'Revisiting Herbert Simon's "Science of Design"', *Design Issues*, 31(2), (Spring), pp. 29–40.

Jarry, A. (1968) *The Ubu Plays*. Trans. C. Connolly and S. Watson Taylor. London: Methuen.

Jarry, A. [1911] (1965) *Selected Works of Alfred Jarry*. Trans. R. Shattuck and S. Watson Taylor. London: Methuen.

Julier, G. (2017) *Economies of Design*. London: Sage.

Kay, J. (2013) 'Circular Thinking', *RSA Journal*, 159(5556), pp. 10–15.

Kercel, S. (2007) 'Entailment of Ambiguity', *Chemistry and Biodiversity*, 4(10), pp. 2369–2385.

Lambert, R. and Flood, R. (2018) *Undertaking Design-Intensive Innovation: A Literature Review*. London: The Design Council.

Latour, B. (2008) 'A Cautious Prometheus? A Few Steps Toward a Philosophy of Design (with Special Attention to Peter Sloterdijk)', in Hackney, F., Glynne, J. and Minton, V. (eds.), *Networks of Design. Proceedings of the International Conference of the Design History Society*. Boca Raton, FL: Universal-Publishers, pp. 2–10.

Lloyd Morgan, C. (1927) *Emergent Evolution. The Gifford Lectures Delivered in the University of St. Andrews in the Year 1922*. 2nd edn. London: Williams and Norgate.

Lucretius [55BCE] (2007) *The Nature of Things*. Trans. A.E. Stallings. London: Penguin Books.

Marenko, B. (2015) 'When Making Becomes Divination: Uncertainty and Contingency in Computational *Glitch-Events*', *Design Studies*, 41(A), (November), pp. 110–125.

Marenko, B. (2014) 'Neo-Animism and Design. A New Paradigm in Object Theory', *Design and Culture*, 6(2), Special Issue: 'Design, Thing Theory and the Lives of Objects', pp. 219–242.

Martin, R. (2009) *The Design of Business: Why Design Thinking is the Next Competitive Advantage*. Boston, MA: Harvard Business Press.

Marx, K. [1856] (1978) 'Speech at the Anniversary of the People's Paper', in Tucker, R.C. (ed.), *The Marx-Engels Reader*. 2nd edn. London: Norton, pp. 577–578.

Massumi, B. (2010) 'The Future Birth of the Affective Fact: The Political Ontology of Threat', in Gregg, M. and Seigworth, G.J. (eds.), *The Affect Theory Reader*. Durham, NC: Duke University Press.

Miller, R. (2011) 'Being without Existing: The Futures Community at a Turning Point? A Comment on Jay Ogilvy's "Facing the Fold"', *Foresight*, 13(4), pp. 22–34.

Miller, W. (2011) *The Darker Side of Western Modernity: Global Futures, Decolonial Options*. Durham, NC: Duke University Press.

Nadin, M. (2011) 'The Anticipatory Profile. An Attempt to Describe Anticipation as Process', *International Journal of General Systems*. doi:10.1080/03081079.2011.622093.

Nadin, M. (2010) 'Anticipation and Dynamics: Rosen's Anticipation in the Perspective of Time', *International Journal of General Systems*, 39(1), pp. 3–33.

Napoleon Dynamite (2004) Directed by J. Hess. [Feature Film]. Los Angeles, CA: Fox Searchlight Pictures.

Nietzsche, F. (1997) *Untimely Meditations*. Ed. D. Breazeale. Trans. R.J. Hollingdale. Cambridge: Cambridge University Press.

Orwell, G. [1949] (2000) *Nineteen Eighty-Four*. London: Penguin Books Ltd.

Poli, R. (2017) *Introduction to Anticipation Studies*. 'Anticipation Science' Series, vol. 1. Berlin: Springer.

Poli, R. (2011) 'Steps Toward an Explicit Ontology of the Future', *Journal of Futures Studies*, 16(1), pp. 67–78.

Poli, R. (2010a) 'The Many Aspects of Anticipation', *Foresight*, 12(3), pp. 7–17.

Poli, R. (2010b) 'An Introduction to the Ontology of Anticipation', *Futures*, 42(7), pp. 769–776.

Poli, R. and Seibt, J. (eds.) (2010) *Theory and Application of Ontology: Philosophical Perspectives*, vol. 1. Dordrecht: Springer Netherlands.

Renwick, C. (2019) 'Movement, Space and Social Mobility in Early and Mid-Twentieth-Century Britain', *Cultural and Social History*, 16(1), pp. 13–28.

Rosen, R. (2012) *Anticipatory Systems: Philosophical, Mathematical and Methodological Foundations*. 2nd edn. New York, NY: Springer.

Rosen, R. (2000) *Essays on Life Itself*. New York, NY and Chichester, W. Sussex: Columbia University Press.

Rosen, R. (1991) *Life Itself. A Comprehensive Inquiry Into the Nature, Origin, and Fabrication of Life*. 'Complexity in Ecological Systems' Series. New York, NY and Chichester, W. Sussex: Columbia University Press.

Serres, M. (1982) 'On the Origin of Language: Biology, Information Theory, and Thermodynamics', in Serres, M., *Hermes. Literature, Science, Philosophy*. Baltimore, MD: Johns Hopkins University Press, pp. 71–73.

Serres, M. (1977) *La Naissance de la physique dans le texte de Lucrèce. Fleuves et turbulences*. Collection: Critique. Paris: Les Editions de Minuit.

Simon, H. (1988) 'The Science of Design: Creating the Artificial', *Design Issues*, 4(1/2), Special Issue: 'Designing the Immaterial Society', pp. 67–82.

Simondon, G. [1964/1989] (2013) *L'individuation à la lumière des notions de forme et d'information*. Paris: Éditions Jérôme Million.

Spinoza, B. de [1677] (1996) *Ethics*. Trans. E. Curley. London: Penguin Classics

Stengers, I. [2002] (2011) *Thinking with Whitehead. A Free and Wild Creation of Concepts*. Trans. M. Chase. Boston, MA: Harvard University Press.

Stengers, I. (2008) 'A Constructivist Reading of *Process and Reality*', *Theory, Culture and Society*, 25(4), pp. 91–110.

Stengers, I. (2005) 'Deleuze and Guattari's Last Enigmatic Message', *Angelaki*, 10(2), pp. 151–167.

Stengers, I. (1997) *Cosmopolitiques. Tome 6. La Vie et l'artifice: Visages de l'émergence*. Paris: Éditions La Découverte.

The Dark Knight Rises (2012) Directed by C. Nolan. [Feature Film]. Burbank, CA: Warner Bros. Pictures.

Tsoukas, H. (1998) 'Chaos, Complexity, and Organization Theory', *Organization*, 5(3), pp. 291–313.

Tsoukas, H. and Chia, R. (2002) 'On Organizational Becoming: Rethinking Organizational Change', *Organization Science*, 13(2), pp. 567–582.

Tushman, M.L. and O'Reilly III, C.A. (1996) 'The Ambidextrous Organization: Managing Evolutionary and Revolutionary Change', *California Management Review*, 38, pp. 1–23.

Verganti, R. (2009) *Design Driven Innovation: Changing the Rules of Competition by Radically Innovating What Things Mean*. Boston, MA: Harvard University Press.

Whitehead, A.N. [1927–28] (1978) *Process and Reality. An Essay in Cosmology, Corrected Edition*. Eds. D.R. Griffin and D.W. Sherburne. New York, NY: The Free Press.

Whitehead, A.N. [1938] (1968) *Modes of Thought*. New York, NY: The Free Press.

Whitehead, A.N. [1925] (1967a) *Science and the Modern World*. New York, NY: The Free Press.

Whitehead, A.N. [1933] (1967b) *Adventures of Ideas*. New York, NY: The Free Press.

Yelavich, S. and Adams, B. (eds.) (2014) *Design as Future-Making*. London and New York, NY: Bloomsbury.

2 Anticipation, creativity and picture perception

Mark Donoghue

Introduction

Anticipation plays an important role in visual cognition as only a limited amount of information from visual sensation can be processed due to the physiology of the visual system. Therefore, in order to make efficient use of this information, in making judgements and guiding action, the brain needs to represent the probability of various events and objects being the cause of visual sensations. In short, we perceive by constantly making inferences based on visual sensations, that is, by anticipating.

In recent years there has been a growing body of theoretical and scientific work that suggests Bayesian Inference can offer a framework to understand the role of inference in visual perception (Knill and Pouget, 2004; Friston, 2009, 2010; Seriès and Seitz, 2013).[1] Bayesian Inference is a way to describe the probability of an event based on prior knowledge of conditions related to the event. In short, this theory supposes the mind generates models that describe the probability of events causing visual sensations. This forms the basis of inferences about their worldly causes. Inference also plays an essential role in picture perception, as the information in a picture is necessarily limited due to the constraints of it being a two-dimensional plane. This chapter aims to highlight how the processes of inference that underpin perception are related to anticipation and how inference plays an even more pronounced role in picture perception. Although the principles apply equally to moving film images, for simplicity only static pictures will be examined here.

The notion of an anticipatory system was initially developed by theoretical biologist Robert Rosen (1934–1998) to describe how biological systems anticipate their environment. Rosen (2012, p. 313) defines an anticipatory system as: 'a system containing a predictive model of itself and/or of its environment, which allows it to change state at an instant in accord with the model's predictions pertaining to a later instant'. That is, anticipation refers to the ability of organisms to adapt their actions to a model of potential environmental development in which they are embedded (Poli, 2010, p. 770). Bayesian Inference can provide a method to explain how systems can build models to guide behaviour, by applying it in a hierarchical fashion in an architecture that generates models to predict changes in the organism and environment at increasing levels of abstraction.[2] The crux of this chapter focuses on the relationship between prediction and creativity. These may initially appear antithetical with the novelty of creativity seemingly at odds with the fixity of prediction. This tension is a recurrent theme in the philosophy of William James (1842–1910). According to James (1987, p. 1033):

> The percepts are singulars that change incessantly and never return exactly as they were before. This brings an element of concrete novelty into our experience. This novelty finds no representation in the conceptual method, for concepts are abstracted from experiences already seen or given, and he who uses them to divine the new can never do so but in ready-made and ancient terms.

For James, perceivers are constantly parsing experience in order to make judgements and guide action, but this mechanism inevitably anchors the flux of experience. This is also an important theme in Henri Bergson's (1859–1941) philosophy.[3] The first section of this chapter examines how Bergson's metaphysics provides a useful way to consider how the past affects the future, how anticipation structures our notion of the future and how creativity plays a role in experience. Next, the connection between Bergson's metaphysics and James's Pragmatism is explored, while Bayesian probability theory is introduced to provide a formal treatment of how current information is combined with prior knowledge to make judgements. The following section involves a discussion of how Bayesian probability facilitates perceptual inference through the hierarchical architecture of predictive processing (PP). This chapter concludes by examining how PP facilitates picture perception and how Bergson and James's philosophies add to this.

Anticipation, Bergson and the ontology of the possible

As noted, there is a tension, in Bergson's philosophy, between both the novelty of experience and how experience is analysed to make predictions. Bayesian probability, through PP, gives a mechanism for this parsing but there is always an element of experience that escapes these models generated to predict future events. Bergson's work, however, demonstrates why this is the case: because analysing experience changes its processual nature and this cannot be recovered once disjointed into discrete events. Thus, as creativity also describes the novelty of experience that escapes such analysis, I will examine this issue of creativity first by discussing Bayesian probability and PP. It is only by understanding how PP is always open to future change to refine its predictions that we can understand how predictions are always accompanied by, and conditioned though, the novelty of creativity. I will argue that PP's future orientation, through its openness to the creativity of experience, puts it in keeping with anticipatory systems. Ultimately, creativity's role in picture perception rests with the way it conditions the generation of models that make perceptual inference possible. This inference is vital for picture perception because, being a two-dimensional surface, picture information is limited and perceivers have to infer depicted objects from pictorial cues. Additionally, creativity may play a role in pictures that are artworks, specifically as a capacity to incite the novelty of experience.

Rosen's work on anticipatory systems, drawing from relational biology, works against the presumptions of Newtonian mechanics that 'never allow future states of the system to affect the present changes of state' (Fuller, 2017, p. 42). According to Rosen (2012, p. 415):

> Anticipation [. . .] involves *selective* response to possible futures represented in the present. The very neatly predictable mechanistic world, where 'the future' is imagined as the one and only possible outcome of the past, cannot be preserved if there are systems that anticipate, for anticipation involves symbolizing multiple unrealized possibilities and selecting from those choices.
>
> (Original emphasis)

This critique of the notion of the future as pre-determined by the past is also an important feature of Henri Bergson's work and formed the impetus for a reconception of time as 'duration' (*la durée*), a process of qualitative rather than quantitative change.

Humans are constantly making forecasts about the future. However, despite 'the fact that our rational mind and especially its use in natural sciences has made it possible to predict events at least to a certain degree of probability' (Jancsary, 2019, p. 63), there is always an unpredictable component to experience that escapes prediction. It is tempting to think these unexpected events are a product of the limitations of the system and that with better scientific or computational tools one could account for this unpredictability.[4] Likewise, with psychological predictions, it is easy to consider the unforeseeable component of experience as a consequence of the limitations of cognitive abilities or prior knowledge. Could a hypothetical ideal observer with an unlimited predictive mind, armed with all the relevant knowledge, be capable of completely predicting upcoming events?

At the end of the nineteenth century, during the height of confidence in Newtonian mechanics, this tendency was even stronger than it is today. Bergson would receive his philosophical and mathematical training in this period, but he would go on to develop a concept of time that countered the contemporary prevailing trends. For Bergson (2001, pp. 219–220), the act of making predictions involves parsing time but:

> By the very fact of breaking up concrete time we set out its moments in homogeneous space; in place of the doing we put the already done; and, as we have begun by, so to speak, stereotyping the activity of the self, we

> see spontaneity settle down into inertia and freedom into necessity. Thus, any positive definition of freedom will ensure the victory of determinism.

In other words, in order to understand upcoming events our analytical mind can only conceptualize our world in rigid and fixed terms, but this activity inevitably cannot account for the processual nature of real phenomena (Jancsary, 2019, p. 72). In *Creative Evolution* (1944) Bergson describes the essential processual nature of reality as the *élan vital* (vital impetus). However, because we are surrounded in an environment where objects generally undergo change very slowly, the nature of reality as an unfolding process is masked and we conceive of the world as consisting mostly of unchanging and static objects. For Bergson, philosophical problems arise when we apply this concrete viewpoint to time.

Bergson (1944, pp. 335–340; 1991, pp. 190–193) demonstrates these problems by referring to the paradoxes of pre-Socratic Greek philosopher Zeno of Elea (c. 495–c. 430 BCE). In one such paradox, Achilles is in a race with a tortoise and gives the tortoise a head start. If the tortoise has a head start, to catch up Achilles will take a certain amount of time to reach the tortoise's position from his starting position. During this time the tortoise will have progressed a much shorter distance, but Achilles still has an additional shorter distance to run in order to catch up with the tortoise. However, even once he reaches the advanced position, the tortoise has proceeded even further. Thus, Achilles is constantly catching up with the tortoise, and although the distance between them is constantly diminishing, he is never able to catch up with or overtake the tortoise. The paradox is essentially a restating of the Dichotomy Paradox where an object in motion before reaching a point must travel half the distance. But before travelling half the distance it must travel one-quarter the distance. But before travelling one-quarter the distance it must travel one-eighth the distance, and so on. The distance to be travelled halves each time tending towards zero but as the distance can be divided infinitely it seems impossible that the object can be in motion at all. Although there are mathematical proofs around these problems, for Bergson these overlook the metaphysical implications of Zeno's paradoxes (Ansell-Pearson, 2002, p. 26). For Bergson, this is a consequence of failing to recognize the distinction between quantitative and qualitative differences. The former refers to how much there is of something by difference-in-degree and the latter refers to the quality of phenomena by difference-in-kind. Therefore, the problems evident in Zeno's paradoxes are a consequence of our tendency to conceive of time in terms of extensive space. Bergson (2001, p. 95) writes: 'space is what enables us to distinguish a number of identical and simultaneous sensations from one another; it is thus a principle of differentiation other than that of qualitative differentiation, and consequently it is a reality with no quality'. It is common to preconceive of time in terms of a timeline where events are positioned (before, after, etc.) on this line. But to consider time as a line means transforming it to a one-dimensional space. This space is composed of an infinity of discrete instances. When Zeno tells of

Achilles running, he does so in terms of a succession of discrete instances, each of which represents a frozen snapshot. In this way, change does not transform time itself, time is like a container in which change occurs. Bergson (2001, p. 90) writes further that when we describe time, 'we generally think of a homogeneous medium in which our conscious states are ranged alongside one another as in space, so as to form a discrete multiplicity'. Our space affects us negatively though, by applying the homogenization implicit in this symbolic representation of space to all concepts (Guerlac, 2006, pp. 68–69).

Bergson argues this commonplace conception of time is a consequence of our mind analysing the world for action, but this inevitably disguises the nature of change in the world. This is due to our failure to distinguish discrete and continuous multiplicities. For example, Bergson (2001, p. 77) describes how when counting sheep in a flock, we treat each of the sheep as an individual in a larger set so we can attribute a number to them. Although there may be differences in the sheep, each sheep is included in an instance of a category. To count these instances their qualitative differences are ignored and the instances are treated as identical and unchanging. However, the sheep in reality do change qualitatively, as each one is unique. Rather than the flock of sheep being a container including a quantity of instances, the flock is a whole that changes qualitatively from sheep to sheep.

Bergson takes this distinction between discrete and continuous multiplicities, and the distinction between qualitative and quantitative change, and extends it. Instead of time being conceived as a distinct quantitative change, Bergson asserts time's true nature is an indistinct qualitative change. Bergson describes this alternative conception of time as duration. When Achilles is running along, the quantitative change in terms of measurable velocity and distance is underpinned by a qualitative change. What we perceive as changes in Achilles' discrete position is founded on a qualitative change in the whole world, which includes Achilles, the tortoise, the road, etc., not as discrete entities but as qualitative associations. In this way, Achilles's steps have the quality of 'overtaking-tortoise' and the tortoise's steps have the quality of 'being-overtaken-by-Achilles'. It is only when our concepts slice these qualities up into an infinity of instances that Zeno's paradoxes occur.

We need to recognize duration as the 'qualitative element of time' (Guerlac, 2006, p. 68). Understanding time this way means change is not what happens between instances, but a continuous multiplicity.[5] Time as duration is a non-metric qualitative varying continuity of the whole in constant flux. In this way, duration is: 'nothing but a succession of qualitative changes, which melt into and permeate one another, without precise outlines, without any tendency to externalize themselves in relation to one another, without any affiliation with number: it would be pure heterogeneity' (Bergson, 2001, p. 104). Elsewhere, Bergson (1991, p. 197) describes how duration is a '*moving continuity* [. . .] in which everything changes and yet remains' (original emphasis). From this we 'dissociate the two terms, permanence and change, and then represent permanence by bodies and change by *homogeneous movements* in space' Bergson (1991,

p. 197). It is too simple to consider duration as a simple indivisible whole, rather it is something whose nature is transformed through division, so that even by assembling the divided components it does not restore the whole's continuous nature (Ansell-Pearson, 2002, p. 72). The immobile nature presupposed by the process of division remains even when all the components are reassembled. In this way, by conceiving of time as a succession of instances, the flux of time is ignored and we resort to counting 'simultaneities' (Bergson, 1944, p. 367). Therefore, if duration is a concept of time that differs from the spatialized quantified succession of instances, and it is a continuous qualitative change, then how do past, present and future differ? In short, past, present and future are distinguished because of the qualitative differences between them. For Bergson (1991, p. 144), the past 'has exhausted its possible action'. According to philosopher Gilles Deleuze (1925–1995), whose work was heavily influenced by Bergson: 'The past [. . .] has ceased to act or to be useful. But it has not ceased to be. Useless and inactive, impassive, it IS, in the full sense of the word: It is identical with being in itself' (Deleuze, 1988, p. 55). The past then is not merely instances of time that have ceased to be, the past continues as 'the form under which being is preserved in itself' (Deleuze, 1988, p. 55) but has qualitatively transformed having expended its potential.

In contrast, the future differs qualitatively from the past as 'an impending action, in an energy not yet spent' (Bergson, 1991, p. 144). As such, the future is the condition for the production of the new. The present then is the becoming in which the future loses its potentiality and becomes an inert 'is' (Deleuze, 1988, p. 55). Duration acts as a universal 'principle of bifurcation' (Mullarkey, 1999, p. 261) between the novelty of the new with the power to act and the inert past. Bergson's critique of discrete spatialized metric time is that it replaces the active progression of events, the actual happening of things, with a static geometric model (Guerlac, 2006, pp. 84–85). This model of time cannot capture the processual nature of duration because in space dimensions can be traversed in either direction, but duration only unfolds in one direction (from impending future to exhausted past potential) as the past differs qualitatively from the future and is irreversible.[6]

However, considering the future as the condition for the production of the new, this could be construed as claiming what is to become already exists in some way, merely needing to be realized. This is an issue Bergson recognized and considered as another example of our tendency to spatialize time. Bergson illustrates this by recalling an episode from the First World War. He says at that time many periodicals were concerned with what literature would be like in the future, after the end of the war. He writes (2017, online):

> One day someone came to ask how I pictured it. I declared, somewhat confused, that I did not picture it. 'Don't you perceive at least,' the questioner continued, 'certain possible directions? Let's grant that the details can't be foreseen; at the very least you, as a philosopher, have some idea of the whole. How do you conceive, for example, the great dramatic works

> of tomorrow?' I will always remember the surprise of my interlocutor when I responded: 'If I knew what the great dramatic works of tomorrow will be, I would write them.'

Bergson goes on to describe how the journalist considered the future as already pre-given and had thought Bergson, being the great philosopher, would have some special insight to allow him to predict this. He explains further that such future works are not yet possible. When the journalist challenged him by stating they must be possible for they will be realized, Bergson says then we can say 'at most, that they will have been possible' (Bergson, 2017, online). When asked to elaborate on this, Bergson explicates that we project the past into the future, reshaping our notions of what is to come according to what has passed. In this way:

> The possible is thus the mirage of the present in the past; and since we know that the future will wind up being the present, as the mirage effect continues relentlessly, we say to ourselves that in this present, which tomorrow will be the past, tomorrow's image is already contained, although we can't quite grasp it.
>
> (Bergson, 2017, online)

However, by projecting past possibilities into the future like this we perform the same sort of operation that masks the processual nature of duration. Instead of dividing time into an infinity of instances, we have divided future into an infinity of discrete possibilities. In effect, '*we perceive only the past*, the pure present being the invisible progress of the past gnawing into the future' (Bergson, 1991, p. 150; original emphasis). If this is the case then it raises the issue of how we can think about the future as the condition of the possible, without the future being conceived as a space of predetermined possibilities. Bergson (1991, p. 242) writes that we need to avoid thinking of possibilities in terms of 'ready-made things superposed the one on the other' but as existing virtually.

For Deleuze (1994, pp. 168–221), instead of thinking about the virtual as a space of predetermined possibilities, it should be thought as undetermined but determinable and that it becomes progressively determined through a process of actualization. Deleuze (1990, p. 114) illustrates this process by referring to the way that different forms are determined through geometric groups. For example, under the projective geometry group, all conic sections (the set of curves including circles, ellipses, parabolas and hyperbolas) are all considered equivalent. Under the affine group the types of curves (circles, ellipses, parabolas hyperbolas, etc.) are considered distinct but the different sizes of the same types of curves are considered equivalent. Under the Euclidean group, different types and sizes of curves are considered distinct. These curves are progressively determined through a sequence of symmetry breaking transitions between groups (DeLanda, 2002, pp. 24–27). In this way, the future as the condition of the possible can be considered as undetermined, like geometric forms under the projective group, but are progressively determined as the future is actualized through duration, as geometric forms being determined through symmetry breaking transitions. This is how we can avoid thinking of possibilities in spatialized terms.

To conclude this section, Bergson proposes a concept of duration as the qualitative change of time. This qualitative change is translated to quantitative change according to our spatialized discrete concept of time. Through this transformation it is possible to project the past onto the future to make predictions but this inevitably maps a space of distinct possibilities onto an undetermined future. Bergson does not doubt the power of prediction but thinks it necessary to recognize this operates like retrospection, only projected onto the future (Guerlac, 2006, p. 86). This transforms the nature of time and the possible, making them into spaces of discrete and determined instances and possibilities. The following section will examine how Bayesian Inference can offer a concrete psychological mechanism for how this happens.

From Pragmatism to Bayesian prediction

In Bergson's work there is a broad opposition between intuition and analysis.[7] This is reflected in his distinction between duration and spatialized linear time. However, this opposition also plays out between sensation and perception. For Bergson (2002, p. 270), perception is pragmatic as it is an 'armature' by which we parse and systematize sensation: that is, how we feel duration. Bergson (2002, p. 270) states: 'Reality flows; we flow with it; and we call true any affirmation which, in guiding us through moving reality, gives us a grip upon it and places us under more favourable conditions for acting.' By positing reality as a flow (duration) onto which we impose a framework to guide action, Bergson's philosophy has much in common with his contemporary William James. The philosophical focus of this section turns towards James, as the mechanism by which we parse experience forms a focus of his epistemology – a philosophical principle he calls Pragmatism.

The close parallels between Bergson and James's philosophies are evident when James (1987, p. 1061) describes perception as 'discrete composition' extracted from the flow of experience; thus:

> We either perceive nothing, or something already there in sensible amount. This fact is what in psychology is known as the law of the 'threshold.' Either your experience is of no content, or it is of a perceptible amount of content or change. Your acquaintance with reality grows literally by buds or drops of perception. Intellectually and on reflection you can divide these into components, but as immediately given, they come totally or not at all.

The perceptual flow, as an alternative expression of duration, is a qualitative change of reflected light (that we see as colour) but is decomposed into discrete objects and occurrences. In this way, the flux of experience 'no sooner comes than it tends to fill itself with emphases, and these salient parts become identified and fixed and abstracted' (James, 1987, p. 783). Therefore, we start from a 'field of pure experience' that is a 'quasi-chaos of latent structure' that forms the ground of perception, and perception is a mechanism that decomposes this field along lines of the qualitative properties intrinsic to the field (Heft, 2001, p. 38). In short, James's philosophy deals with the nature of this field of experience and how this field is decomposed into concepts – as radical empiricism and Pragmatism respectively. In this way:

> Without the pragmatic tool of conceptualization, it would be impossible to master the concreteness and fluidity of the world of perception. But unless conceptual schemes are continually referred back to their locus of origin, the dynamism of the life of consciousness may be transformed into closed and static systems.
>
> (Stevens, 1974, p. 22)

James points out that our perceptual system transforms the 'dynamic relations' of visual flux to the static relations of concepts and consequently our 'conceptual scheme' consists of 'discontinuous terms' that only 'cover the perceptual flux in spots and incompletely' (James, 1987, pp. 1022–1023). For him, the way the perceptual system performs this transformation is a 'method' (Pragmatism) and a 'genetic theory of what is meant by truth' (James, 1987, p. 515). Therefore, truth is a consequence of a process that analyses the flux of experience. James (1987, p. 398) describes how true beliefs have 'cash-value' and that we need to consider what 'practical difference would result from one alternative or the other being true'. Truths are, therefore, not eternal but produced by a process that continuously evaluates the practical implications of holding beliefs.

> The truth of an idea is not a stagnant property inherent in it. Truth *happens* to an idea. It *becomes* true, is *made* true by events. Its verity *is* in fact an event, a process: the process namely of its verifying itself, its veri-*fication*. Its validity is the process of its valid-*ation*.
>
> (James, 1987, p. 574; original emphases)

Therefore, Pragmatism's key features are that it: emphasizes that knowledge is rooted in practice; insists 'that the concepts employed in our thinking should be operable, that we should be able to do something with them'; and acknowledges that 'there is a degree of fallibilism inherent in [. . .] the human knower' (Chang, 2012, p. 197).

Regarding perception specifically, truth is a process that constantly verifies the causes of visual sensations. The 'cash-value' of perceptual concepts is how adequate these are at guiding our actions in the world and these malleable concepts are constantly being refined and reverified accordingly. Bayesian

Inference, when applied in a hierarchical predictive processing model, can offer a concrete implementation of James's pragmatic method of perception because such a model is constantly being adjusted to improve predictions and, therefore, accords with James's notion of verity as a process.

A principal problem in visual perception is how to understand a three-dimensional spatial world when light that is reflected off these objects is refracted onto the two-dimensional retinal surface. The crux of the problem is that:

> light that reaches the eye from any scene conflates the contributions of reflectance, illumination and transmittance, as well as a host of subsidiary factors that affect these primary physical parameters. Even more important [. . .] spatial properties such as the size, distance and orientation of physical objects are also conflated in light stimuli.
>
> (Howe and Purves, 2010, p. 1)

This is known as the 'inverse optical problem'. In short, because information from distal objects is conflated on the retina, perceivers have to infer its causes based on knowledge, essentially making well informed guesses as to the causes of visual sensations.[8]

The three shapes in Figure 2.1 have the same projection on the image plane. Not only this but theoretically any projection could be produced by an infinite number of possible shapes. At this stage, in Bergson's and Deleuze's metaphysics, we could consider any shape that causes these to be virtual, in that it is undetermined but determinable. In principle, we cannot recover the three-dimensional shape from the information given solely in the two-dimensional projection. However, in practice, perceivers always have prior knowledge to rely on to resolve this problem. If we know one shape is more likely, then we can predict that it is the probable cause of the projection. This is the case with the retinal projection. Importantly, it is also the case with a projection in the form of a picture. Helmholtz (1885, pp. 281–282) writes:

> The means we possess in this case [of visual perception] are just the same as those the painter can employ in order to give the objects represented on his canvas the appearance of being solid objects, and of standing at different distances from the spectator.

Consequently, because the same problem applies to both retinal and picture projections, how perceivers solve this for lived perception is also informative regarding picture perception. Instead of inferring possible causes of retinal sensation, picture perceivers have to infer the intended object in the picture space. Artists guide this inference by including, as picture cues, the same kinds of cues we use in lived perception to infer the causes of visual sensation.

In practice, perceivers generally have a wealth of prior knowledge to rely on when interpreting a visual scene. Considering the figures with equivalent

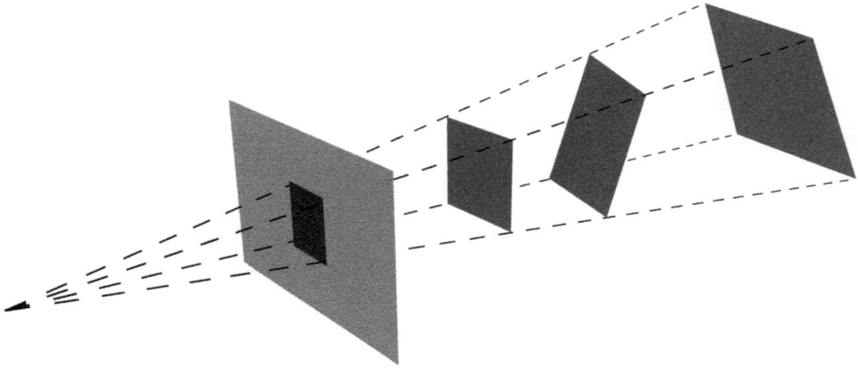

Figure 2.1 These shapes produce the same two-dimensional projection so, when a perceiver only has access to the projection, how do they know the causal shape? In short, they infer the most probable shape based on prior knowledge. © Mark Donoghue

projection (Figure 2.1), suppose perceivers know from experience that one of these is far more common in the environment than the others, it would be natural to assume the projection was likely caused by the most common shape. Bayesian Inference is a statistical technique that allows us to formalize judging the probability of a projection being caused by an object, by considering the information we have from visual sensation and prior knowledge about the environment.

Formally, Bayes's Theorem can be given as:

$$P(H \mid D) = \frac{P(D \mid H) \cdot P(H)}{P(D)}$$

The term $P(H|D)$ refers to the probability of the hypothesis (H) given the data (D) and is called the *posterior probability*. Applied to visual perception, this refers specifically to the probability of an object causing the projection given the visual sensation. $P(D|H)$ refers to the probability of the data (D) given the hypothesis (H) and is called the *likelihood*. In the case of visual perception, it refers to the probability of the projection being caused given a specific object. In the above example, this is equal for all the shapes because they cause the same projection. The term $P(H)$ refers to the probability of hypothesis (H) and is called the *prior probability*. Regarding visual perception it refers to the probability of the hypothesized cause of a visual sensation existing there. (For example, if you see a kangaroo, the probability of there being a kangaroo is much higher if you are in Australia.) In the case of the above example, this refers to our prior knowledge about how probable each hypothesized causal shape is. The term $P(D)$ refers to the probability of the projection and is called the *evidence*. For simplicity, we do not need to concern ourselves with this as it normalizes the posterior probability $P(H|D)$ as probabilities sum to one. In other words, we need only concern ourselves with the relative sizes of posterior probabilities and can omit $P(D)$ to give:

$$P(H \mid D) \propto P(D \mid H) \cdot P(H)$$

Putting this together with the example of equivalent projections above, the projections of each shape look identical, the *likelihood* $P(D|H)$ is the same in each case. Therefore, the highest probable cause of the projection – the *posterior probability* $P(H|D)$ – is solely determined by our previously acquired knowledge of the probability of encountering the shapes: the *prior probability* $P(H)$.[9]

Compare this to an opposite case. Suppose we have three shapes, a cube, cone and sphere, hidden under three cups. The cups are shuffled, so we do not know which shape is under which cup. When one of the cups is lifted to reveal the hidden shape, because it could only be one of the possible three shapes, and each is equally probable, the *prior probability* $P(H)$ for each is the same (one-third). In this case, the *posterior probability* $P(H|D)$ – that tells the perceiver which shape caused the sensation – is determined by the *likelihood* $P(D|H)$. The *likelihood* is the information the perceiver has from the visual scene at hand. Suppose the cup is lifted revealing a sphere and this is projected forming an image (either on the retina or page). The *likelihood* for a circle projection being caused by a sphere is far greater than the other two shapes. The only other possible circle projection could be from the base of the cone, while the sphere from any angle will produce a circle. Therefore, the perceiver will conclude the shape is a sphere.

Occasionally *posterior probabilities* will be equally likely and consequently the perceiver will be confused. This is the case with this bistable corner illusion (Figure 2.2). The cube can be interpreted either as receding from the viewer, or as protruding towards the viewer. On viewing the image, it will appear to flip periodically between the two interpretations. This image cannot be resolved into a single interpretation because the *likelihood* ($P(D|H)$) and *prior*

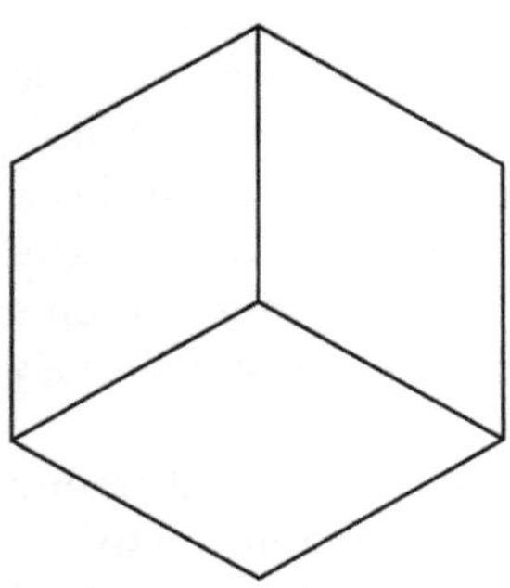

Figure 2.2 Bistable cube illusion where the shape can appear as both concave viewed from above, or convex viewed from below. © Mark Donoghue

probability P(H) are the same for each of the possible interpretations. Therefore, both interpretations are equally probable.

These probabilities are in the form of probability functions and can be visualized in graph form. Viewing probabilities in this way makes it easier to understand Bayesian Inference. In the example of the bistable image (Figure 2.3), the horizontal axis represents possible hypothesis, while the vertical axis represents the probability of these. However, suppose we add lighting to the scene (Figure 2.4). In this case, either it is the corner of a cube advancing towards the viewer with the bottom plane in shadow, or the corner of a cube shape receding from the viewer with the bottom plane in shadow. The *likelihood* of each is again equal because the image is the same. But the *prior probability* for the second condition is now much less probable because, when we take into account prior knowledge about lighting conditions, it is very improbable the bottom plane would be in shadow as lighting generally comes from above. Therefore, as the *posterior* is proportional to the *likelihood* multiplied by the *prior*, the first interpretation comes out to be more probable.

Now, it may seem that Bayesian Inference is at odds with the principles of anticipatory systems. According to Poli (2010, p. 770), in anticipatory systems the '[f]uture states may determine present changes of state'. This is in contrast to classical Newtonian systems where 'the dynamics of the system depends crucially on present and past states of the system. No future information is ever allowed to play any role whatever' (Poli, 2010, p. 770). Consequently, because Bayesian Inference depends on present and past information (*likelihood* and *prior probability*) it seems to comply with the definition of a Newtonian system. However, I contend that any possible implementation of Bayesian Inference can explain how future states affect present systems. I shall, therefore, reconsider Bayesian Inference along the lines of an anticipatory system in the following section.

Hierarchical predictive processing

To consider Bayesian Inference as an anticipatory system requires examination of the origin of prior knowledge in the form of *prior probability* functions. After all if, according to Bayes's theorem *P(H)* are just taken as a given, how do perceivers learn from experience? In other words, '[w]ithout direct input, how does image-independent knowledge of the world get put into the visual system?' (Kersten et al., 2004, p. 285). In brief, because the visual system engages in a process of 'iterative estimation', the origins of prior predictions become less of an issue, as one can begin with an arbitrary value that is progressively adjusted to better predict the input (Clark, 2013, p. 183).

The visual cortex is 'traditionally viewed as a hierarchy of neural feature detectors, with neural population responses being driven by bottom-up stimulus features' (Egner et al., 2010, p. 16,601). This bottom-up stimulus driven approach to visual processing is considered to engage in feature detection

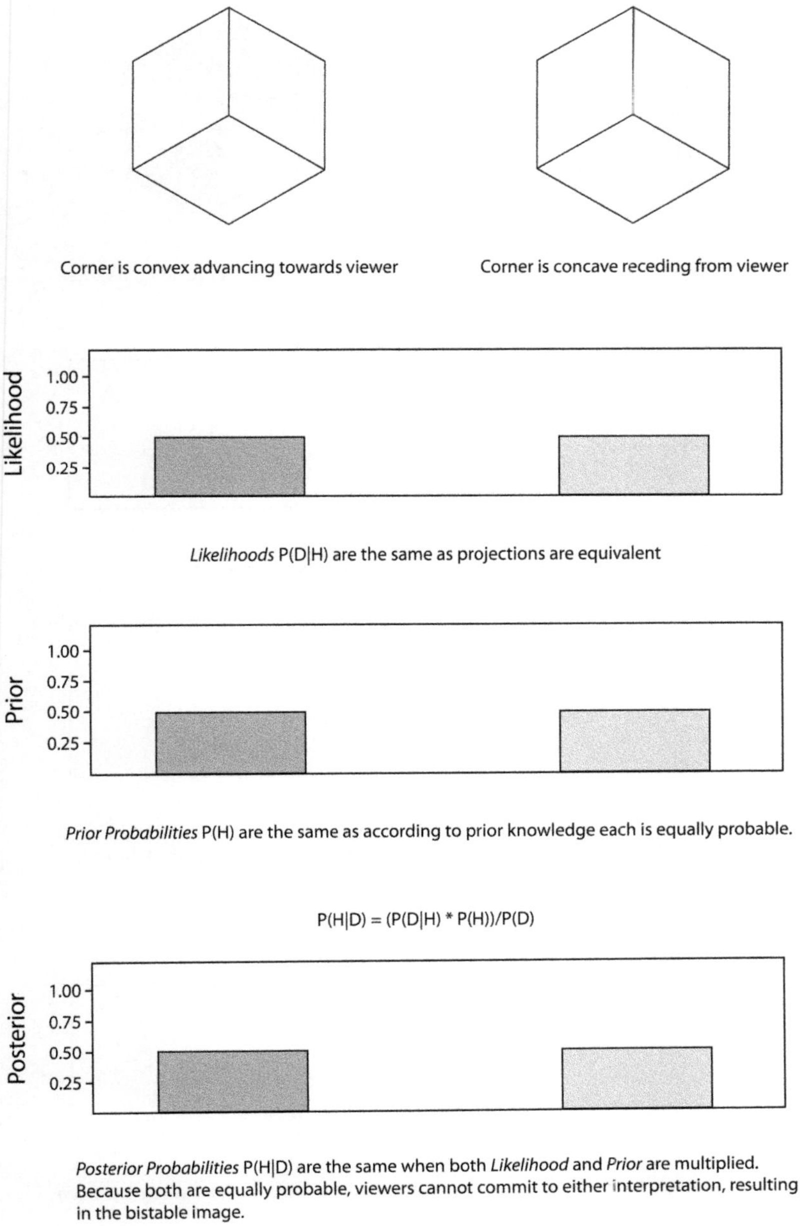

Figure 2.3 Bistable cube illusion illustrated with probability functions. © Mark Donoghue

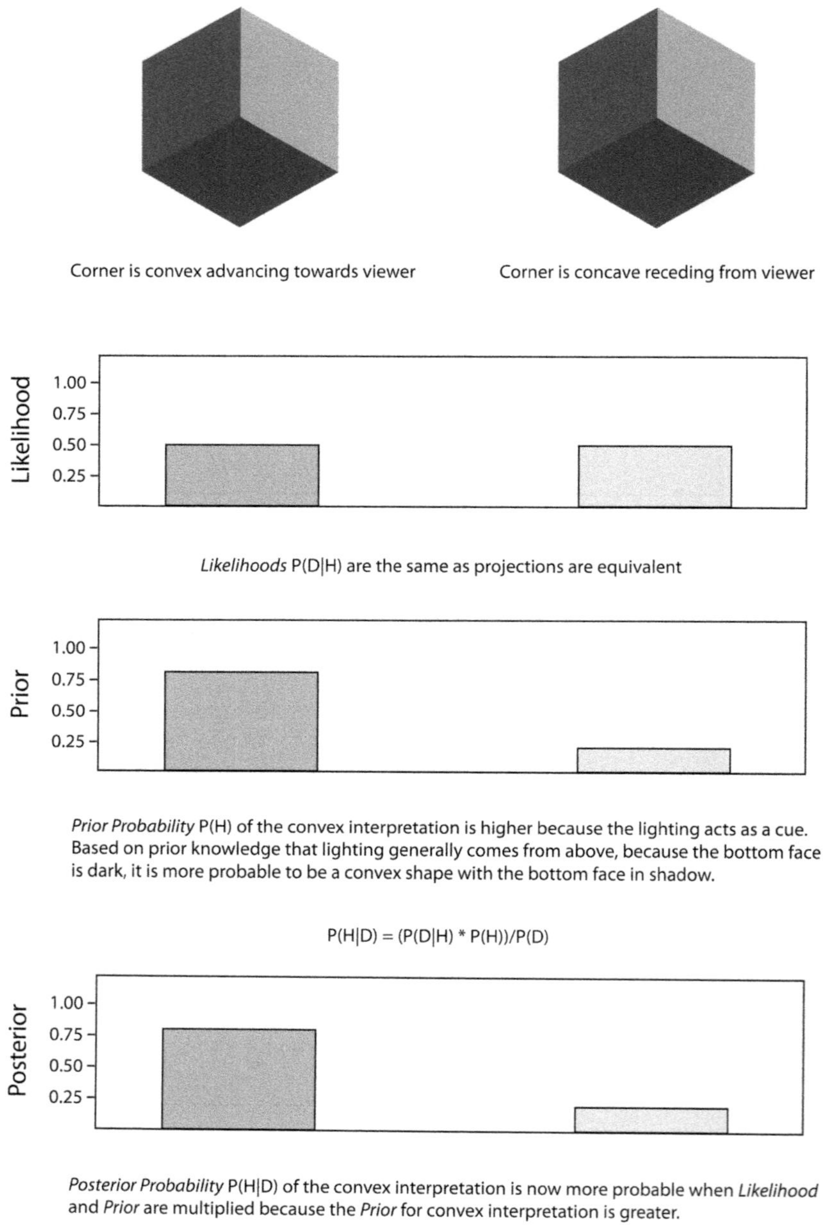

Figure 2.4 Cube image now includes lighting, making the convex interpretation more probable. © Mark Donoghue

where features are extracted and recombined to build up representations, from simple to complex object properties. Where the feature detection approach assumes a process of 'evidence accumulation', the hierarchical predictive process (PP) theory of visual perception inverts this by implementing a hierarchy that involves a 'recurrent cascade of predictions' where each level encodes predictions 'spanning multiple spatial and temporal scales' (Clark, 2014, p. 24; see also: Rao and Ballard, 1999; Lee and Mumford, 2003; Friston, 2005, 2010). Philosopher Andy Clark (2016, p. 21) sums up such a probabilistic theory of vision:

> A good probabilistic generative model for vision would [. . .] seek to capture the ways that lower level visual patterns (ultimately, retinal stimulations) are generated by an inferred interacting web of distal causes. A certain pattern of retinal stimulation, encountered in a given context, might thus be best accounted for using a generative model that (as an admittedly simplistic illustration) combines top-level representations of interacting agents, objects, motives, and motions with multiple intermediate layers capturing the way colours, shapes, textures, and edges combine and temporally evolve. When the combination of such hidden causes (which span many spatial and temporal scales) settles into a coherent whole, the system has self-generated the sensory data using stored knowledge and perceives a meaningful, structured scene.

The PP theory of visual perception essentially implements Bayesian Inference in an iterative procedure where a hypothesis about the cause of a visual sensation is selected and predictions are made about upcoming visual input. This visual input is then compared to the prediction and the difference between the two – the prediction-error – is used to refine future predictions. This process then continues with each iteration theoretically making predictions more accurate.

In principle, the PP framework, rather than detecting features directly from the visual input, develops internal models of the world that can be used to make predictions about upcoming visual sensations. This works through a principle of 'prediction-error minimization', to use the error between model and visual input repeatedly to nudge the model towards greater accuracy. In an ideal case where there is no prediction-error, it can be safely assumed the model completely captures the state of the environment. However, in practice there is always some prediction-error.[10]

In terms of Bayes's Theory, minimizing prediction-error means maximizing the *likelihood* $P(D \mid H)$ because it is the probability of the visual input (D) given a hypothesis (H) about the cause of a visual sensation generated by a cognitive model. In other words, a high probability of the data given the hypothesis means a good fit between model and visual input. Increasing the *likelihood* $P(D \mid H)$ will in turn increase the *posterior probability* $P(H \mid D)$, which will lead to better inferences about the causes of sensations. However, the *posterior probability* $P(H \mid D)$ is not only proportional to the *likelihood* $P(D \mid H)$ but also to the *prior probability* $P(H)$. Where does this *prior probability* $P(H)$ come from? The key feature of this kind of PP framework is that it is a recursive time-extended process in which the *posterior probability* $P(H \mid D)$ at one-step becomes the *prior probability* $P(H)$ in the following step. In short, a *prior* can initially be uninformative about a scene but because at each iteration of the process the *posterior probability* $P(H \mid D)$ becomes the subsequent *prior probability* $P(H)$, this nudges the system towards increasing accuracy as the *priors* gradually become more informative.

This process is implemented in a hierarchy where priors $P(H)$ (which are the probabilities of hypotheses and therefore represent the best guess as to worldly causes of sensation) generated by a model at one level are used to make predictions of the visual input from the preceding level. This is accompanied by prediction-error that is generated by the mismatch between predicted input and actual input being used to adjust the model of the subsequent level to produce more accurate hypotheses. In this way, a 'rich representation of worldly states of affairs is signalled in the top-down predictions of sensory input' and 'prediction-error is propagated up through the system in a bottom-up fashion' (Hohwy, 2013, p. 47).

Each level of the hierarchy is composed of error and representation units (see Figure 2.5). The error units calculate prediction-error by receiving signals from the representation units, which encode the probable causes of input – of the same level and level above – and compare the predicted input with the actual input. The representation units are driven by prediction-error from the same level and level below to update predictions about the input. In this way:

> The generative model therefore maps from causes (e.g., concepts) to consequences (e.g., sensations), while its inversion corresponds to mapping from sensations to concepts or representations. This inversion corresponds to perceptual synthesis, in which the generative model is used to generate predictions.
>
> (Bastos et al., 2012)

The perceptual system is constantly using prior knowledge to predict sensory input to dampen prediction-error by engaging in a two-fold process where predictions are propagated downwards to anticipate input, while input that is not captured by predictions (prediction-error) is propagated upwards to refine models.[11]

Theoretician of anticipatory systems Pierre Rossel (2010, p. 82) draws a distinction between anticipation and prediction, with 'the latter being a representation of a particular future event while the first term refers to a future-oriented action, decision or behaviour based on an implicit or explicit prediction'. My contention is, however, that PP *is* an anticipatory system, in the sense that

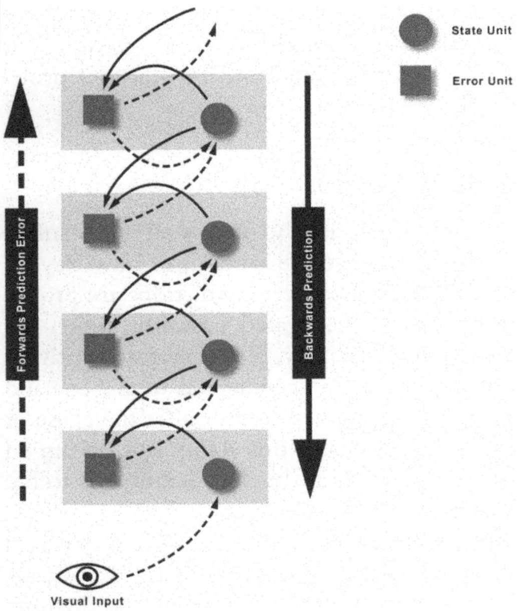

Figure 2.5 Proposed architecture of hierarchical predictive processing. Each level of the hierarchy attempts to predict the signal from the lower level through feedback connections. The difference between the actual signal and expected signal is the prediction-error and is passed back to the higher level via feedforward connections. This prediction-error also amends the prediction model at each level. This process of prediction and model amendment (via prediction-error) occurs concurrently so there is a top-down cascade of prediction and bottom-up propagation of prediction-error. For more detail see: Rao and Ballard (1999); Friston (2009, 2010); Bastos et al. (2012). © Mark Donoghue

it relies on prior knowledge to generate predictions but is ultimately future-oriented, as the system is always open to recalibration by recruiting prediction-error. According to Rosen (2012, p. 314):

> there will be a discrepancy between the behaviors of S [state] and the corresponding behavior predicted by M [model of state], and that this discrepancy will generally grow in time. The character of this discrepancy is, of course, unpredictable from within M, since the basis for it has been abstracted away in the process of the encoding from which M was generated.

In terms of the PP account this discrepancy is prediction-error. The way, according to PP, models are constantly tested and adjusted accordingly fits into James's Pragmatism with the acknowledgement that a perceiver's knowledge is always fallible, and it is by the application of concepts through actions that they are modified. Bergson's philosophy explains why there is always an unpredictable component as the model is an abstraction that changes experience (the qualitative change of duration) into distinct probability distributions.

Conclusion: predictive processing and creativity

According to the PP theory, perception is deeply integrated with inference based on prior knowledge. PP gives a plausible account of how Bayes's Theorem could be implemented in a hierarchy where predictions are propagated downwards, while prediction-error is propagated upwards. Given the essential role inference plays in perception generally, it is therefore relatively unproblematic to draw the connection between perceiving a natural scene and artificial picture perception. In brief, because perceivers can infer the cause of visual sensations based on probability distributions, this ability can be turned to pictures to infer what the picture maker intends the depicted objects to be, according to the picture surface's marks.

Picture perception differs from perception of a natural scene in that it is static. According to PP, perception is produced between a cascade of top-down predictions on the one hand, and a stream of bottom-up prediction-error. However, in the case of picture perception, only top- down prediction occurs because pictures lack the dynamic aspect of actual lived perception. For example, when perceiving a natural scene, if several hypotheses regarding the cause of the visual sensation are probable, then the perceiver can resolve this by interacting with the scene (taking a different viewpoint, etc.) to recruit additional prediction-error to refine predictions and resolve judgement over probable causes. When perceiving a natural scene, prediction-error is constantly adjusting cognitive models to make better predictions, but this is impossible in a static picture because predictions about the pictorial scene are not testable and, therefore, cannot recruit prediction-error to improve predictions. In short, because prediction-error minimization is a time-extended process, a static picture is unable to facilitate this type of error minimization.

Picture perception relies on predictions generated through actual perception to make inferences about a pictorial scene. For example, the three cylinders in Figure 2.6 are exactly the same size. However, if the cylinders are included in a scene with cues to perspective in the form of converging edges, the visual system makes the judgement that the cylinders are receding in space. This has a very tangible effect on perception as it generates an illusion where the cylinders no longer appear the same two-dimensional projection size. In terms of Bayes's Theorem, there are two possible hypotheses regarding the cylinders in the image with depth cues. Hypothesis one (*H*1) is that three identical sized cylinders in the same plane caused three identical sized projections.

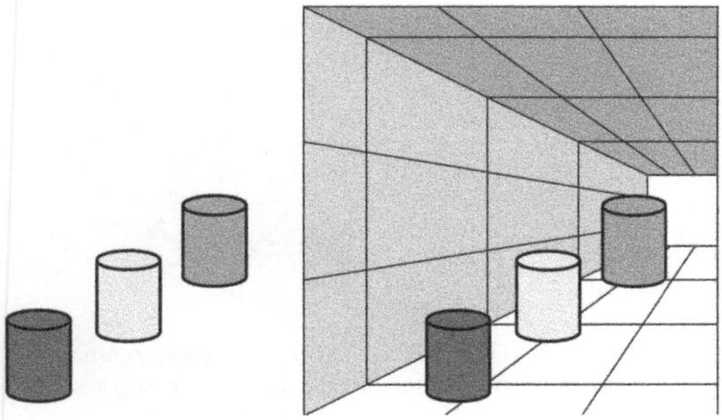

Figure 2.6 Three Cylinders Illusion. © Mark Donoghue

Hypothesis two (*H2*) is that the cylinders are increasing in size but receding in space, therefore maintaining an identical two-dimensional projection size. The visual system can judge between these by implementing Bayes's Theorem:

$$P(H1 \mid D) \propto P(D \mid H1) \cdot P(H1)$$

$$P(H2 \mid D) \propto P(D \mid H2) \cdot P(H2)$$

The *likelihood* for hypothesis one and hypothesis two is equal – $P(D \mid H1) = P(D \mid H2)$ – because the data (the image with the cylinders of equal projection size) could be caused by either. However, the *prior probability* for hypothesis two is greater than that for hypothesis one – $P(H2) > P(H1)$. This is because the perspective cues come into play and the perceiver, according to prior experience, assumes the cylinders are sitting on the ground plane receding into the picture space. Technically, the projection could be caused by three identically sized cylinders in the same plane but that would assume the cylinders were floating in mid-air and such an interpretation would go against prior experience and the pictorial cues. Overall, the *posterior probability* of hypothesis two (projections are caused by different sized cylinders receding in space) wins out over hypothesis one (projections are caused by three identically sized cylinders in the same plane) – $P(H2 \mid D) > P(H1 \mid D)$. Moreover, once the perceptual system is committed to the hypothesis, this concretely affects our visual perception causing the cylinder projections to appear to be different sizes (despite being the same two-dimensional size). In the case of the left image of the cylinders on the white background without any perspective cues, it does not force the perceptual system to commit to the differing sized cylinders hypothesis so we can more accurately judge the projection size. However, when this is juxtaposed with the right image that includes the perspective cues it is apparent how top-down knowledge can distort bottom-up visual input (see Figure 2.7).

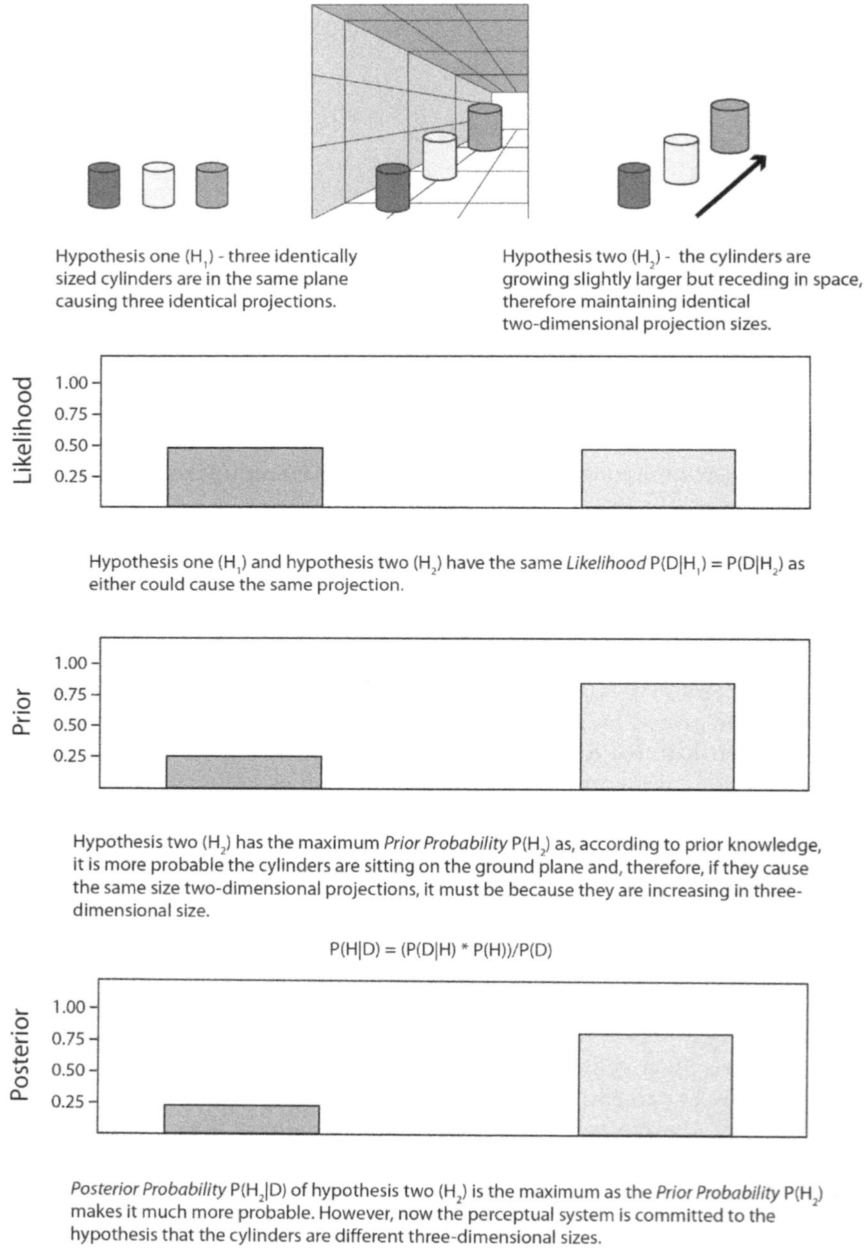

Figure 2.7 Probability functions illustration for the three cylinders illusion. © Mark Donoghue

The three cylinders illusion demonstrates the kinds of judgements perceivers are constantly making when viewing pictures and highlights how perceivers are predisposed to apply prior knowledge (in the form of probability distributions) to pictorial forms. After all, a picture is not caused by the depicted objects, in the same way actual objects cause projections on the retina, but perceivers still judge pictorial forms according to the kinds of probable causes generated through PP. Where picture perception differs from actual perception is that it cannot recruit prediction-error to refine predictions as pictures are static and non-dynamic.[12] Therefore, pictures rely on models generated in actual perception to facilitate inferences about depicted objects based on pictorial forms. Without any prior knowledge, it would be impossible to see a depicted object as infinitely many three-dimensional forms could cause a single projection. All perceivers can do is make judgements about the most probable depicted object in the pictorial scene based on established probability distributions.

Recall that, for Bergson and James, we segment the dynamism of duration or experience and this fundamentally changes its nature so that, even after these segments have been reassembled, we cannot recover this essential dynamism. Throughout Bergson's oeuvre, he argues the pragmatic processes of perception work to regulate the flux of duration and, although this is an effective method for guiding actions, this generates a predilection in Western thought to consider things in terms of space as discrete units. This is echoed by James who describes a process of 'vicious abstractionism' where:

> We conceive a concrete situation by singling out some salient or important feature in it, and classing it under that; then, instead of adding to its previous characters all the positive consequences which the new way of conceiving it may bring, we proceed to use our concept privatively; reducing the originally rich phenomenon to the naked suggestions of that name abstractly taken, treating it as a case of 'nothing but' that concept, and acting as if all the other characters from out of which the concept is abstracted were expunged.
>
> (James, 1987, p. 951)

It is possible to argue the PP account of perception falls into a similar trap with concepts being defined as models that encode probability distributions. There are interpretations of PP that conceive of perception as strictly indirect which falls into James's 'vicious abstractionism'. According to Frith (2007, p. 135; see also Hohwy, 2007):

> Our brains build models of the world and continuously modify these models on the basis of the signals that reach our senses. So, what we actually perceive are our brain's models of the world. They are not the world itself, but, for us, they are as good as. You could say that our perceptions are fantasies that coincide with reality.

Clark (2016, p. 170) critiques these accounts of the PP framework that claim perception is best thought of as 'a kind of hypothesis, model, fantasy, or virtual reality' and suggests this introduces 'a kind of representational veil between agent and world'. Instead, it is the 'structured probabilistic know-how' that enables perception of a world of distal causes beyond the 'veil of surface statistics' through a strategy of 'prediction-driven learning' (Clark, 2016, p. 170). In addition, such an account fails 'to take sufficient account of the role of action [. . .] in both selecting and constantly testing the ongoing stream of prediction itself' (Clark, 2016, p. 171).[13]

Therefore, Clark (2016, p. 171) concludes that perceivers are not stuck behind a veil of probabilistic inference but, rather: 'prediction-driven learning delivers a grip upon affordances: the possibilities for action and intervention that the environment makes available to a given agent'. Such a claim can be enhanced by referring to Bergson and James's ontologies, however, as there is always an element of novelty that escapes conceptualization. For example, Bergson (2017) describes how 'philosophy has never frankly admitted this continual creation of unforeseeable newness' which is reiterated by James (1987, p. 1057) who states '[t]he same returns not, save to bring the different'. For both, this element of essential novelty always escapes conceptualization because this process changes the nature of the experiential flow.

If experience does inevitably involve this element of novelty, in terms of the PP account, we can conclude there is a limit to how developed the cognitive models' predictions can become as there is always prediction-error to be minimized to adapt the models. Consequently, this process can never reach an ultimate conclusion where visual input is fully predicted. At this point, as the relationship between Bergson and James's philosophies and the PP account of perception has been explored, it is possible to approach the issue of creativity.

We generally think of creativity as a faculty that enables novel thinking. However, the PP account paints cognition as dependent on prior knowledge. According to psychologist Christopher Frith (2007, p. 137):

> The imagination is utterly uncreative. It has no predictions to make and no errors to resolve. We don't create in our heads. We create by externalizing our thoughts with sketches and doodles and rough drafts so that we can benefit from the unexpectedness of reality. It is this continual unexpectedness that makes interacting with the real world such a joy.

Imagination in this description is an application of top-down cognitive models which are divorced from actual perception and therefore lack bottom-up prediction-error. We cannot imagine something only to discover, through introspection, that our imagining is inaccurate – there is nothing to which we can compare the cognitive model. In short, imagination cannot create something truly novel because that is the very quality that escapes conceptualization

(Hinton, 2007; Clark, 2014). Therefore, the genuinely creative relies on the essential novelty of the world.

As was pointed out above, picture perception lacks the dynamism of actual perception and therefore, perhaps, pictures also lack creativity as they are derivative of pre-established probability distributions. This issue can be addressed, however, by referring to Bergson's philosophy and its relevance to art. This focuses on the dichotomy between intuition and analysis. Intuition involves the encounter with duration (or experience in James's terms), while analysis involves the conceptualization (and inevitable spatialization) of duration. Bergson (1946, p. 189) writes:

> an absolute can only be given in *an intuition*, while all the rest has to do with *analysis*. We call intuition here the sympathy by which one is transported into the interior of an object in order to coincide with what there is unique and consequently inexpressible in it. Analysis, on the contrary, is the operation which reduces the object to elements already known, that is, common to that object and to others. Analyzing then consists in expressing a thing in terms of what is not it. (Original emphases.)

For Bergson, there is a relationship between the two, but this is strictly in the direction of intuition to analysis and, once analysed, the intuitive sympathy cannot be recovered because, as his critique of the spatialization of time makes clear, this process transforms qualitative into quantitative change. One could conclude picture making involves a kind of geometric analysis that, likewise, sacrifices the intuitive sympathy. This seems to be Bergson's position when he uses an analogy of making a sequence of sketches of Paris to highlight the distinction between intuition and analysis. He writes:

> Now at the bottom of all the sketches made in Paris the stranger will probably write 'Paris' by way of reminder. And as he has really seen Paris, he will be able, by descending from the original intuition of the whole, to place his sketches in it and thus arrange them in relation to one another. But there is no way of performing the opposite operation; even with an infinity of sketches as exact as you like, even with the word 'Paris' to indicate that they must bear close connection, it is impossible to travel back to an intuition one has not had, and gain the impression of Paris if one has never seen Paris.
>
> (Bergson, 1946, p. 200)

In short, the intuitive encounter with Paris cannot be recovered, regardless of how many sketches are produced because the act of sketching involves an analysis that transforms the experience. It is possible to consider this in terms of the PP account, as the projections of the Paris environment are reliant on probability distributions generated through prediction-error processes conditioned in actual perception.

On the one hand, art is often treated as the expression of higher order principles, with the quality of a work determined by how well they are expressed. Thus, '[p]rinciples of [. . .] order are logically prior to their instances and independent of them. Therefore, they allow for prediction: the same principles are expected to govern further events' (Lorand, 1999, p. 404). However, if this were the case, it would fall into Bergson's critique of predetermined possibilities and relegate works of art to the position of merely realizing pre-existing immaterial forms. On the other, we often consider novelty as essential to an artwork's originality. In short, there is a tension in artistic value between the expression of higher artistic principles that, being logically prior, are in principle predictable, and the unpredictable novelty of an artwork. However, if intuition cannot be recovered from analysis, how can an artwork that is the realization of higher artistic principles (products of analysis) be novel? For it is intuition that enables an encounter with duration and duration itself that guarantees novelty as qualitative change. In terms of a picture, perceivers encounter a projection, a realization of a geometric order, which suggests an analytical process that dismisses an intuitive sympathy with the depicted object.

Therefore, in Bergson's account, 'the intellect can exercise via analysis its power over the products of intuition, but intuition is unable to reciprocate – it cannot construct a whole out of conceptual components' (Lorand, 1999, p. 408). Overall, in Bergson's work there is an attitude that analysis corrupts the intuition of duration and once fractured cannot be recovered. As he writes: 'From intuition one can pass on to analysis, but not from analysis to intuition' (Bergson, 1946, p. 212). However, philosopher Ruth Lorand (1999, p. 409) argues this is not the case and Bergson's account needs to be adapted to accept 'positive interactions of both sides', so that 'the two orders are interdependent and that no order is prior to or independent of the other'. The justification for this move comes from the commonplace understanding of art as generating an intuitive sympathy by incorporating novelty. For example, a film is not merely motion broken into a sequence of static frames but genuinely conveys change as duration (Lorand, 1999, pp. 408–409).[14] In terms of two-dimensional pictures – returning to Bergson's example of Paris sketches – this means that perceivers can, in fact, recover an intuitive grasp of Paris from sketches that are a product of analysis. In this way, creativity is injected into picture making by introducing an intuitive sympathy; which all seems contrary to Bergson's own arguments.

Exactly how an artwork generates an intuitive sympathy cannot be reduced to a formula or rule, as this would merely be the application of an analysis. Given how intuition is pointed in the direction of duration and therefore entails genuine quantitative change, it is impossible to provide a general principle for how art enables intuition. Therefore, one can only examine the method of intuition as it is applied in an artwork on a case-by-case basis. One could argue the skill of art criticism exemplifies this examination.

There is not space here to examine thoroughly the issue of art criticism. So, I will use the familiar example of Van Gogh's (1888) *Le Café de nuit* (The Night Café) to illustrate this intuitive method (see also: Ward, 1976; Harris, 1999). Van

Gogh's painting goes beyond an analysis of the scene as it (the painting) cannot be reduced to a simple geometric projection. Although it uses geometric elements to convey the interior space, it does not strictly adhere to the axioms of projective geometry. In this way, Van Gogh is attempting to go beyond the analytical visual perception of the scene to recover an intuitive visual sensation. The exaggerated perspective seems to propel the viewer into the space of the painting. The same could be said about the colour as it is not a simple reduplication of luminance values but tries to convey the feeling of a warm and bright café interior at night.

I suggest there is a distinction between picture making to communicate information (visual communication) and pictures as artworks. Where pictures are constructed to communicate information, they can simply act as prompts to perceptual inference and therefore engage solely in analysis. In contrast, where pictures are also artworks, they still act as prompts to perceptual inference but in a way that opens perceptual models to intuition and the possibility of creative production. This conclusion may seem at odds with the PP account of perception, especially given the important role probabilistic inference plays in it. After all, how can a picture generate a genuinely novel experience if it is divorced from the prediction-error of actual perception? However, psychologists Schwartenbeck, FitzGerald, Dolan and Friston (2013) argue that in order to minimize prediction-error in the long-term, perceivers may acquire perceptual policies that value new experiences in order to recruit short-term prediction-error to improve cognitive models. Therefore, prediction-error minimization 'by no means precludes agents from active exploration or appreciating novelty but rather explicitly predicts that this is an important factor in guiding our behavior' (Schwartenbeck et al., 2013, p. 4). Clark (2016, p. 267), developing this insight, claims art might profit from such novelty-seeking policies and artists may design such works to 'support and encourage [. . .] open-ended forms of exploration and novelty-seeking'. This novelty-seeking activity loosens the grasp of predictive models, as a model that is too prescriptive will ultimately prove inaccurate in the long run. Bergson and James's philosophies spell out this as an intuitive sympathy that discloses the nature of duration or experience as the genuine novelty that underpins our perceptual and conceptual faculties.

After this exploration of Bergson and James's philosophies and how they relate to the PP account of perception it is possible to consider the role creativity plays in picture making. On the one hand, creativity, as the essential novelty of duration or experience, conditions predictive models by making prediction-error possible. It is these models that enable picture perception by providing the prior knowledge needed to make inferences about depicted objects from pictorial cues. On the other hand, creativity works in pictorial artworks as a method for novelty-seeking. This novelty-seeking takes advantage of perceptual policies that value new experiences as the source of improved predictions. Artworks provide a way to engage in novelty-seeking in a simulated environment but, given how novelty always escapes the concepts that it conditions, the ways in which artworks perform this depends on the artwork in question, because a general principle for this cannot be created.

Notes

1 The importance of the role that inference plays in perception can be traced back to the origins of psychology in the work of Hermann von Helmholtz (1821–1894). See: Westheimer (2008).
2 There are other mathematical techniques to describe anticipation. For example, Rosen's (2012) work refers extensively to category theory. Regarding models, see: Haraway (2016, p. 63).
3 James and Bergson recognized the consonance in their respective philosophies in their own lifetimes. See: James (1987, pp. 731–781) and Bergson (2002).
4 The French polymath Pierre-Simon Laplace (1749–1827), developing an idea from Gottfried Leibniz (1646–1716), postulated a being of super intelligence that, if they knew the position, velocity and force on every particle in the universe at one time, they could know this at every time through extrapolation. This is known as Laplace's Demon. See: Green (1995).
5 On the concept of multiplicity, see Deleuze's (1994) *Difference and Repetition*.
6 Bergson (2001, p. 181) writes: 'If I glance over a road marked on the map and follow it up to a certain point, there is nothing to prevent my turning back and trying to find out whether it branches off anywhere. But time is not a line along which one can pass again.' See also: Alfred North Whitehead's *Process and Reality* (1978) and *The Concept of Nature* (1964) for a complementary treatment of time.
7 Guerlac (2006, p. 109) describes this in terms of 'presence' and 'representation'.
8 This inferential approach in perception has its roots in the origin of psychology with Helmholtz (2000) and went on to influence the works of Mackay (1956), Neisser (1967, 1978) and Gregory (1970, 1980), becoming known as the 'analysis-by-synthesis' psychological tradition (for review see Yuille and Kersten, 2006).
9 For summaries of this, see: Frisby and Stone (2010, pp. 307–323); Stone (2012, pp. 155–178); and Hohwy (2013, pp. 32–38).
10 Predictive processing began as a method of image compression to improve computing bandwidth; see Shi and Sun (2019) for more detail.
11 For a more detailed treatment see: Rao and Ballard (1999); Friston (2005, 2010, 2012); and Bastos et al. (2012).
12 Although cinema is not static it is non-dynamic and the constraints of a static picture still applies. In lived perception we can generate hypotheses about the causes of sensations and test these, recruiting prediction-error to improve perceptual models, but despite the pictures moving in cinema, because the perceiver has no control over the camera, they cannot test perceptual models and therefore are still dependent on models generated in lived perception to perceive cinema pictures. In contrast, computer games are a kind of moving pictures that is dynamic so perhaps it is possible to generate perceptual models of a world simulated in a computer. However, most games are designed with the perceptual and physical constraints of the actual world in mind, so it is easy for the player to adapt to the game world. In this way players apply the models generated through lived perception to the virtual game world.
13 For more on this, see: Clark (1997, 2008); and Clark and Chalmers's (1998) previous work on 'enactive perception'.
14 For more on this, see: Bergson (1946); and Deleuze's (1986, 1989) work on cinema.

References

Ansell-Pearson, K. (2002) *Philosophy and the Adventure of the Virtual: Bergson and the Time of Life*. London and New York, NY: Routledge.

Bastos, A.M., Usrey, W.M., Adams, R.A., Mangun, G.R., Fries, P. and Friston, K.J. (2012) 'Canonical Microcircuits for Predictive Coding', *Neuron*, 76(4), pp. 695–711.

Bergson, H. (2017) *The Possible and the Real*. Available at: http://bergsonian.org/the-possible-and-the-real/ (Accessed: 15 December 2019).

Bergson, H. (2002) 'On the Pragmatism of William James: Truth and Reality', in Ansell-Pearson, K. and Mullarkey, J. (eds.), *Henri Bergson: Key Writings*. London and New York, NY: Continuum, pp. 267–273.

Bergson, H. (2001) *Time and Free Will: An Essay on the Immediate Data of Consciousness*. Trans. F. Pogson. New York, NY: Dover Publications.

Bergson, H. (1991) *Matter and Memory*. Trans. N.M. Paul and W.S. Palmer. New York, NY: Zone Books.

Bergson, H. (1946) *The Creative Mind*. Trans. M.L. Andison. New York, NY: Philosophical Library.

Bergson, H. (1944) *Creative Evolution*. Trans. A. Mitchell. New York, NY: Modern Library.

Chang, H. (2012) *Is Water H2O?: Evidence, Realism and Pluralism*. New York, NY: Springer.

Clark, A. (2016) *Surfing Uncertainty: Prediction, Action, and the Embodied Mind*. Oxford: Oxford University Press.

Clark, A. (2014) 'Perceiving as Predicting', in Stokes, D., Matthen, M., and Biggs, S. (eds.), *Perception and its Modalities*. Oxford: Oxford University Press, pp. 23–43.

Clark, A. (2013) 'Whatever Next? Predictive Brains, Situated Agents, and the Future of Cognitive Science', *The Behavioral and Brain Sciences*, 36(3), pp. 181–204.

Clark, A. (2008) *Supersizing the Mind: Embodiment, Action, and Cognitive Extension*. Oxford: Oxford University Press.

Clark, A. (1997) *Being There: Putting Brain, Body, and World Together Again*. Cambridge, MA: MIT Press.

Clark, A. and Chalmers, D.J. (1998) 'The Extended Mind', *Analysis*, 58(1), pp. 7–19.

DeLanda, M. (2002) *Intensive Science and Virtual Philosophy*. London and New York, NY: Continuum.

Deleuze, G. (1994) *Difference and Repetition*. Trans. P. Patton. New York, NY: Columbia University Press.

Deleuze, G. (1990) *The Logic of Sense*. Trans. M. Lester with C. Stivale. Ed. C.V. Boundas. London: The Athlone Press.

Deleuze, G. (1989) *Cinema 2: The Time*-Image. Trans. H. Tomlinson and R. Galeta. London: The Athlone Press.

Deleuze, G. (1988) Bergsonism. Trans. H. Tomlinson and B. Habberjam. New York, NY: Zone Books.

Deleuze, G. (1986) *Cinema 1: The Movement Image*. Trans. H. Tomlinson and B. Habberjam. London: The Athlone Press.

Egner, T., Monti, J.M. and Summerfield, C. (2010) 'Expectation and Surprise Determine Neural Population Responses in the Ventral Visual Stream', *Journal of Neuroscience*, 30(49), pp. 16601–16608.

Frisby, J.P. and Stone, J.V. (2010) *Seeing: The Computational Approach to Biological Vision*. Cambridge, MA and London: The MIT Press.

Friston, K. (2012) 'Prediction, Perception and Agency', *International Journal of Psychophysiology*, 83(2), pp. 248–252.

Friston, K. (2010) 'The Free-Energy Principle: A Unified Brain Theory?', *Nature Reviews Neuroscience*, 11(2), pp. 127–138.

Friston, K. (2009) 'The Free-Energy Principle: A Rough Guide to the Brain?', *Trends in Cognitive Sciences*, 13(7), pp. 293–301.

Friston, K. (2005) 'A Theory of Cortical Responses', *Philosophical Transactions of the Royal Society of London. Series B: Biological Sciences*, 360(1456), pp. 815–836.

Frith, C. (2007) *How the Brain Creates Our Mental World*. Chichester: Wiley-Blackwell.

Fuller, T. (2017) 'Anxious Relationships: The Unmarked Futures for Post-Normal Scenarios in Anticipatory Systems', *Technological Forecasting and Social Change*, 124, pp. 41–50.

Green, R. (1995). *The Thwarting of Laplace's Demon*. London: Palgrave Macmillan UK.

Gregory, R.L. (1980) 'Perceptions as Hypotheses', *Philosophical Transactions of the Royal Society of London. Series B, Biological Sciences*, 290(1038), pp. 181–197.

Gregory, R.L. (1970) *The Intelligent Eye*. London: Weidenfeld & Nicolson.

Guerlac, S. (2006) *Thinking in Time: An Introduction to Henri Bergson*. Ithaca, NY: Cornell University Press.

Haraway, D.J. (2016) *Staying with the Trouble. Making Kin in the Chthulucene*. Durham, NC and London: Duke University Press.

Harris, N. (1999) *The Masterworks of Van Gogh*. Surrey: Colour Library Direct.

Heft, H. (2001) *Ecological Psychology in Context: James Gibson, Roger Barker, and the Legacy of William James's Radical Empiricism*. Hillsdale, NJ: L. Erlbaum Associates.

Helmholtz, H.L.F. von (2000) *Helmholtz's Treatise on Physiological Optics. The Perceptions of Vision*, vol. 3. Trans. J.P.C. Southall. Bristol: Thoemmes.

Helmholtz, H.L.F. von (1885) *Popular Lectures on Scientific Subjects*. Trans. E. Atkinson. New York, NY: D. Appleton and Company.

Hinton, G.E. (2007) 'Learning Multiple Layers of Representation', *Trends in Cognitive Sciences*, 11(10), pp. 428–434.

Hohwy, J. (2013) *The Predictive Mind*. Oxford: Oxford University Press.

Hohwy, J. (2007) 'Functional Integration and the Mind', *Synthese*, 159(3), pp. 315–328.

Howe, C.Q. and Purves, D. (2010) *Perceiving Geometry: Geometrical Illusions Explained by Natural Scene Statistics*. New York, NY: Springer.

James, W. (1987) *Writings 1902–1910*. New York, NY: Literary Classics of the United States.

Jancsary, J. (2019) 'The Future as an Undefined and Open Time: A Bergsonian Approach', *Axiomathes*, 29(1), pp. 61–80.

Kersten, D., Mamassian, P. and Yuille, A. (2004) 'Object Perception as Bayesian Inference', *Annual Review of Psychology*, 55(1), pp. 271–304.

Knill, D.C. and Pouget, A. (2004) 'The Bayesian Brain: The Role of Uncertainty in Neural Coding and Computation', *Trends in Neurosciences*, 27(12), pp. 712–719.

Lee, T.S. and Mumford, D. (2003) 'Hierarchical Bayesian Inference in the Visual Cortex', *Journal of the Optical Society of America. A: Optics, Image Science, and Vision*, 20(7), pp. 1434–1448.

Lorand, R. (1999) 'Bergson's Concept of Art', *The British Journal of Aesthetics*, 39(4), pp. 400–415.

Mackay, D. (1956) 'The Epistemological Problem for Automata', in Shannon, E. and McCarthy, J. (eds.), *Automata Studies*. Princeton, NJ: Princeton University Press, pp. 235–251.

Mullarkey, J. (1999) *Bergson and Philosophy*. Edinburgh: Edinburgh University Press.

Neisser, U. (1967) *Cognitive Psychology*. New York, NY: Appleton-Century-Crofts.

Neisser, U. (1978) *Cognition and Reality: Principles and Implications of Cognitive Psychology*. San Francisco, CA: W.H. Freeman and Company.

Poli, R. (2010) 'An Introduction to the Ontology of Anticipation', *Futures*, 42(7), pp. 769–776.

Rao, R.P.N. and Ballard, D.H. (1999) 'Predictive Coding in the Visual Cortex: A Functional Interpretation of Some Extra-Classical Receptive-field Effects', *Nature Neuroscience*, 2(1), pp. 79–87.

Rosen, R. (2012) *Anticipatory Systems: Philosophical, Mathematical, and Methodological Foundations*. 2nd edn. New York, NY: Springer.

Rossel, P. (2010) 'Making Anticipatory Systems More Robust', *Foresight*, 12(3), pp. 72–85.

Schwartenbeck, P., FitzGerald, T., Dolan, R.J. and Friston, K. (2013) 'Exploration, Novelty, Surprise, and Free Energy Minimization', *Frontiers in Psychology*, 4(710). Available at: https://www.ncbi.nlm.nih.gov/pmc/articles/PMC3791848/ (Accessed: 28 January 2020).

Seriès, P. and Seitz, A.R. (2013) 'Learning What to Expect (in Visual Perception)', *Frontiers in Human Neuroscience*, 7, pp. 1–14.

Shi, Y.Q. and Sun, H. (2019) *Image and Video Compression for Multimedia Engineering: Fundamentals, Algorithms, and Standards*. Boca Raton, FL: CRC Press.

Stevens, R. (1974) *James and Husserl: The Foundations of Meaning*. The Hague: Martinus Nijhoff.

Stone, J.V. (2012) *Vision and Brain: How We Perceive the World*. Cambridge, MA and London: The MIT Press.

Van Gogh, V. (1888) *Le Café de nuit* (The Night Café) [Oil on Canvas]. New Haven, CT: Yale University Art Gallery. Available at: https://artgallery.yale.edu/collections/objects/12507 (Accessed: 4 August 2020).

Ward, J.L. (1976) 'A Reexamination of Van Gogh's Pictorial Space', *The Art Bulletin*, 58(4), pp. 593–604.

Westheimer, G. (2008) 'Was Helmholtz a Bayesian?', *Perception*, 37(5), pp. 642–650.

Whitehead, A.N. (1964) *The Concept of Nature*. Cambridge: Cambridge University Press.

Whitehead, A.N. (1978) *Process and Reality: An Essay in Cosmology*. Eds. D.R. Griffin and D.W. Sherburne. New York, NY: Free Press.

Yuille, A. and Kersten, D. (2006) 'Vision as Bayesian Inference: Analysis by Synthesis?', *Trends in Cognitive Sciences*, 10(7), pp. 301–308.

3 Flowing or frozen anticipation?

Runes and the creativity of time

Anne Marchais-Roubelat

Introduction

As creativity relates to every time we try to conceive, produce or react to something, it may be studied as an individual or social activity happening within time, unless it is a characteristic of time itself. While the assessment of the usefulness of creativity over time may not be a question in the general context of futures studies it should be considered when these are linked to implementation, as is the case for anticipation (Fuller, 2017). Implementation implies organization, as 'the attempt to order the intrinsic flux of human action, to channel it towards certain ends by generalizing and institutionalizing particular cognitive representations [. . .] [and as] a pattern that is constituted, shaped, and emerging from change' (Tsoukas and Chia, 2002, p. 567). Questioning the nature of time underlying the conception of organizational creativity – and so innovation, then change – challenges its relation to anticipation. In the organizational field, creativity is justified by its contribution to innovation and, more largely, change (Woodman et al., 1993), but the standard conception of time in innovation management does not take into account the move of change (Chia, 1999). From the contrast between standard time, as frozen, and flowing time, the purpose of this chapter is to suggest a way for anticipation to think and use creativity as a characteristic of flowing time.

From this definition of creativity, the first part focuses on organizational creativity and its assessment over time in relation with organizational change. It stresses that research on organizational creativity adopts the standard conception of time in the research field of organizational change (Chia, 1999; Tsoukas and Chia, 2002) to which it belongs, and that has 'little attempt to understand the nature of change *on its own terms*' (Chia, 1999, p. 210; original emphasis). By contrast, anticipation questions change itself as a movement, insofar as anticipation is linked to implementation (Fuller, 2017). This leads us to look for a temporality that could allow us to think movement differently and for which the Heideggerian comparison between the 'flowing' time of History and the frozen one of natural sciences acts as a guide. This will be examined in the second part of this chapter.

Even if, for Heidegger (2007, p. 68), the goal of history is 'to present the ensemble of the effectivity and development of objectifications of human life in all their singularity and uniqueness as they are understandable in their relation to cultural values', the *flowing time* he highlights in history cannot be directly transposed to futures studies, as 'the historical object, as historical, is always past: in the strict sense it no longer exists' (Heidegger, 2007, p. 68). In order to conceive a *flowing time* in futures studies, and more specially in anticipation, one has to look for inspiration elsewhere, in space or in time. This supposes a heterogeneous, fluctuating matter, which makes it possible to imagine alternative relationships over time, on which scholars are forced to speculate, no interpretation being sure nor fixed because it evolves over time and accumulated knowledge, the same as for the future – but which is also capable of being analysed according to a Western scientific thought. This is the case of ancient runes, as both writing and magical practices. Their number, order and shape changed over time and space, they are studied by the disciplines of linguistics and archaeology, and their mysterious origins and uses constitute an inexhaustible source of speculation.

The third part will imagine what could be a *flowing time* in futures studies emerging from the runes and their study, as well as from a trace of their complex relation to time as found in literature, especially in the Poetic Edda of the early Nordic settlers of Iceland, which compiles – using poetry and magic – the ancient Scandinavian mythology. The complexity of the subject matter here means I have broken this part into smaller sub-sections, in order to fully explicate this conception of flowing time. I will also explore the assessment of creativity, and the role people play in future-oriented flowing time, as well as the challenges the world of runes launches in designing flowing time in futures studies and anticipation.

This discussion decentralizes the action from those who participate, in order to highlight the transformation the use of runes brings to ways of thinking, practising and being on which they rely. Such a 'lay' conception of time, so to speak, will be shown to be useful for future research, opening new horizons for creativity and anticipation.

Creativity, organisation and the paradox of anticipation

Creativity is difficult to define and yet very rich, as evidenced by the multiple variants of its definition and the research it generates. Beyond these debates, organizational creativity appears to be critical for futures studies that imagines tomorrow in a useful way for today; that is, useful for anticipation. Within futures studies however, the directly utilitarian character of anticipation questions the relation of creativity to time.

This section will first discuss the relation of the scientific definition of creativity to time and its consequences on the way it is used in organization studies as well as in futures studies, including the field of anticipation. As anticipation is directly linked to implementation (Fuller, 2017), then to future action, such a use of creativity raises a temporal paradox.

The role of 'when' in assessing creativity: an open debate

Although the definition of creativity is 'still hotly debated, [. . .] standard definitions point to originality and effectiveness requirements' (Runco and Beghetto, 2019, p. 7), so that creativity is commonly defined as 'the ability to produce novel and useful responses' (Zabelina et al., 2019, p. 47; who make reference to Runco and Jaeger's (2012) 'standard definition of creativity'). Even if there is no consensus, definitions are commonly based on the conjugation of an ability to produce – or the product implemented by this ability – and of the assessment of this product in terms of novelty as well as usefulness. Creativity may also produce immaterial objects as behaviours or ideas, when for example 'defined as the generation of new ideas as well as the assessment of those ideas to be included in a domain' (Simon et al., 2018, p. 444). Whatever the definition, creativity's usefulness has to be assessed; and this assessment still raises questions and debates.

One of the main debates is about 'for whom', as well as 'where *and* when something is creative' (Runco and Beghetto, 2019, p. 7; original emphasis). The question of 'for whom' has been considered in a proposition of consensual definition of creativity proposed by Teresa Amabile (1982, p. 1001):

> A product or response is creative to the extent that appropriate observers independently agree it is creative. Appropriate observers are those familiar with the domain in which the product was created or the response articulated. Thus, creativity can be regarded as the quality of products or responses judged to be creative by appropriate observers, and it can also be regarded as the process by which something so judged is produced.

For Amabile here, the assessment of something's creativity is the result of a social convention: it is the judgement of recognized experts in the field in which the response is given. This is not simply the observation of obvious success by everyone; this could never exist as a criterion for assessing creativity, because the relative weights of people's distinct or implicit theories of creativity differ (Loewenstein and Mueller, 2016). Even in the case of a radical innovation, the assessment of its creativity changes depending upon both who assesses (for whom) and the time this assessment is made (when). An example can show this at work.

At the beginnings of French nuclear power generation in the early 1950s, the creativity of the engineers working on the project is clear for *Electricité de France* (EDF, the French electricity utility). The generation of electricity from this product is both new and useful: in 1953 it is not certain whether producing electricity from nuclear energy is possible; and French economic growth needs a growing production of energy for which nuclear power could be a solution. For EDF this is clearly a creative activity. At the same time, for the *Commissariat à l'Energie Atomique* (CEA) – France's atomic energy agency, whose mission is to produce nuclear energy, and which is collaborating with EDF on the project – this creativity takes on another meaning, as its novelty and usefulness lie in the production of material for nuclear bombs. The difference in the usefulness assessments is what justifies the collaboration of the two institutions (and highlights one of the spaces for their disagreements) for this radically innovative project (Marchais-Roubelat, 2000, 2016).

The French nuclear programme is developing in a doubly conflicting context: first mainly between the industrial actors during the period of technological change from Natural Uranium-Graphite-Gas to Pressurised Water Reactor (Barth, 1981; Marchais-Roubelat, 2000), then with the development of anti-nuclear protest (Barth, 1981; Beltran, Picard and Bungener, 1985). After the 1973 oil crisis, nuclear energy provides for EDF – winner of the first conflict – a way out of possible economic catastrophe at the scale of the country. Since the 1970s, the relationship between nuclear power and economic and ecological development has not been so clear cut: the assessment of nuclear power's creativity is therefore relative to an organization's purpose, its social, political, economic and environmental context and its evolution over time. The perception of, and consensus around, creativity depends on the goals pursued by the various participants in the action. Furthermore, it depends on both the historical circumstances of and the time scale chosen by these same participants: two aspects of 'when' that deserve to be explored.

The role of time in debates about creativity

In their 'review of the literature on who determines what counts as creative and how it should be studied', the cognitive psychologist Mark Runco and creativity expert Ronald Beghetto (2019, p. 7) highlight two main approaches: the 'personal view' of creativity based on individual interpretations; and the 'social view' based on social judgements. The 'personal view' focuses on cognitive mechanisms and subjectivity. 'This includes the claim that something need only be original and effective for the individual creator to be considered creative' (Runco and Beghetto, 2019, p. 7). By contrast, in the 'social view', 'it is especially clear in the claim that there is no creativity without some sort of social recognition or consensus' (Runco and Beghetto, 2019, p. 8).

Rather than a blended option, Runco and Beghetto's model of primary and secondary creativity proposes a differentiated one. This model is composed of two successive phases: in the first one, personal creativity produces a creative outcome. It needs to meet an audience to become a social process, producing a creative outcome that is 'a unique interpretation and experience of primary outcomes by an external audience' (Runco and Beghetto, 2019, p. 8). However, even in that model, the questions of 'when' and 'for whom' are not resolved, as the circumstances in which the second phase follows the first may play an important role in its development. Furthermore, the circumstances in which the output is experienced socially may influence any assessment of creativity that this group may give.

In a paper published in 1993 in the *Academy of Management Review*, Richard Woodman, John Sawyer and Ricky Griffin propose an interactionist model adding to individual and group creativity a third organizational one, locating the then little studied concept of organizational creativity in the developing field of organizational change and innovation. They propose to define organizational creativity as: 'the creation of a valuable, useful new product, service, idea, procedure, or process by individuals working together in a complex social system' (Woodman et al., 1993, p. 293). As the authors explain:

> It is, therefore, the commonly accepted definition of creative behavior, or the product of such behaviors [. . .] placed within an organizational context. We frame the definition of organizational creativity as a subset of the broader domain of innovation. Innovation is then characterized as a subset of an even broader construct of organizational change. [. . .] It is our belief that the organizational sciences can benefit from systematic investigations of creative behaviors in complex social systems. Creativity for individuals and organizations [. . .] represents a dramatic aspect of organizational change that may provide a key to understanding change phenomena and, ultimately, organizational effectiveness and survival.
>
> (Woodman et al., 1993, pp. 293–294)

If it is asserted that researches on organizational creativity are both positioned and justified by innovation and change, as well as organization duration (through survival), their relation is not assessed. The authors have adopted an interactionist approach, focusing on the factors of creativity – individual, in groups and in organization. The only reference to time relates to one of the individual factors ('antecedent conditions', Woodman et al., 1993, p. 297) and its link to the past.

Studies in the organizational field associate creativity with an answer to the uncertainty of future under many forms: for example, innovation (Baer, 2012), competition, world or organizational transformation (Weik, 2012; Teodoridis et al., 2019); and problem solving or enacting (Hagtvedt et al., 2019). Postulating that creativity 'fuel[s] competitive advantage' (Loewenstein and Mueller, 2016, p. 321), that it is 'one of the critical success factors for organizations in today's rapidly changing business environment' (Wang et al., 2018, p. 1443), or that 'organizational creativity is crucial to solving problems in uncertain conditions' (de Vasconcellos et al., 2019, p. 440), there is much focus on creativity in order to improve its functionality. In the field of organizational creativity, Teresa Amabile proposes for example to define creativity as 'the production of novel and useful ideas by an individual or small groups of individals working together', creativity constituting a component of organizational innovation defined as 'the successful implementation of creative ideas within an organisation'; 'implementation' encompassing, in this context, 'elements of developing ideas and putting them to use' (Amabile, 1988, p. 126). Thus, she separates the utility of creativity from its transformation in organizational utility (the evaluation of which is then transferred to innovation's implementation).

Few critical studies on organizational creativity offer new paths for research. The work by organizational behaviour scholars Jeffrey Loewenstein and Jennifer Mueller (2016) therefore stands out. They write: 'much of the creativity literature has focused on how to foster the generation of creative ideas, with less emphasis on examining how to recognise creative ideas' (Loewenstein and Mueller, 2016, p. 340). Furthermore, and of particular importance to this chapter, research on the relation of creativity to time is little questioned in the organization field. Phase models are proposed regarding the creativity process or its insertion in other processes, such as innovation. Taking up the differentiation made by Amabile between creativity creating ideas and innovation implementing new and useful ideas, a group of psychologists and management specialists – respectively: Kathrin Rosing and James Farr; Ronald Bledow, Michael Frese, Nataliya Baytalskaya and Johanna Johnson Lascano – working on team innovation propose a model integrating the two existing theoretical perspectives, the linear and the complex, in order to link the temporal pattern of innovation activities of a team with the outcome it produces.

> The linear perspective assumes a sequence of distinct phases that presumably follow each other step by step from creativity to implementation [. . .] whereas the complexity perspective assumes that creativity and implementation are intertwined and unfold in a cyclical and chaotic manner.
>
> (Rosing et al., 2018, p. 799)

Their 'results show that teams engage in creativity throughout the entire life cycle of team projects; however, innovative teams refrain from focusing on implementation in early time frames and increase their focus on implementation over the course of the project' (Rosing et al., 2018, p. 798). The creative process, for them, continues right through the project, but to enhance innovation, concerns about implementation need to be removed from the initial stages of an innovation practice. This research regards creativity and implementation as two temporally differentiated processes, introducing the idea of slowing down any rush to implementation. From a temporal perspective, during the course of an implementation, it would be then useful to differentiate the interlaced creative and innovative subprocesses, such that modifying their relative paces could differently orient these processes relative to any future utility. From this the strategic question for organization becomes: when should implementation be considered?

Other approaches use time as a scale for making comparisons. For example, regarding the relative creative performances of specialists and generalists, recognition of a difference in innovation pace, as discussed above, leads researchers to recontextualize previous innovation models. On this, innovation, strategy and creativity scholars Florenta Teodoridis, Michaël Bikard and Keyvan Vakili (2019, p. 920) write that the

> recombination [of knowledge] based on newly emerging knowledge in a faster-paced domain is not as trivial as prior literature suggests. It involves

> important skills to identify those opportunities early on and to exploit them efficiently. The role of those skills for creative success has received little attention to date.

In this case, time is a means to propose a typology of innovation (fast- and slow-paced domains), explaining the conditions under which specialists or generalists perform creative tasks better, or not, than each other.

Whenever creativity is examined in relation to uncertainty – and therefore with a future that it helps to manage – time is poorly studied in the field of creativity and organizational creativity. The 'when' is almost non-existent; the 'how' and the 'why' predominate. When it is taken into account, time is considered only as a tool providing a scale of measurement. We see this happening even in the typologies of pace of innovation. However, time ought to be studied further in organizational creativity because, in this context, it is necessary to assess the value of creativity for organizations over time. This raises the question of the nature of time in organizational conceptions of change. Philosophically minded innovation management scholar, Robert Chia (1999, p. 211), explains that the

> dominant approach to the analysis of change continues to view the latter as something 'exceptional' rather than as a sine qua non of all living systems [. . .]. There has been little attempt to understand the nature of change on its own terms and to treat stability, order and organization as exceptional states.

By contrast to this dominant Parmenidian approach Chia, writing with Process Organization theorist Haridimos Tsoukas, conceive of organization as 'the attempt to order the intrinsic flux of human action, to channel it towards certain ends by generalising and institutionalising particular cognitive representations' and as 'a pattern that is constituted, shaped, and emerging from change' (Tsoukas and Chia, 2002, p. 567). In this view, organizations take on a signification with reference to goals to be achieved and which consequently guide the evaluation of their usefulness, and they participate in a movement of Heraclitean change which goes beyond them and encompasses them. The consequence is that 'although an organisation fixes the definition of its representations (generic cognitive categories) for certain purposes, it does not have total definitional control over them [. . .]. Definitional control is compromised because of organizational interactions with the outside world' (Tsoukas and Chia, 2002, p. 573), because 'humans have the intrinsic ability to interact with their own thoughts and, therefore, to draw new distinctions, imagine new things, and employ metaphor, metonymy, and mental imagery' (Tsoukas and Chia, 2002 p. 574). This questions the supposed role of organizational creativity in innovation and change if change is not considered in a Parmenidian view, as a voluntary transition from one state to another (Chia, 1999), but in a Heraclitean way as 'an ongoing process in organization' (Tsoukas and

Chia, 2002, p. 580). One could imagine, then, that this processual approach to organization interweaves movements whose different paces insert multiple futures in the present in becoming.

Time and the instrumentalization of creativity in futures studies

Insofar as it brings together novelty and usefulness, creativity is a matter of time because it cannot be conceived without the idea of becoming as an emerging movement intertwined with a multiplicity of other movements with varied rhythms and directions. These movements, over time, provide to creativity both its novelty and its usefulness within a larger movement that encompasses and exceeds it, but also feeds on it. As such it is at the heart of futures studies, that 'think about and study the "not yet" of the future' (Bell, 2005, p. 432). If anticipation is part of futures studies, its link to implementation creates additional problems: when and in which direction can implementation be undertaken to orient the spread of the effects of creativity over time in particular directions? When this orientation is directed towards achieving objectives, when does anticipation become an organizational question too?

As an ability, creativity resides in power, out of time. This is in contrast with its manifestation in the product of the creative act and the responses that this manifestation produces; both of which can be dated in terms of their two attributes – novelty and usefulness. To be novel, a response must be compared to a former one to which it is supposed to add something previously non-existent. Therefore, any creative outcome (as a result, or as a process of production) should be different from those produced from the past to the present. To be useful, any creative outcome will be assessed as such when produced.

Futures studies produces a creative result as words 'speaking about the future' (Voros, 2005, p. 87) whose usefulness is to shed light on the action to come without prejudging what to do. The creative outcome may be a scenario whose design process or whose existence is sufficient to make it useful.

This is visible, for example, in the way that designers work, who, Jamie Brassett (2016, p. 173) notes, 'when inserting the future in the crannies of the present [they take] future specks to grow the present as a crystal in the material medium and environment that already resonate with future and present possibilities'; even if we do not know when it happens or for what it will be useful. More broadly, the consequences of creativity as an act of the future without prejudging the future are threefold: first, it promotes the search for innovative ideas without necessarily having to seek an implementation; second, it encourages a focus on creativity tools and methods (Jaoui, 1970; de Brabandere and Iny, 2010; Montuori, 2011; Van Vliet et al., 2012); and third, it emphasizes the role of creativity management in considering both a future organization's general survival and any specific crisis management it may have to encounter (Williamson, 2001; Sommer and Pearson, 2007). However, anticipation is more than an addition to the already multiple existing designations of future studies (Sardar, 2010) fitting into a tendency to gain in practical utility (Gibbs and Flotemersch, 2019; Slaughter, 2020). As it is directly associated with an implementation (Fuller, 2017; Groves, 2017), anticipation brings also its specificity: as it has to be useful also *for* the implementation to come or the purpose it serves. So that, because anticipation entails an upcoming implementation, any assessment of its nature – made during the act of anticipating – has to stay valid for an indefinite future. The estimation of its plausibility (Ramírez and Selin, 2014), for example, expresses the social acceptability at the time this estimate of the creative result of anticipation is made. Such an estimate is likely to facilitate an implementation, but it does not assess the future utility of this implementation, which cannot be known until it has been done. So three movements are intertwined over time.

First, there is the conception and actualization of *temporal process* in future studies – which includes creativity and which has its own effects. Second, the later (or contemporary) movement related to the first, in anticipation in particular, is *implementation* – which follows its own specific temporal dynamics and direction. Third, there is another later (or contemporary) movement in the *materialization* of the implementation's effects during which anticipation's usefulness occurs – and which may evolve over time as the effects continue to occur and influence the future course of events.

These three interlaced movements can happen simultaneously or in an offset way, their durations are not necessarily equal and their paces, constraints and directions are not necessarily the same. Thus, in anticipation, creative novelty emerges from the past from which it differs, but its usefulness will continue to develop in the future as it depends also on the two other movements: the implementatation and the materialization of effects. Both of these movements take place in a future world different from that of anticipation itself, as 'the consequences of our actions will occur in a world completely different from that in which we will have prepared them' (Berger, 1964, p. 222). From this point of view there is a heterogeneity of time. To consider that creativity is useful at the time of the anticipation reduces this heterogeneity for the benefit of a homogeneous time, as in this specific case usefulness assumes a continuity between the present and the future. This depends on a present assessment of the usefulness of the future outcome of its implementation while this outcome will be assessed for a future specific goal different from what we can imagine in the present. Among futures studies, the specificity of anticipation – as it is linked to implementation and therefore to a future within reach – is that the assessment of a future creative outcome depends not only on the creativity itself, but also on the effects of the implementation over time. From this discussion two main questions must be answered: is the time of creativity homogeneous or heterogeneous? What is the nature of the event in a homogeneous or heterogeneous time? The next part of this chapter will examine both of these questions in relation to the philosophy of Martin Heidegger (1889–1976).

Frozen time of creativity or flowing creativity of time: Heidegger and the nature of time in science

As we have seen above, insofar as anticipation implies implementation in futures studies, this implementation introduces an organizational dimension. From this point of view, the standard conception of time in organizational change has to be questioned. Heidegger's thinking on the nature of time in sciences from a comparison between history and physics provides a basis to conceptualize flowing heterogeneous time in contrast to the standard frozen homogeneous one. As Heidegger's comparison only applies to History, flowing time has to be examined from the perspectives of organization studies and futures studies before assessing the appropriateness of its adaptation to thinking the place of creativity in anticipation.

Heidegger on scientific time

In his lecture on 27 July 1915, Heidegger (2007) differentiates between concepts of time depending on their function in natural sciences and in History, which appears to be the only science in counterpoint to all others. The function of time in natural sciences is to enable measurement based on laws that are universal, in a frozen time, as 'the flow [of time] freezes, becomes a segment; and only as a segment can it be measured' (Heidegger, 2007, p. 66). Furthermore, on the scale of frozen time, 'events get their determinate place in time and are thus fixed *historically*' (Heidegger, 2007, p. 67; original emphasis). An example could be a cycle, as for example an alternation of demographic growth and famine. Every famine must respect this model, wherever and whenever it occurs. Therefore, scientific phase models belong to frozen time as scientific frozen time is concerned with the quantitative measurement of duration and not with time itself. Such a frozen time has a quantitatively determinable, mathematical character and is homogeneous; here, events – like famine – may replicate as elements belonging to the same class.

By contrast, flowing historical time is characterized by a discontinuity expressed by periods, as 'each period manifests in the ensemble of its creations and expressions a characteristic that distinguishes it from another' (Heidegger, 2007, p. 70). In this heterogeneous time, an event as a historical concept is a 'specific individual' within the flow of time, it is unique and singular. 'The famine in Fulda in the year 750' is a historical concept. It cannot be generalized. Philosopher Francisco Naishtat (2018) examines Heidegger's lecture in order to clarify a controversy over the concept of time between 'the early Heidegger and the young Benjamin'. He writes:

> the function of time in history, according to Heidegger, is not a measurement, but a fully qualitative notion, which shall be understood through the idea of meaning, since the time of history allows carving out an event as a meaningful historical individuality. When we inscribe an event on a

> certain date, we are not measuring, but rather placing that event in a unit of meaning in relation to a 'before' and an 'after'.
>
> (Naishtat, 2018, p. 441)

As Naishtat explains, in Heidegger's conception of historical flowing time the event does not contain its full meaning in itself: it belongs to a larger unit of meaning which includes with it a before and an after. This conception is essential for anticipation because any future meaning of events partly depends on others, which themselves may influence and be influenced by any present implementation as an anticipatory, creative act.

In this heterogeneous time, periods are durations that give unities of interpretation to the specificity of events in historical contexts. Events chosen as starting points of these periods are marks that may be discussed as they first depend on the historical significance given to them, however this significance is not unique. To state that the beginning of the Renaissance aligns with the discovery of America in 1492, or the fall of Constantinople in 1453, is less an issue about placing a date on a time scale than the expression of a difference in the significance that is attributed to the specific periods in which these events are related. Similarly, the beginning of the production of nuclear electricity in France may be given as late 1953 when the decision was made to add a small electricity recovery unit to the two CEA reactors; or as 1956, when the first protocol of agreement between EDF and the CEA was signed. Or, an earlier institution of the nuclear in France could be with the research of Pierre Curie and Marie Skłodowska Curie, for example. Changing the beginning of a period changes the historical context that gives it – the period and its context – sense and therefore alters the meaning of the events that happen during its duration. In such a flowing time, the meaning of an event may change depending on the period as well as the circumstances taken into account, but as events are specific (as 'the famine in Fulda in the year 750'), they never replicate (as another famine in Fulda would be different from the one of 750).

Organizational frozen and flowing time: consequences for anticipation

From a standard conception of change in organization, wondering whether the goal pursued today will be different in another period is out of scope, because questioning in this way would assume time to be heterogeneous (as we just saw for anticipation). In homogeneous time, anticipation can be confused with planning or risk management. This is the case in project management, for example, when Program (sometimes Project) Evaluation and Review Technique (PERT) or Gantt charts are used within set time frames and organizational scope, in relation to a fixed goal with recognized, critical path dependencies. However, since the rise of the neo-institutionalist conception of organization, the idea that 'organizations tend to disappear as distinct and bounded units' and evolve according to 'myths' (Meyer and Rowan, 1977,

p. 346) has spread and the concept of risk has tended to take over all corporate futures-thinking. This also becomes insufficient, because risk assumes a frozen time, since it deals with the probability of the occurrence of an event whose effects are not desired.

Traces of flowing time can be found in organization studies, however, especially through the notion of 'fluid world' which assumes that each event is unique, as 'in a fluid world, wise people know that they don't fully understand what is happening right now, because they have never seen precisely this event before' (Weick, 1993, p. 641). To illustrate this, organizational theorist Karl E. Weick (1993, p. 642) evokes the pre-hunt divinatory ritual of the 'Naskapi' native First Nation tribe of northern Canada:

> any attempt to hunt for caribou is both a new experience and an old experience. It is new in the sense that time has elapsed, the composition of the hunter band has changed, the caribou have learned new things, and so forth. But the hunt is also old in the sense that if you've seen one hunt, you've seen them all: there are always hunters, weapons, stealth, decoys, tracks, odors, and winds.

Here Weick superimposes frozen time (the old experience: 'if you've seen one hunt, you've seen them all') and flowing time (a new experience as the circumstances have changed). Depending on the underlying nature of time, the hunts constitute repeating or unique events. In this example, the divinatory ritual is effective in creating the ability to produce novel and useful responses to a repetitive problem (the hunt) as neither experience of the past (that would lead to deplete the stock of animals) nor patterning in human choices (that would enable hunted animals to learn and take evasive actions) would be creative enough to allow hunters to return sustainably with game. Time is not a measurement of duration for human action (the repetition of events called hunts), it has its own, flowing nature that together changes and maintains the world. Future events will never be an exact repetition of past ones and the paradox of novelty/usefulness of creativity dissolves in the flowing time, as time itself becomes creativity whose usefulness depends on the human ability to recognize and to use it. The flow of time constitutes the reference of novelty, as future will be different from past. Yet the effectiveness of this novelty is contingent as it depends on the participants' abilities to see, use, and transform it in each event. This ability varies according to the participants and the organizations through which they act. It may also depend on the period during which events happen. Although attractive, the idea of a 'fluid world' has not opened up new avenues of research in organization theory; nevertheless it creates a bridge with anticipation in a flowing time.

To approach better the relationship between anticipation and flowing time, we need to leave the standard approaches of change and the nature of time that supports them and turn to the much more speculative, heterogeneous approach of a 'fluid world' where events are unique and humans (as well as animals and things) are embedded in a changing world that exceeds them. This is a conception of world that approaches a neo-animist paradigm (see Marenko, 2014, for this in a design context), where humans seek to anticipate, for themselves, to achieve their own moving and evolving objectives.

Such an encompassing, dynamic world seems to have something magical about it, so we first suggest imagining a conception of frozen and flowing times, through the language of runes and their magical relationship to time. If runes are certainly not the only way to perceive flowing time, they are explicitly linked to a complex notion of time that we can only imagine today and from which we can find some traces, especially through the Poetic Edda, the ancient Norse poems of Iceland.

Runes and the *Poetic Edda*: a complex conception of time

By their origins, their pronunciations, their meanings or their uses, runes continue to be a source of speculation for scholars. The specialist in historical linguistics Theo Vennemann writes about the *futhark* – broadly, the name of the Old Norse 'alphabet', based upon the first six letters in its list: 'f', 'u', 'th', 'a', 'r', 'k' – that it 'can only be understood if it is investigated not as a synchronic given but as a historical object, the result of centuries of change, in part well-studied sound changes, in part manipulations whose motivations still have to be discovered' (Vennemann, 2012, p. 592). I will first discuss the complex conception of time associated with runes from two perspectives. Firstly, the meaning and the value of the runes themselves force us to question their values, contexts and meaning in relation to past scenarios. Secondly, runes stress different relations to time through magic. This is evident in a passage from *Hávamál*, one of the poems that comprise the *Poetic Edda*, the collection of anonymous, Old Norse poems. In this passage, Óðinn appropriates the runes in an initiation rite that has caused a lot of speculation (Fleck, 1971a, b). I will return to this below to examine the implications of such a conception of time for creativity, as well as the difficulties it raises.

Runes and conceptions of time

There is no single, stable runic alphabet over space and time, as the order of the runes, their shapes, pronunciation and even number, have changed; Nordeide and Edwards (2019, p. 14) explain: 'the runic alphabet changed from twenty four runes in the early Iron Age [Elder *futhark*] to sixteen runes (the so-called Younger *futhark*) in the late Iron Age, including the Viking Age'. These evolutions have accompanied changes in their contexts, use and meaning, with different constructions in Anglo-Saxon (*futhorc*, see Table 3.1), the Viking Age (Younger *futhark*, see Table 3.2a) and the largely Swedish variant ('short twig' *futhark* or *Rök* runes, see Table 3.2b) – all of which occur through the late Iron Age (see, for example: Osborn and Longland, 1980; Nordeide and Edwards, 2019).

Runes have names and meanings: for example, in the *Younger Futhark*, h 'hagall' (hail), n 'nauðr' (need, longing; see Mitchell, 2008 for a valuable discussion of this rune and its etymological possibilities), i 'íss' (ice) (Wrenn, 1932, p. 32) – these become especially important when thinking about rune magic, of which more below. They can turn into other forms (for example, in the Anglo-Saxon *futhorc* the 15th inverted becomes the 29th; see Table 3.1); or they can be combined with other runes like letters to form words.

Table 3.1 Anglo-Saxon *futhorc*. The Dickins (sometimes known as Dickins-Page) system of transliteration

1	2	3	4	5	6	7	8:	9	10	11	12	13	14	15	16:
ᚠ	ᚢ	ᚦ	ᚩ	ᚱ	ᚳ	ᚷ	ᚹ	ᚻ	ᚾ	ᛁ	ᛄ	ᛇ	ᛈ	ᛉ	ᛋ
f	u	þ	o	r	c	g	w:	h	n	i	j	ʒ	p	(x)	s:

17	18	19	20	21	22	23	24:	25	26	27	28	29	30	31.
ᛏ	ᛒ	ᛖ	ᛗ	ᛚ	ᛝ	ᛟ	ᛞ	ᚪ	ᚫ	ᚣ	ᛠ	ᛣ	ᛤ	ᚸ
t	b	e	m	l	ŋ	œ	d:	a	æ	y	êa	k	k̄	ḡ

Source: Page (1998, p. 249).

Table 3.2a Two main variations of the Viking Age runic alphabet

1	2	3	4	5	6	7	8	9	10	11	12	13	14	15	16
ᚠ	ᚢ	ᚦ	ᚬ	ᚱ	ᚴ	ᚼ	ᚾ	ᛁ	ᛅ	ᛋ ᛋ	ᛏ	ᛒ	ᛘ	ᛚ	ᛦ
f	u	þ	ą o	r	k	h	n	i	a	s	t	b	m	l	R

Table 3.2b

1	2	3	4	5	6	7	8	9	10	11	12	13	14	15	16
ᚠ	ᚢ	ᚦ	ᚭ ᚭ	ᚱ	ᚴ	ᚽ	ᚿ	ᛁ	ᛆ ᛆ	ᛌ	ᛐ	ᛓ ᛓ	ᛙ	ᛚ	ᛧ
f	u	þ	ą o	r	k	h	n	i	a	s	t	b	m	l	R

Source: Nordeide and Edwards (2019, p. 14).

We can see, also, across all three alphabetic forms shown here, not only a reduction in rune numbers (from Anglo-Saxon *futhorc* to *Younger futhark*), but some fairly radical changes in form. These differences not only characterize the geographic contexts of their use, but their temporalities too.

Against the argument that runic writing develops from a direct origin in Latin or Greek, Vennemann defends the controversial thesis that it derives 'directly from the Phoenician writing system of the Carthaginians who dominated the Atlantic coasts from the fifth to the end of the third century' (Vennemann, 2012, p. 529; see also: Mailhammer and Vennemann, 2019). For Vennemann, this thesis explains a number of specificities of runic writing, including: 'the form of those runes for which Latin offers no model', along with 'various aspects of the order of the older futhark'. He continues (2012, p. 529):

> The Phoenician thesis explains the concentration of the oldest runic finds in the Scandinavian regions between the German and the Baltic Seas rather than close to the Roman Empire and answers the question of the reasons for the extraordinarily early availability of an alphabetic writing system of its own in northern Europe.

However we may arrange ourselves in these debates, as soon as we question the origins and changes – across time and in space – of the runic alphabet, we encounter concerns about the organization of the world in which it appeared. The theses that scholars defend propose alternative stories, different ways of thinking and creating space and time; and, in certain ways, this approach is comparable to the alternative scenarios one could propose in futures studies.

Another important debate concerns the magical character of runes. As magical incantations, runes were used for many different purposes: protecting, treating illnesses, making curses or detaching shackles, for example (McKinnell et al., 2004; Mitchell, 2008). Generally, 'conclusions about runic practice – and about rune magic – [have] largely been drawn on the Scandinavian material' (Page, 1998, p. 106), especially in the series of references to the magical use of runes in the *Poetic Edda*.

Historian R.I. Page is opposed to the theory that considers 'all early runic inscription as magical, no matter what their apparent meaning' (Page, 1998, p. 105). However, he notes their relation to magic in the Eddic Poems; as well as in some few cases – among them an amulet of a series of seven 'a-', three 'R-' and three 'n-runes' – where they appear as magic symbols. He explains:

> That even a sceptic must admit evidence of the occasional magical use of runes in early Scandinavia is important to the present argument, for the Scandinavian material is needed to support the, in my opinion, weak evidence from Anglo-Saxon England.
>
> (Page, 1998, p. 105)

Page's overall scholarly position in this field is that we should not extrapolate meanings from Scandinavian examples into the Anglo-Saxon context; the changes between these peoples and locations are too dramatic. While Scandinavian magic may have been bound up with runes, he argues, this is not necessarily so in Anglo-Saxon England (Tornaghi (2010) argues otherwise).

There is, then, uncertainty about the conceptions of the world, or worlds, which leads scholars to interpret different meanings of events as linked to runes. The specialist of Nordic culture and literature Stephen Mitchell examines the n-rune's role in Nordic charm magic, asking: 'how [. . .] did practitioners understand their activities helping them? Within what mentalities did the operator function? And what power was assumed, believed in, called out, conjured?' (Mitchell, 2008, p. 221). He continues:

> modern pagans have invested their own interpretations into the n-rune [. . .] suggesting that it embodies the ideas of blockage, the distress implicit in something called a need-rune. No one will object to participants in a modern syncretic religious revival imposing whatever meaning they want on such objects, but such understandings do not say much about what a twelfth-century Swede or a sixteenth-century Icelander thought, much less the views of earlier carvers.
>
> (Mitchell, 2008, p. 222)

Examining his corpus, Mitchell notices that the charm operators call on deities, diabolical entities and even the physical world (for example, the sun), but not 'abstract concept[s], such as happiness, depression, or despair' (Mitchell, 2008, p. 222; see also: McKinnell et al., 2004). The question then arises, that:

> while we may confidently push aside the New Age pagan reading of the need-rune as irrelevant to the understanding earlier periods would have possessed, is it possible that scholarly views are in a similar state? Just because we possess a present-day ability to read so very much entertaining meaning into the so-called 'need-rune', does it necessarily follow that we truly understand its historic value?
>
> (Mitchell, 2008, p. 223)

This question illustrates another side of the heterogeneity of time, that which corresponds to anachronism and concerns both the future and the past, with historians and New Age pagans both reading-in value to runes, even while their motives may differ. Mitchell (2008, p. 227), once more, concludes:

> At some point, certainly by, for example, the time of the various rune poems in Old English, Old Icelandic, and Old Norwegian, the meanings that we today associate with 'ned' has supplemented, or supplanted, its original supernatural meaning, so that many generations of sorcerers must

> have used the term without having any sense of how it related to the giants and gods with which it was so often surrounded.

Mitchell reminds us that for many scholars, the social production of magic appears to be based on two components: tradition, as its institutional component; and the actions of the sorcerer (speaking, carving and so on) as its individual component. For Mitchell, the individual action itself is also a double one: for 'the carved inscription connotes two additional related but separate performances: both the physical deed of carving or writing, that is, the making of the inscription, which at the same time echoes the recently orally performed words of the charm, i.e., "I write. . ." or "I carve. . ."' (Mitchell, 2008, p. 223).

As event, the magical use of rune would be twice double, so that its evolution over time can refer to multiple time frames: the time of tradition spanning several generations; the individual time of the act of producing magic; and the time of individual act as part of the temporal context of the tradition which it contributes to, makes last or evolves. Furthermore, the individual act creates two categories of effect that also fit into separate and complementary time frames: the one of carving, whose effects will last as long as the record, its image, remains even if its meaning is lost; and that of the creation and duration of the charm. These temporalities complement each other to create the full meaning of the event, which has an objective part that stays over time (the carved rune) and a complementary symbolic one which evolves over time (Marchais-Roubelat, 2012a),[1] depending on multiple time frames that are separate and evolve at different rates, creating a permanent movement.

In ancient times, the use of these runic symbols helped the Germanic peoples, Vikings and Anglo Saxons who used them to face the changing and ambiguous events that chart the creativity of time. We do not know much about runes as many of the materials on which they were carved (skin, wood, bone) have perished over the centuries. The runic inscriptions that remain, carved into weapons or stone, are the subject of conjecture; and while related cannot necessarily be treated homogeneously – as we have seen (Page, 1998; Mitchell, 2008; Vennemann, 2012). In such a moving universe, one cannot use a standard representation of runes.

Beyond the debates on the theory of rune-magic, scholars agree on the magic aspect of the runes in a poem of the *Poetic Edda*: the *Hávamál* ('Sayings of the High One [Óðinn/Odin]'). While there is no critical consensus about the origins of *Hávamál*, John McKinnell (2014, p. 79) – an expert in mythological and heroic poetry – explains that it is not always necessary to hunt for origins, as 'two important studies [. . .] have chosen to ignore the problem of origins and to attempt reading of the text simply as we have it in the Codex Regius'. In this vein, we can focus on aspects of the poem that highlight a temporal attitude. One passage in particular sheds light on the relation of runes to time, through the sacrifice of Óðinn:

> I know that I hung on the windswept tree for nine whole nights, pierced by the spear, given to Óðinn, myself given to myself, on that tree whose roots no man knows. They refreshed me neither with bread nor with drink from the horn. I peered down, I took up (? learnt) runes, howling I took them up, and then fell back.
>
> (Page, 1998, p. 107)

Óðinn learns sorcery and the runes as a result of a mystical and paradoxical sacrifice of himself to himself. In their magical use, runes link divination to destiny through the power of the high priestesses of magic who serve the three maidens, the Norns. Weavers of fate and casters of chance, the Norns shape destiny by carving runes into the trunk of Yggdrasil, the great Earth tree. The strands of our discussion – of time, anticipation, divination and creativity – are beginning to come together.

Through the use of the runes, we can propose at least three structures of time: first, a linear time oriented from past to future, where letters are written one after the other and symbolized by the three Norns – one is a girl, the second a beautiful young adult, the third one an old woman. The time of life cycles belongs here. Yggdrasil is an ash tree, its leaves appear, they grow, wither and disappear, and the cycle repeats over time. Second, a frozen time of Is, a suspended (vertical) time, where Óðinn is hanging onto Yggdrasil's branches by his own will and is learning about rune magic, told from the linear time of the Norns but also outside of it. This is the time of thought, of analysis and modelling; and, more generally, abstraction. Third, a flowing time of Yggdrasil's sap circulation – the global water cycle – the time of life and action. This life cycle is doubled by the life cycle of the world itself: 'The Ash Yggdrasil is a symbol of the world; it is the tree of life and fate. Its appearance and condition reflect at all times the state of the world' (Flom, 1939, p. 148). In *Grímnismál*, from the *Poetic Edda* – but also quoted from by Snorri Sturluson, writer of the *Prose Edda* (Jónatansdóttir, 2020, p. 411) – it is said about Yggdrasil:

> Yggdrasil's Ash suffers anguish more
> Than mortal has never known;
> A hart gnaws it above, it rots at the side,
> And Niðhöggr rends it beneath.
> [. . .]
> More serpents lie under Yggdrasil's Ash
> Than a witless fool would ween.
>
> (Flom, 1939, pp. 154–155)

The time of action is multi-layered as separated time frames complement each other and evolve at different rates. This is where the Norns exert influence over the course of destiny by carving runes into Yggdrasil's trunk and where Yggdrasil develops while at the same time the great dragon Niðhöggr undermines it by biting its roots. This is a time of contradictions, multiplicity, even swarming, which is very much like the time people and organizations live.

These three temporalities are completely differentiated, as their paces and rhythms, as well as their rules are different: during the frozen time, Óðinn thinks but does not drink or eat, he is as if dead and no longer belongs to the flowing time. However, linear time continues, as Óðinn knows he hung for nine whole nights. During flowing time, Óðinn acts physically, taking up and howling. And yet the three separate temporalities are braided together, just as the meaning of the magic runes may combine or transform into each other.

We have noted above that i *íss* is 'ice', but it can also mean 'will', 'frozen'; and ✱ *hagall* 'hail', can mean 'crystal', 'snowflake' or 'ash tree'. This last giving us a reference to Yggdrasil. *Íss* hangs without moving or living, like Óðinn's willing thought, in a frozen time. But *íss* may unfreeze – its state changes in a hailstone or a snowflake – and may become *hagall* and so, like Yggdrasil, teeming with life. Isolating a moment to immobilize a time interval leads to error, as a shape as well as the meanings attached do not stay unchanged. However, frozen time and flowing time continually move into and out of each other, from a universe of will and thought to a teeming world of intuition. Conversely, Yggdrasil's sap may freeze and lead to a suspended time of abstraction, removed from the living world even if linear time goes on flowing. Runes, as we have been developing them here, are events, their meanings are manifold and they evolve, transforming and combining over time.

Despite the little knowledge we have, runes allow us to propose a flowing-living time, without denying either linear or abstract-frozen time, with which it combines. On a plane where the chronology would be on the horizontal axis, we could imagine frozen time as a vertical which would move along it, while horizontal segments would represent intersections between the plane and the worlds of flowing time. These worlds essentially belong to other dimensions from a symbolic universe, they cannot be represented on the plane without losing their creative value. In order to highlight them, we must use various modes of creativity (Marchais-Roubelat, 2012b).

Conclusion: from magical divination to anticipation in futures studies

As magic is very far from today's conception of time in futures studies – that is dominated by attempts to develop it as 'science' – we need to identify the characteristics of future-adapted flowing time in order to make the imaginative connection from ancient, creative thoughts to a more contemporary, anticipatory approach. If, as futures practitioner Joseph Coates (2010, p. 1430) writes, 'the study of the future is not and cannot be a science', then 'the essential nature of FS [Futures Studies] is about attempting to make useful statements about the future so that our present actions are better informed' (Voros, 2005, p. 88). Until now, there has been no place for divination or magic in such a world, as futurists make 'efforts to satisfy the yearning for "scientific" consideration of the future', even if 'scientific is defined here in a rather minimalist fashion as analysis based on explicit and open methods that test hypotheses pertaining to a particular subject through inter-subjective evaluations' (Miller, 2007, p. 342). The function of proving authenticity – which, for Heidegger, explains the qualitative structure of time in science of history – is not suitable for questioning creativity in a logic of anticipation. If flowing time allows us to conceive of becoming, more concepts need to be used in order to fashion its relation to anticipation. Two difficulties have to be overcome: first, the transition from a magic flowing time to a future-oriented one, where periods may be distinguished in the otherwise heterogeneous flow; and, second, to research into the transformation of action over time through events previously symbolized by magic runes.

Before approaching these two problems, it is necessary to consider what the flow of heterogeneous time may be in a future-oriented attitude – in contrast to the usual, frozen time of futures studies – in order to define whether, for anticipation, the time of creativity is homogeneous or heterogeneous.

Anticipation has to deal with unknown futures as well as the implementation into the present of such a creative act. Thus, it has to take into account the temporality of an action that is not a human activity (either as an individual intentional act or a collective action) but a process within which human activities occur. In other words, human activity and will is not at the centre of the action, even though it does participate in it. Under these conditions, action may be conceived as a process, as philosopher and futurist Bertrand de Jouvenel (2012, p. 113) defined it: 'the evolution of a phenomenon which is not a *goal* chosen by a human will but the effect of a complex concurrence of actions not consciously aimed at the effect' (original emphasis). Thus, action is a historical process of becoming. It is a fluid world, as things are incessantly changing under the influence of human actions as well as independently, so that 'wise people know that they don't fully understand what is happening right now, because they have never seen precisely this event before' (Weick, 1993, p. 641).

As future events are unknown, they may be replaced in heterogeneous flowing time by rules which give unity of meaning to the periods during which they can be observed or anticipated. A rule is a constraint on behaviour or a relation between variables which operates during a phase (Marchais-Roubelat, 2000): it gives a unity of meaning to a period that is taking place and influences acts, but it is also changed by acts and then a new period appears. So, when events no longer follow particular rules, a rule can change even if others carry on; and when rules change, the periods also change, creating heterogeneity in the time of action. Phases in flowing time are periods that give time a qualitative structure and function, be it past or to come (Marchais-Roubelat and Roubelat, 2008). A phase is not a period defined *a priori*, neither is it a time scale for measuring a succession of events. It produces a qualitative context giving a unity of meaning to recognize, anticipate or react to the effects of time's creativity; and this context is valuable only as long as we observe that the rule works, whenever we cannot guess which particular event will happen. Thus, one has to go back-and-forth between the frozen time of action, modelling through the rules and the flowing time during which the rules work. With the rules then allowing differentiation of intertwined processes whose directions and paces differ. When the evolution of the phenomenal part of events no longer corresponds to the rule, the phase changes. The problem to solve, as well as the novelty and the usefulness of the given response, change too, even if no significant event occurs. Therefore, creativity evolves over time, as do the goals that it was meant to serve during the phase when it was first perceived, with rules made only to recognize 'when' things are changing. Anticipation – as an aspect of futures study and in its present implementation – goes back and forth between the two temporalities, as time is homogeneous inside the phase and heterogeneous outside.

Several rules – and therefore phases – coexist during the same action, each one moving at its own rhythm, pace and scale, and all contributing to action. Actors play, disregard or create rules. The usefulness of their acts depends not solely on their will, but also on the circumstances created by the rule of the phase during which their acts and their effects occur. A phase change creates an irreversibility that may disrupt other rules working in parallel. At this scale of action time is flowing and one can perceive it as heterogeneous through the change of phases.

In frozen time, an event is a phenomenon that appears; the problem for anticipation is then to imagine the value of its future effects. In flowing time, an event must be completed to acquire its full meaning. It is a visible sign, a phenomenon, whose complementary part we either do not know yet, or whose complementary part we know, but which may change depending upon the contingencies of a particular time or observer. Then any change of meaning over time possibly creates irreversibilities. The event therefore brings together the objective part of a phenomenon and the symbolic part of its interpretation (Marchais-Roubelat, 2012a), and they become process in Jouvenal's sense (1967). When the flow of time freezes, the event also freezes into an object that one may place on a time scale, measure and locate.

As frozen and flowing time are differentiated in action, they may be used together for anticipation as the time paradox of creativity in anticipation only exists in frozen time. Flowing time *is* creativity, as it introduces novelty even in repetitive activities; with its usefulness depending upon 'when' it is assessed and 'who' is assessing. Frozen time, therefore, serves as a guide for an assessor to measure time's flow and to gauge when, how, how long and for what this creativity may be used. In this way, flowing time does not replace frozen time, it completes it and opens new research fields and applications for anticipation.

If creativity is at the heart of futures thinking, it also raises the problem of the conception of becoming. In a homogeneous, frozen time, creativity may be conceived as an individual or organizational ability to create innovative and useful responses to problems. As long as the response it produces is in itself useful, it may be considered as a tool for future thinking. For anticipation, however, such a conception of creativity is not sufficient as its implementation raises temporal paradoxes. So to anticipate, one must conceive of the nature of time in which creativity is thought and driven.

Through the runes and the *Poetic Edda*, we can design an interlacing of frozen and flowing temporalities. The nature of the problem changes as the question is not to enhance human creativity, but to appreciate time's creativity sufficiently early to be able to use it; as well as to recognize when a rule governing an action is changing, as it will change the usefulness of creativity.

The runes' association with mystical functions questions the value and validity of making future-oriented studies 'scientific'. In today's debates on creativity, as we have seen, frozen time is used as a scale for measuring human activity. By contrast, flowing time suggests that we can design the creativity of time as temporal flows made of processes in becoming, that the human may use but never totally control. Events, therefore, may be seen as symbols composed of two complementary parts: a factual part and a part that completes its meaning or its work according to its becoming over time. As eventful meanings evolve diversely according to stakeholders, their contexts and temporalities, they allow a high anticipatory creativity in both dimensions of novelty and usefulness.

Notes

1 The idea of combining the symbolic part of an event with Heidegger's conception of flowing time in history in relation to anticipation is proposed by Bettina Bouchayer (Conservatoire National des Arts et Métiers/Technische Universität Berlin) in her forthcoming PhD thesis, *Does the event make the process?*

References

Amabile, T.M. (1988) 'A Model of Creativity and Innovation in Organizations', *Research in Organizational Behavior*, 10, pp. 123–167.

Amabile, T.M. (1982) 'Social Psychology of Creativity: A Consensual Assessment Technique', *Journal of Personality and Social Psychology*, 43(5), pp. 997–1013.

Baer, M. (2012) 'Putting Creativity to Work: The Implementation of creative Ideas in Organizations', *Academy of Management Journal*, 55(5), pp. 1102–1119.

Barth, J. (1981) *Quarante ans d'énergie nucléaire dans le monde*. Palaiseau, France: SOFEDIR.

Bell, W. (2005) 'Creativity, Skepticism, and Visioning the Future', *Futures*, 37, pp. 429–432.

Beltran, A., Picard, J.F. and Bungener, M. (1985) *Histoire(s) de l'EDF*. Paris: Dunod.

Berger, G. (1964) 'Sciences humaines et prévision', in Berger, G., *Phénoménologie du temps et prospective*. Paris: Presses Universitaires de France, pp. 218–226.

Brassett, J. (2016) 'Speculative Machines and Technical Mentalities: A Philosophical Approach to Designing the Future', *Digital Creativity*, 27(2), pp.163–176.

Chia, R. (1999) 'A "Rhizomic" Model of Organizational Change and Transformation: Perspective from a Metaphysics of Change', *British Journal of Management*, 10, pp. 209–227.

Coates, J.F. (2010) 'The Future of Foresight – A US Perspective', *Technological Forecasting and Social Change*, 77, pp. 1428–1437.

De Brabandere, L. and Iny, A. (2010) 'Scenarios and Creativity: Thinking in New Boxes', *Technological Forecasting and Social Change*, 77, pp. 1506–1512.
De Vasconcellos, S.L., Lapuente Garrido, I., and Couto Parentec, R. (2019) 'Organizational Creativity as a Crucial Resource for Building International Business Competence', *International Business Review*, 28, pp. 438–449.
Fleck, J. (1971a) 'Ódinn's Self-Sacrifice – A New Interpretation: I: The Ritual Inversion', *Scandinavian Studies*, 43(2), pp. 119–142.
Fleck, J. (1971b) 'Ódinn's Self-Sacrifice – A New Interpretation: I: The Ritual Landscape', *Scandinavian Studies*, 43(4), pp. 385–413.
Flom, G.T. (1939) 'The Drama of Norse Mythology', *Scandinavian Studies and Notes*, 15(5), pp. 135–157.
Fuller, T. (2017) 'Anxious Relationships: The Unmarked Futures for Post-Normal Scenarios in Anticipatory Systems', *Technological Forecasting and Social Change*, 124, pp. 41–50.
Gibbs, D.A. and Flotemersch, J. (2019) 'How Environmental Futures can Inform Decision Making: A Review', *Futures*, 108, pp. 37–52.
Groves, C. (2017) 'Emptying the Future: On the Environmental Politics of Anticipation', *Futures*, 92, pp. 29–38.
Hagtvedt, L.P., Dossinger, K., Harrison, S.H. and Huang, L. (2019) 'Curiosity made the Cat More Creative: Specific Curiosity as a Driver of Creativity', *Organizational Behavior and Human Decision Processes*, 150, pp. 1–13.
Heidegger, M. (2007) 'The Concept of Time in the Science of History', in Kisiel, T. and Sheehan, T. (eds.), *Becoming Heidegger. On the Trail of His Early Occasional Writings, 1910–1927*. Evanston, IL: Northwestern University Press, pp. 61–72.
Jaoui, H. (1970) 'Creativity Methods in Technological Forecasting', *Futures*, 2(4), pp. 373–375.
Jónatansdóttir, K. (2020) 'Snorri Sturluson, The *Prose Edda* (ca. 1220–40)', in McNabb, C.H. (ed.), *Medieval Disability Sourcebook. Western Europe*. Goleta, CA: Punctum Books, pp. 411–415.
Jouvenel, B. de [1967] (2012) *The Art of Conjecture*. Trans. N. Lary. New York, NY: Routledge.
Loewenstein, J. and Mueller, J. (2016) 'Implicit Theories of Creative Ideas: How Culture Guides Creativity Assessments', *Academy of Management Discoveries*, 2(4), pp. 320–348.
Mailhammer, R. and Vennemann, T. (2019) *The Carthaginian North: Semitic Influence on Early Germanic. A Linguistic and Cultural Study*. Amsterdam: John Benjamins Publishing Company.
Marchais-Roubelat, A. (2016) 'Innovation, action des faits futurs et relations de dominance. Le cas de la gouvernance de l'électronucléaire en France', *Prospective et Stratégie*, 7, pp. 79–95.
Marchais-Roubelat, A. (2012a) 'Contracts to Frame Sustainable Futures: The Rational and Symbolic Sides of Contract Functions and Dysfunctions', *Society and Business Review*, 7(1), pp. 50–64.
Marchais-Roubelat, A. (2012b) *La décision. Figures, symboles et mythes*. Bourges: Apors Editions.
Marchais-Roubelat, A. (2000) *De la décision à l'action. Essai de stratégie et de tactique*. Paris: Economica.
Marchais-Roubelat, A. and Roubelat, F. (2008) 'Designing Action-Based Scenarios', *Futures*, 40, pp. 25–33.

Marenko, B. (2014) 'Neo-Animism and Design. A New Paradigm in Object Theory', *Design and Culture*, 6(2), pp. 219–242.

McKinnell, J. (2014) *Essays on Eddic Poetry*. Eds. D. Kick and J. Shafer. 'Toronto Old Norse-Icelandic' Series. Toronto: University of Toronto Press.

McKinnell, J., Simek, R. and Düwel, K. (2004) *Runes, Magic and Religion: A Sourcebook*. Vienna: Fassbaender.

Meyer, J. and Rowan, B. (1977) 'Institutionalized Organizations: Formal Structures as Myth and Ceremony', *American Journal of Sociology*, 83(2), pp. 340–363.

Miller, R. (2007) 'Futures Literacy: A Hybrid Strategic Scenario Method', *Futures*, 39(4), pp. 341–362.

Mitchell, S. (2008) 'n-Rune and Nordic Charm Magic', in Magnúsdóttir, A.G., Janson, H., Johansson, K.G., Malm, M. and Rogström, L. (eds.), *'Viska alla vara välkomna!' Nordiska studier tillägnade Kristinn Jóhannesson*. Göteborg: Göteborg University, pp. 219–229.

Montuori, A. (2011) 'Beyond Postnormal Times: The Future of Creativity and the Creativity of the Future', *Futures*, 43(2), pp. 221–227.

Naishtat, F. (2018) 'The Crisis of Historical Time at the Beginning of the Twentieth Century: An Early Counterpoint Between Benjamin and Heidegger as a Crucial Issue for Thinking Modernity, Globalization and its Historical Space', in Roldán, C., Brauer, D. and Rohbeck, J. (eds.), *Philosophy of Globalization*. Berlin: De Gruyter, pp. 435–447.

Nordeide, S.W. and Edwards, K.J. (2019) *The Vikings*. Amsterdam: Arc Humanities Press.

Osborn, M. and Longland, S. (1980) 'A Celtic Intruder in the Old English "Rune Poem"', *Neuphilologische Mitteilungen*, 81(4), pp. 385–387.

Page, R.I. (1998) *Runes and Runic Inscriptions. Collected Essays on Anglo-Saxon and Viking Runes*. 2nd edn. Woodbridge, NJ: The Boydel Press.

Ramírez, R. and Selin, C. (2014) 'Plausibility and Probability in Scenario Planning', *Foresight*, 16(1), pp. 54–74.

Rosing, K., Bledow, R., Frese, M., Baytalskaya, N., Johnson Lascano, J. and Farr, J.J. (2018) 'The Temporal Pattern of Creativity and Implementation in Teams', *Journal of Occupational and Organizational Psychology*, 91, pp. 798–822.

Rowlinson, M., Hassard, J. and Decker, S. (2014) 'Research Strategies for Organizational History: A Dialogue between Historical Theory and Organization Theory', *Academy of Management Review*, 39(3), pp. 250–274.

Runco, M.A. and Beghetto, R.A. (2019) 'Primary and Secondary Creativity', *Current Opinion in Behavioral Sciences*, 27, pp. 7–10.

Runco, M.A. and Jaeger, G.J. (2012) 'The Standard Definition of Creativity', *Creativity Research Journal*, 24, pp. 92–96.

Sardar, Z. (2010) 'The Namesake: Futures; Futures Studies; Futurology; Futuristic; Foresight – What's in a Name?', *Futures*, 42(1), pp. 177–184.

Simon, F., Allix-Desfautaux, C., Khelil, N. and Le Nadant, A.L. (2018) 'Creativity within Boundaries: Social Identity and the Development of New Ideas in Franchise Systems', *Creativity and Innovation Management*, 27, pp. 444–457.

Slaughter, R.A. (2020) 'Farewell Alternative Futures?', *Futures*, 121, pp. 1–14.

Sommer, A. and Pearson, C.M. (2007) 'Antecedents of Creative Decision Making in Organizational Crisis: A Team-Based Simulation', *Technological Forecasting and Social Change*, 74, pp. 1234–1251.

Teodoridis, F., Bikard, M. and Vakili, K. (2019) 'Creativity at the Knowledge Frontier: The Impact of Specialization in Fast- and Slow-paced Domains', *Administrative Science Quarterly*, 64(4), pp. 894–927.

Tornaghi, P. (2010) 'Anglo-Saxon Charms and the Language of Magic', *Aevum: Rassegna di scienze storiche linguistiche e filologiche*, 84(2), pp. 439–464.

Tsoukas, H. and Chia, R. (2002) 'On Organizational Becoming: Rethinking Organizational Change', *Organization Science*, 13(5), pp. 567–582.

Van Vliet, M., Kok, K., Veldamp, A. and Sarkki, S. (2012) 'Structure in Creativity: An Exploratory Study to Analyse the Effects of Structuring Tools on Scenario Workshop Results', *Futures*, 44, pp. 746–760.

Vennemann, T. (2012) *Germania Semitica*. Series: 'Trends in Linguistics. Studies and Monographs, no. 259', series ed. P.N. Aziz Hanna. Berlin and New York, NY: Mouton de Gruyter.

Voros, J. (2005) 'Speaking about the Future: "Pro-Vocation" and "Ante-Diction"', *Futures*, 37, pp. 87–96.

Wang, S., Liu, Y. and Shalley, C.E. (2018) 'Idiosyncratic Deals and Employee Creativity: The Mediating Role of Creative Self-Efficacy', *Human Resources Management*, 18(57), pp. 1443–1453.

Weick, K.E. (1993) 'The Collapse of Sensemaking in Organizations: The Mann Gulch Disaster', *Administrative Science Quarterly*, 38, pp. 628–652.

Weik, E. (2012) 'Introducing "The Creativity of Action" into Institutionalist Theory', *M@n@gement*, 15(5), pp. 563–581.

Williamson, B. (2001) 'Creativity, the Corporate Curriculum and the Future: A Case Study', *Futures*, 33, pp. 541–555.

Woodman, R.W., Sawyer, J.E. and Griffin, R.W. (1993) 'Toward a Theory of Organizational Creativity', *Academy of Management Review*, 18(2), pp. 293–321.

Wrenn, C.L. (1932) 'Late Old English Rune-Names', *Medium Ævum*, 1(1), pp. 24–34.

Zabelina, D.L., Friedman, N.P. and Andrews-Hanna, J. (2019) 'Unity and Diversity of Executive Functions in Creativity', *Consciousness and Cognition*, 68, pp. 47–56.

4 Ernst Bloch's ontology of not-yet being

Intuiting the possibility of anticipation's fulfilment

Nathaniel J.P. Barron

Introduction

For futurists Roberto Poli and Riel Miller (2010, p. 1) the study of anticipation 'is an effective way to unpack the future'. In other words, by analysing the category of anticipation we might come to learn something of how the future is structured, if it is structured at all. Design presents itself as an appropriate site to explore this connection between anticipation and what is not yet. The gamut of design practices and modes ideation have been said to unite in a common tendency, namely, in anticipation of the future (Dunne and Raby, 2013). At one remove from considering design *as* anticipation – as 'futurology' (Dunne and Raby, 2013, p. 3) – is the philosophical question of what drives and renders possible design's anticipatory projections; this is a question that has risen to prominence in recent anticipation studies, and one that is said to require an ontological orientation in thought (Poli, 2010). Jamie Brassett (2015, p. 33), a philosopher of design, has adopted such an orientation, couching design's futural directedness as 'a process that engages with possibilities and the dynamic and complex ways that they may actualise'. At a fundamental level design can simply be construed as an articulation of possibility and actuality. As old metaphysical by-words for being itself, the ancient metaphysicians continuously cast actuality as preceding possibility 'in both concept and substance', as Brentano (1975, p. 29) has stated of the Aristotelian doctrine. The ancient tongue of metaphysics expressed the ontology of the world and all within it as essentially actualized, then, as possessing no incompletion whatsoever.[1] Ostensibly, design adopts a full reversal of this ancient premise – its profoundly anticipatory gesture is overturning this closure of what being *could* be.[2] Thus, we find that what makes anticipation possible is the ontological antecedence of possibility itself – design thrives and indeed *exists* at all thanks to possibility's antecedence. We can go one step further, however, and say that this antecedence of possibility is *exemplified* by design in its embodying the pulse of a broader tendency of the human animal in both its individual and collective life: that of imagining another world – perhaps a better one – and helping that new, preferred future to emerge (Rodgers and Bremner, 2018; Poli, 2019). In response to the question posed above, we might ponder the following trajectory of thought: is design's anticipatoriness not driven and made possible by a disjunction in being (there is a certain untimeliness to design) and, moreover, by the possibility of a *plus ultra* (that which is capable of being bettered) of disjunction?

Framed in this way, design lands us squarely with the encyclopaedist of hopes Ernst Bloch (1885–1977), whose utopian thought supplies us a phrase for this double acknowledgement: arguably, design carries within itself 'the intuition of a possible better being' (Bloch, 1985a, p. 116).[3] Yet despite proponents of the burgeoning field of anticipation studies recognizing that Bloch's work avails us of rich tools to categorize anticipation (see: Poli, 2010, 2017, 2019), there so far exists insufficient engagement on the part of philosophy of design with this thinker who was so acutely aware of the significance of anticipation. Bloch is – it must be noted – equally reticent of design. Of the few comments he does provide his general tenor remains one of suspicion of and contempt for the evident complicity of industrial design with capitalism; geared as industrial design has often been towards high and fast profits, it lacked the alleviation of labour that it first sold as a promise. Design has, Bloch therefore wrote, a 'pervasive destruction of the imagination on its conscience' (Bloch, 2000, p. 11). A real anticipation of the future had dropped off design's scale; and yet this could only have borne down on design's conscience if it had been seen to betray its anticipatory potential. In what follows, I will explore how Bloch's utopian philosophy can reconnect design with this (lost?) potential, a potential that Bloch inextricably associates with a peculiarly conceived utopian category of fulfilment.[4]

To shed light on what he can contribute to design and anticipation studies, in what follows I will focus on how Bloch's ontology of not-yet being closely mirrors Roberto Poli's (2010) notion that an anticipatory system is self-referential and impredicative; although, as I also hope to show, if the reason for why there is anticipation is to be approximated, this conception must be supplemented by the possibility of fulfilment – which Poli precludes. The nub of Bloch's utopian philosophy, as we will see, is that fulfilment constitutes a real possibility within incomplete being. My argument will be that Bloch accesses this truth via an immediate experience – an intuition, a direct seeing – of one's own existence, an experience which Bloch categorizes as 'the darkness of the just lived moment' (Bloch, 1985b, p. 147). Reading this category serves to shed light on Bloch's approach to anticipation as a necessary feature of human existence in a world that remains incomplete in its core of being, a world, indeed, whose very incompletion can be disclosed via an intuitive sense of ourselves. My discussion below will help clarify, within the horizon of Bloch's philosophy, the inextricable connection of what Poli (2010) has distinguished as the human capacity to anticipate from what renders anticipation ontologically possible. The possibility of anticipation is united, for Bloch, with the necessity of a *possible* fulfilment.[5]

In the first section of this chapter, I will discuss the key categories that make up Bloch's ontology of not-yet being, while in the second I will briefly outline Poli's theorization of matter, in which Poli's categorization of anticipation is firmly embedded. I will critique Poli's categorization on the basis of its explicit abandonment of the possibility of fulfilment. In the third section, I will tie these discussions together, attending to how Bloch's self-referential and impredicative ontology articulates a unity between the human capacity to anticipate and the ontological correlate that makes anticipation possible. This site of ontological unity stands as the possibility of a not-yet actual fulfilment. As I have already mentioned, Bloch discloses this possibility of fulfilment by way of 'the darkness of the just lived moment' (Bloch, 1985a, p. 147). I will then conclude with some brief comments on what impact Bloch's utopian intuition may have for design specifically, and for anticipation studies more broadly.

Bloch's ontology of not-yet being

In *The Principle of Hope*, his magnum opus, Bloch goes about hyperbolically assailing all former philosophy for its putative failure to treat sufficiently the realest dimension of being: 'the future tense', Bloch (1995, p. 6) proclaims, 'philosophically [. . .] has still not been adequately considered to this day'. Once the future is rejected as philosophically insignificant, then the notion that reality is ontologically open is discounted as totally irrelevant; subsequently we are left to make peace with this intolerable reality of ours. But Bloch wants to crack open this closed notion of the real. Indeed, for Bloch, it is the future that gives 'reality its real dimension' (Bloch, 1995, p. 285). The 'sustaining intuition' (Siebers, 2018, p. 51) of Bloch's philosophical vision is of an existence that remains unfinished; if the future is being's realest dimension, then what is traditionally referred to via the signifier of 'being' is in truth an incomplete referent, and with this incompletion there is still work to be done to render it concluded. Present existence possesses an as of yet undetermined surplus of itself, then, and Bloch's life's work was to render this surplus, this residue of un-actuality, recognizable in the myriad layers of its existence – from colportage to art to knowledge to being itself. Bloch therefore shares with design the same overarching intention: to 'materialis[e] the "not yet" now' (Brassett and Marenko, 2015, p. 12). As Bloch writes:

> to find the *right thing*, for which it is worthy to live [. . .]: *that is why we go*, why we cut new, metaphysically constitutive paths, summon what is not, build into the blue, and build ourselves into the blue [. . .] *incipit vita nova*.
> (Bloch, 2000, p. 3; emphases added)

The importance of the category of fulfilment for Bloch's 'ontology of not-yet being', which I mentioned at the beginning of this chapter, is clearly perceptible in this famous passage from the *Spirit of Utopia*. The present existential unavailability of the 'right thing' (it does not yet exist) is the reason why we anticipate, why we create, and why in the process of anticipating and creating we re-create and re-anticipate ourselves – it is because of this unavailability that we build ourselves into the blue. I will return to these points in more detail when I broach Poli's work, but first let me clarify some of the key categories of Bloch's ontology, which are all dimly present in the passage just quoted.

Anticipation is possible – it only makes sense as a category – in a world in which possibility possesses an ontological ascendency over actuality. I have already highlighted this point, and Bloch wholeheartedly agrees. From the outset, Bloch's ontology is totally constructed within an anti-Eleatic train of thought. Compared to 'the land of actuality, the ocean of possibility is much greater' (Bloch, 1978, p. 356). Bloch's (2019) prioritization of possibility cuts against the grain of orthodox Aristotelianism; it holds that when being is considered incomplete, that which is *un-actual* comes first in the order of being. Bloch neither philosophically prioritizes nor commences from already existent actualizations (although of course he does not jettison their importance). Nor does he begin from an already realized pure actuality (*actus purus*), as many of his counterparts in the history of philosophy have tended to do.[6] Instead, Bloch begins from the aphotic conditions of possibility. His point of departure is being deprived of completion, and so precedes ontology if the latter is understood as preoccupied with originary being or with what exists actually (see: Poli, 2006; Scognamiglio, 2010). The core of being is un-actual, is 'still night, incognito, ferment', 'around which everyone, everything, and every work is still built' (Bloch, 2000, p. 173).

It is decisive to note that this elevation of possibility to the status of precedency does not, however, preclude Bloch's ontology from concerning itself with the creative and anticipative transgression of this condition of un-actuality. Indeed, the un-actual arrives first in the order of being for another reason besides being's incompletion: the not-yet existent goal of anticipation that is always aimed at, is what propels incomplete being forward. Bloch suggests that incomplete being is invariantly directed to an overcoming of its own unfinished state; in the darkness of possibility's ocean there is said to be a real possibility of transgressing un-actuality and, moreover, that all of present existence is in some sense disposed towards this real possibility (Bloch, 1975, p. 30). Crucially, however, by the lights of Bloch's own philosophical commencement point, this disposition towards fulfilment[7] is not to be characterized by transcendence (despite Bloch's often close association with the theological tradition), but by transcending *minus* transcendence (Vilmar, 1965; Agar, 2014). This move intensifies the critical function of Bloch's philosophy in its casting present being as intolerable, for no evasion towards a pre-empted place beyond present being is feasible in the sphere of utopian philosophy.[8] For Bloch, to be sure, there is no pre-given truth which incomplete immanence could point to as its actual transcendent ground or destination; hence, there can be nothing but immanence to incomplete being's tending, since what it is directed towards is a not yet actual fulfilment of itself. The fulfilment sought for – what is ultimately anticipated – is itself *of* immanence, then.[9] This process of immanence can be best described as promulgating a strictly *utopian* mode of anticipation;

for the contents of its fulfilment are constituted in their process and therefore, thus far, remain 'undecided' (Bloch, 1995, p. 193). What is ultimately sought for ('the right thing') takes place within immanence's *self-relation* – there is no beyond this relation. This expresses Bloch's notion of 'the incomplete system' of his philosophy having its correlate in the 'objectively real incompletion' at the heart of being (Bloch, 1985a, p. 471), a *shared* incompletion that renders anticipation possible and, crucially, fundamental to the very nature of philosophy itself. Indeed, it is simultaneously this being dispossessed of an un-actual fulfilled being and the possibility of this dispossession being overhauled that inaugurates and augments creative anticipation. It is on this score that Bloch's ontology accords with and in fact to an extent inspires Poli's (2010, p. 12) suggestion that 'anticipatory systems are self-referential or impredicative systems'. Self-referential insofar as the creative anticipations of immanence are born from and always refer to the incompletion of immanence, whose fulfilment would be *its own* fulfilment; impredicative by virtue of Bloch's transcendence-free ontology: the fulfilled contents of this incompleteness are not pre-modelled but are anticipated as not-yet.[10]

It is in this Blochian thought-context that I can now offer a general definition of creative anticipation, which can be defined as a manifestation of 'the unresolved utopian tension constantly undermining everything shaped' (Bloch, 2000, p. 228). The already formed existents that surround us and that we ourselves are, are pervaded by a tension of immanence incomplete with itself yet striving for an utter *novelty of itself* that is actualized only in the process of a creative anticipation of its arrival. As Bloch writes of the utopia category on this score:

> At the very beginning Thomas More designated utopia as a place, an island in the distant South Seas. This designation underwent changes later so that it left space and entered time. Indeed, the Utopians, especially those of the eighteenth and nineteenth centuries, transposed the wishland more into the future. In other words, there is a transformation of the topos from space to time. With Thomas More the wishland was still ready, on a distant island, but I am not there. On the other hand, when it is transposed into the future, not only am I not there, but utopia itself is also not with itself. This island does not even exist. But it is not something like nonsense or absolute fancy; rather it is not *yet* in the sense of a possibility; *that* it could be there if we could only do something for it. Not only if we travel there, but *in that* we travel there the island utopia rises out of the sea of the possible – utopia, but with new contents.
>
> (Bloch, 1988, p. 2; original emphasis)

In this light, it becomes clear why Bloch's project of constructing a new metaphysics has been said to be one of erecting a metaphysics *of the new* – a metaphysics of a utopia that rises out of the process of its own anticipation (Siebers, 2018). Fulfilled being essentially amounts to a not-yet Novum (singular, not plural, for being is what unites, not divides, multiplicity), and this Novum constitutes 'the *unum necessarium*' (Bloch, quoted in Claussen, 2008, p. 272), the one thing necessary; the 'right thing' that would hush the constant undercutting tension of incomplete immanence. As Siebers (2013, p. 63) contends, at work here is the radical idea that 'novelty and ultimacy belong together'. Bloch advances, then, a philosophical vision in which being's incompletion tends towards a potential novelty of itself as the ultimate ingredient of its own consummation, an Ultimum whose contents remain unresolved and will continue to be unresolved unless there is a creative anticipation of fulfilment's arrival (Bloch, 1995, p. 202). As Bloch states rather cryptically of this, utopian philosophy is not guided by the notion that the '*idea* exists', for we do not know what this fulfilment could *be*, since its contents remain un-actual and emerge only through their own constitutive anticipation (thus its impredicativity); but rather by the notion that the 'idea *exists*', for we do know that it *could* be when being is radically open by dint of its own incompleteness (Bloch, 2000, p. 177). What we have here, again, is an ultimate idea of fulfilled being that exists as a possibility preceding its actual contents; it is only known that this idea is *not* not possible, is not *im*possible. This litotic idea of fulfilled being is therefore 'utopianly real' (Bloch, 2000, p. 178); its u-topic reality cannot by definition be 'based on a declaration of facticity' (Bloch, quoted in Löwy, 1976, p. 37), and thus the anticipation of its arrival serves a critical function with regards to present reality (Catalani, 2020). Not only would this as of yet non-present 'right thing' mark a novelty in being (thus it is non-reducible to what has so far become), but it also stands as facticity's condition of possibility as such (it is responsible for what has thus far arrived), for facticity is nothing less than previous spent anticipations – failed transcendings, as it were – of the ontological disposition towards the Novum that would be the Ultimum. Such is Bloch's ontology of not-yet being in broad outline.[11]

Roberto Poli's anticipation in relation to Bloch

Poli's work is of note not only because it has attended to the category of anticipation with rigour and interdisciplinarity, and not merely because it places great stock on the future and anticipation in the context of ontological speculation, as does Bloch. Poli's work is also significant insofar as it casts an eye to utopia in its categorization of anticipation (Poli, 2017, pp. 65–67). To his merit, then, Poli does not depreciate this category entirely. Below I want to outline briefly how Poli's theorization of anticipation, which stands at the forefront of this new form of studies, is embedded in a conception of matter, before then highlighting, from a Blochian standpoint, a flaw of its application to the category of utopia. In doing so, I hope to lend to the category of fulfilment a higher status within the emerging field of anticipation studies. At the very least, I hope to show that fulfilment needs must figure in our ontological understanding of anticipation's possibility. Such is Bloch's lesson to contemporary anticipation studies.

It must be said at the outset that Poli and Bloch carry markedly different styles. On the one hand we have Poli, who approaches the category of anticipation scientifically and analytically; on the other hand, Bloch, who assumes a more expressive and affective tone. Their differences in this respect notwithstanding, both Poli and Bloch, who equally share the mantle of enhancing our understanding of anticipation, share commonalities in their approach; an uncontentious point that is, as we will see, recognized by Poli himself. While I will not be able to do the full measure of Poli's interventions on anticipation justice here (his body of work is extensive), a number of points are worth making before I juxtapose his with Bloch's position on the score of utopia and fulfilment. I have already mentioned how Poli's notion of an anticipatory system being self-referential and impredicative maps neatly onto Bloch's ontology of not-yet being. More would need to be done to fully analyse this apparent convergence, however, which is not my aim in this chapter. Differences do begin to emerge between Bloch and Poli, however, precisely at a moment when they seem to most closely align: namely, in Poli's quite interesting categorization of matter and what, as I will argue, is its subsequent misapplication to the category of utopia, and this by its own logic.

Briefly put, Poli lucidly elaborates Aristotle's notion of essence (what it *was* to be that thing) as the *formal* aspect of an entity which marks its past traces: 'the entity's past history is retrievable as far as it has left traces in the entity's actual configuration' (Poli, 2006, p. 75; emphasis removed). The *material* aspect of that entity, its matter, on the other hand, marks its 'indeterminability', that is, its '*principle of openness*' to what it may become in the future (Poli, 2006, p. 75; original emphasis). Within this material aspect of an entity Poli makes a further sub-division between what he will call the entity's possibility and the same entity's potentiality:

> upon deeper reflection, matter shows two different components. The first [. . .] is the reflection of form around the present. The same rules used to obtain the actual from the past can be used to project what is actually given in the future [. . .] The problem [. . .] is that all the possibilities should be calculated in advance. Which is unlikely, to say the less [sic], and simply unfeasible as far as complexity issues are considered. The second and truly demanding component of matter is embedded in the creativity of causality, that is, permanent openness towards novelty [. . .] To mark the difference between the two components of matter, I shall distinguish between possibility and potentiality.
>
> (Poli, 2006, p. 76)

In other words, anticipating the future of an entity is split two ways, between anticipating its future as a variance of repeatability, 'the continuation of the past in the future' (Poli, 2006, p. 76), that is, its material *possibility*, and anticipating its future as a real novelty of itself, that is, its material *potentiality*. I suspect that

Bloch would remain in likely agreement with Poli at this point, having himself developed a similar approach, although he employs contrasting terminology:

> The real possible has two sides, a reverse side as it were, on which the measures of the *respectively* Possible are written, and a front side on which the Totum of the *finally* Possible indicates that it is still open.
>
> (Bloch, 1995, p. 206; original emphases)

But in fact, Bloch's articulation of what appears to be a similar conception of the material aspect of an entity is what sets him apart from Poli, as I will now show. Indeed, against this backdrop it seems erroneous of Poli to subsequently declare, both in the light of Bloch's ontology *and* in view of his own distinction between possibility and potentiality, that 'no serious scholar of utopia would claim that utopias are aimed at the perfect society' (Poli, 2019, p. 7), and, nevertheless, that perhaps the most serious scholar of utopia – Bloch – is one of a few philosophers to have 'conducted the most extensive categorical analyses of the future' (Poli, 2017, p. 88; see also: Poli, 2006). This is misguided because, as we have seen, Bloch is a utopian scholar who holds *open* the possibility of fulfilment precisely insofar as reality is itself radically open. Poli seems to miss the association of the Novum with fulfilment, with the Ultimum. But this is a note of the score that cannot be discarded from Bloch's categorization of the future, lest incomplete being is anticipated as an ocean without a shore, even if that shore is not yet determined, but created through a process of its anticipation (Bloch, 2000). Here, and more importantly, Poli's position of barring the possibility of fulfilment (potentiality, in his terms) bears a resemblance to Theodor W. Adorno's position in the context of a reasonably well-known disputation with Bloch over the contradictions of utopian longing (see: Bloch, 1988; Truskolaski, 2020). As Bloch (quoted in Claussen, 2008, p. 272) wrote to Adorno on this: 'It is true that the utopian conscience [. . .] has remained alive, quite explicitly, in your rich and successful writings. But the snag [is] the abandonment of the great line, the *unum necessarium*'. In other words, to confine anticipation to process alone, while omitting its what-for, is, for Bloch, to restrict the lifeblood of anticipation.

A similar Adornian stance seals Poli's comments. To let anticipation roam in the domain of potentiality is presumably to side with what Poli (2006, p. 78) terms Bloch's 'mystical theory of [. . .] potentiality', that is, to side with a non-rational element, we can suppose, a charge repeatedly levelled against Bloch due to his penchant for theological ideas (Rehmann, 2020). That a singular desirable state of the future – not-yet actualized – might abolish further need for anticipation is a potentiality that Poli cannot countenance; he cannot entertain the prospect that an arrival of what is new could fundamentally re-configure any previous understanding of what *anticipation is itself* and thus *what anticipation can do*, to echo a phrase of Spinoza's (Poli, 2010). It should be noted that this accusation of mysticism against Bloch's category of possible fulfilment turns on Poli's (2019, p. 7) claim that, in opposition to utopias of order, '[u]topias of freedom are [...] things that by definition are never achievable', and, moreover, that 'it is unlikely that human societies will ever achieve perfect justice'. The problematic nature of such claims, from a Blochian standpoint and in light of Poli's own distinction between possibility and potentiality, is that they close down the *potentiality* of anticipating and experimenting with the idea of, in this case, perfect justice. Both statements presume to know the contents of perfect justice and freedom – they remain regulative, and thus abstractly static, marred by traces of the past – which are subsequently, in a Kantian manoeuvre, calculated in advance as entirely unrealisable.[12] 'Nevertheless, the idea of justice acts as a motivating force', argues Poli (2019, p. 7). How, then, if its goal is rendered impossible, is this force motivated, if every anticipation is consigned to failure before it begins? If reality really is *open*,[13] do Poli's statements not foreclose the expansive field of anticipation to the realm of mere possibility; do they not decidedly confine the very notion of fulfilment 'within the boundaries of the category of possibility', thereby projecting the past onto the future, measuring the future by the rules of the past (Poli, 2006, p. 76)? Do they not thereby steal away the openness of the future from anticipation and the very temporal core of what it aims at? Surely *real* anticipation anticipates not repetition alone (mere possibility) but that which is irreducible to what has come before it (potentiality)? But this is all apparently foreclosed by one fell swoop of the pen. Anticipation is circumscribed to a predictable facticity; but have we not learnt from Poli himself that the future is not a fact?[14]

I think that this is an important critique to level against Poli, since – returning to Bloch – it is the human animal's unique factual location within this nexus of incompletion, novelty and ultimacy which best discloses the possibility of fulfilment, for the human is said to be a peculiar entity that is an intense and (at times) highly erratic anticipatory tensioning. In a letter to the evangelical theologian Paul Tillich (1886–1965), for example, Bloch (1985b, p. 827) states that three categories constitute the most important yet 'most un-thought through' philosophically: 'Front – Novum – Ultimum'. The incompletion of immanence is said to possess a Front in which the actualization of the Novum is anticipated and experimented with most decisively. The place that Bloch assigns to human beings is crucial insofar as we are grasped as that incomplete anticipatory form residing on the Front of the broader ontological process of creative anticipation. We are a unique site of tensioning in incomplete being between what has and what has not yet become. Bloch's suggestion here is that at the level of the human collective, as at the level of individual consciousness, the human animal occupies the 'foremost segment of history' (Bloch, 1995, p. 200). '[H]uman wakefulness', Bloch (2000, p. 167) argues, constitutes the highest developed form of creative anticipation; albeit, like all other hitherto shaped forms, and like the contents of fulfilment itself, human shapedness remains inconclusive – what Derek Hales (2015, p. 140) terms 'conjectural'.[15] The human animal conjectures at an intense level of Frontal exposure: we are 'surely a leap like no other on the historically-forming Front' (Bloch, 1975, p. 173). It is on the Front of incomplete being, then, where the human is to be found truly anticipating, where fulfilment discloses its possibility most powerfully. This is why Hudson (2019, p. 161) has correctly argued that Bloch's utopian philosophy can be read as 'link[ing] human subjective experience of fulfilment with a finalism of the object'.

In the light of the above, Poli has not satisfactorily demonstrated why, within the context of the potentiality of matter, the Novum precludes fulfilment. Indeed, why should a categorical analysis of what is not, not include within it the possibility of fulfilment? It is Poli himself, after all, who recognizes, in his very denial of the possibility of fulfilment, that nowhere yet has true fulfilment roamed the Earth. On this account, Bloch and Poli are poles apart.

The darkness of the just lived moment

How is this *possibility* of fulfilment (potentiality in Poli's terms) disclosed? I have registered how Bloch conceives of the human animal as occupying the Front of incomplete immanence. It can now be said that Bloch's ontology commences from an intuition of the Front of incomplete immanence, that is, from the human animal's immediate experience.[16] In the light of its thoroughgoing immanence, Bloch's philosophy may be described as developing an 'intra-ontology', then, one for which '[b]eing must be approached from within' (Burke, 2005, p. 185). To discern the possibility of fulfilment we would therefore have to follow this movement into immanence. Indeed, Bloch writes that 'the still present non-congruence of even the best realization with the goal-image' (Bloch, 1995, pp. 192–193) is 'the central question of the world', and this is the question of 'what we ourselves are, what all things are without distance from themselves' (Bloch, 1985a, p. 148). Here Bloch appeals to a concept of fulfilment that is defined by an immediacy without remove from itself. Fulfilment would be an overcoming of this breach within immediacy.[17] Let us turn to immediacy, then.

Bloch, to be sure, opens most of his works with variations of the following introspective statement: 'What then? I am. But I do not yet have myself. We know therefore nowhere what we are, too much is full of something that's missing' (Bloch, 1975, p. 11). Here we are provided a description of Bloch's Frontal departure within anticipative immanence, a commencement point *in* and *of* immanence. In fact, it describes *the most* immanent of all starting points *within incomplete being itself* – with consciousness and thus with 'the front of immediacy' (Bloch, 1985b, p. 147). How, then, is the incompleteness of being and its possibility of fulfilment disclosed on the Front of immediate experience? Both are brought to light, Bloch argues, when the Front of immediacy simply *self-regards* itself – the very operation that Bloch's quotation above undertakes. When this 'darkness of the most *immanent immanence*, in the nearness within', self-regards itself, Bloch (1985a, p. 150) argues, a distance (a disjunction) within immanence opens itself up to our vision.

Firstly, then, the incompletion of being is disclosed in this self-regard of immanence insofar as the disjunction of immanence is seen as homologous with our distance from ourselves; our distance from ourselves is thus inscribed into immanence as such and as a whole. From the simplest and most immediate of human experiences – from the experience of ourselves – Bloch constructs a new ontology, an ontology born of the 'Not-there in the midst of the immediate nearness of occurrence' (Bloch, 1985a, p. 193). Bloch otherwise categorizes this Not-there as 'the darkness of the just lived moment' (Bloch, 1985a, p. 147). Bloch articulates this category with an ocular analogy:

> [what] we humans are, represents only an untrue form, to be considered only provisionally. We have no organ for the I or the We; rather, we are located in our own blind spot, in the darkness of the lived moment, whose darkness is ultimately *our own darkness*, being-unfamiliar-to-ourselves, being-enfolded, being-missing.
>
> (Bloch, 2000, p. 200; original emphasis)

In intuiting the ground of ourselves as being-missing we intuit 'the darkness of the lived moment' (Bloch, 2000, p. 200), whose disjunctive moment belongs to being, too. Being's nearness to itself is intuited as a not-having itself: 'unknowing [is] the ground of the manifestation of this world' (Bloch, 2000, p. 205). This not-having is not an unknowing that operates merely on the plane of epistemology, however, rather it assumes the stakes of an ontological problem. As Bloch (1995, p. 7) writes: 'intention towards possibility that has still not become [. . .] is not only a basic feature of human consciousness, but [. . .] a basic determination within objective reality'. However, if one is willing to accept the homology that is made here by Bloch of incomplete being and a consciousness that cannot fully see itself, it is quite another thing to accept that the intuition of incomplete being provides an intuition of the possibility of its fulfilment. For Bloch, nevertheless, this intuition of the Front of immediacy just as much bears the possibility of fulfilment as it discloses present disjunction. This can be intimated in the following passage:

> Had the searcher totally nothing, then he would not be able to search. That, what is missing, the Not-had, is in him equally present as Had, otherwise it would not be capable of being sublated into Not-having.
>
> (Bloch, 1978, p. 406)

In other words, I would not know that something is missing were that something not present as a possibility, for were it not present as a possibility, I would not then know that I am incomplete. Indeed, for Bloch, this darkness of our lived now is inextricable from the future and from the real possibility that this not-yet fulfilment constitutes:

> the *future*, the *topos of the unknown within the future*, where alone we occur, where alone, novel and profound, the function of hope also flashes, without the bleak reprise of anamnesis – is itself nothing but *our expanded*

> *darkness, than our darkness in the issue of its own womb, in the expansion of its latency* [. . .] *the very thing-in-itself everywhere is this, which is not yet, which actually stirs in the darkness, the blueness, at the heart of objects.*
>
> (Bloch, 2000, p. 201. Original emphases.)

To close this chapter, I would like to consider an intuitive image that Bloch judges discloses ontological disjunction and the possibility of a *plus ultra* of that disjunction. It is an irony, to be sure, that Bloch's utopian intuition of the incomplete, anticipative ground of the human animal – what Poli refers to as Bloch's mysticism – should surface and be rendered visible – intuited concretely – by the Viennese physicist and philosopher Ernst Mach (1838–1916), a positivist so close to the hardened facts. Mach is said to provide a self-reflection of empirical experience that stumbles upon the incomplete and anticipative correlate of that experience, and thereby renders perceivable both the distance in immanence as well as the possibility of fulfilment. Incidentally, the same youthful Bloch who had read the Kantian thing in itself as an objective phantasy had also corresponded with Mach and, as Pelletier (2018, p. 29) argues, on occasion had sided with Mach's reduction of 'the "objective" world to the sensations we have of it'.[18] In his 1903 letter to Mach, for example, Bloch states that '[a]n extra-psychical existence is evidently meaningless. Reality is intuitability; we never come out of the circle of the optical' (Bloch, 1985b, p. 20). Curiously, 'Machism' combined with the Marxian problematic during the course of the Russian Revolution, particularly in the work of Alexander Bogdanov (Jensen, 1978), whose 'empiriomonism' Vladimir Lenin branded 'solipsism' (Lenin, 1972, pp. 99–100). The 'ego', Mach (1914, p. 13) wrote, 'can be so extended as to embrace the entire world. The ego is not sharply marked off, its limits are very indefinite and arbitrarily displaceable'.

Bloch eventually overcame his youthful fascination with Mach's extreme phenomenalism (Moir, 2020, pp. 40–44). And yet, in a woodcut self-portrait of Mach's, entitled *Self-regarding Ego* [*Selbstanschauung Ich*] (see Figure 4.1) and which appeared in Mach's *Analysis of Sensations* (1914), Bloch espies a self-reverting representation of immediate sense-impression that captures the double-edged truth of the darkness of the just lived moment: disjunction and the possibility of its *plus ultra.* Indeed, *if* being is appearance, as Mach held, then an analysis of appearance exhibits a strange though, for Bloch, fundamentally optative blind spot. This is how Bloch describes Mach's self-portrait:

> It is even more perplexing when the body looks at itself as it really is, without prejudice and without a mirror. It was [. . .] Mach who, in attempting to gain a primal sense-impression of his own person, removed himself from the mediating effects of this device [that is, the mirror]. Mach's *Analysis of Sensations*, in most respects a rather questionable work, nonetheless contains a striking picture: a self-portrait of Mach lying on a couch. This extraordinary drawing is entitled *Self-regarding Ego*. In fact, no image could be freer from the mirror's supplemental effects, and none more immediate

Figure 4.1 'Self-Regarding Ego' by Ernst Mach. © Ernst Mach 1897, *Analysis of Sensations*, Continuum Publishing, an imprint of Bloomsbury Publishing Plc.

> in terms of what is given by the senses. This is how the reclining, self-regarding Mach appears: strands of beard waft over a massive shoulder, and where the head and neck should be joined to that shoulder is nothing but blank space; further down, grotesquely foreshortened, extend the torso, the legs, and feet. [. . .] Furthermore, the self-portrait is not situated at a distance from the observer, as the columns, balustrades, and heavenly ceilings of baroque art are. Singularly enough, [. . .] the subject that occupies the centre of a self-portrait does not appear at all: the man who regards himself is decapitated. Mach's descriptive drawing of the ego is [. . .] amateurishly executed, and the philosophical aim for which it was undertaken is not of the highest order. Nonetheless: the headless drawing of the artist is a novum in the history of the self-portrait.
>
> (Bloch, 1998, pp. 193–194)

Via an immediate intuition of oneself represented sensuously, 'immediate in terms of what is given by the senses' (Bloch, 1998, p. 193), an intuition free from the 'round eye of the mirror' (the past traces of possibility?) (Claudel, quoted in Merleau-Ponty, 2004, p. 300), the incompleteness of one's own ego is revealed.[19] Only a 'primal sense-impression of his own person' (Bloch, 1998, p. 193), a representation turned toward the subject himself (the movement from immanence to immanence), occasions Mach to apperceive an otherwise imperceptible truth: the certainty of the incompleteness of one's self-existence, the absolute poverty of one's own content. To paraphrase Poli (2017, p. 268), prior to his self-portraiture Mach had been 'inattentional' to his blindness, was incapable of perceiving what was 'in plain sight'. In this 'reverting, intuitant activity', however, Mach 'self-objectifies' (Goudeli, 2002, p. 149) himself and appears as pervaded and grounded by an unknown region, thereby accessing, for Bloch, the unity of the seer and seen in mutual incompleteness. The indifference of the white paper *just is* the indifference between subject and world in their unifying incompletion, their background postulate of non-identity and its anticipated overcoming. Again:

> Had the searcher totally nothing, then he would not be able to search. That, what is missing, the Not-had, is in him equally present as Had, otherwise it would not be capable of being sublated into Not-having.
>
> (Bloch, 1978, p. 406)

The uniqueness of this discovery, in which Mach turns upon himself, is that it wins a vision that develops *in praesentia* – before one's very eyes – the incompleteness, the blank space, the gap in which something that is missing appears to us in the stream of our sense-perceptive experience. From an indirect aesthetic intuition, Mach's sketch reveals, for Bloch, the ontological significance of incompleteness as the unitive truth of the reality of subject and object. Mach's portrait, a representation of an analysis of sensations, is a *utopian* intuition, interfusing subject and object in the open future that is their unitive ground. Mach's auto-figurative self-portrait is thus a self-portrait of incomplete being by incomplete being. It is a self-portrait of and by disjunctive being. It is equally a self-portrait of and by the possibility of this disjunction's *plus ultra*.

Conclusion: designing fulfilment

> [W]e have thoughtlessly believed that a design was a tracing, a copy, a second thing.
>
> Maurice Merleau-Ponty, *Maurice Merleau-Ponty: Basic Writings*, 2004, p. 296

In sum, Bloch's ontology of not-yet being posits the correlate of our experience as incomplete and as situated within a future that has not yet been determined, but that harbours the real possibility of a fulfilment. This ontological condition of incompletion, and this real possibility of fulfilment, can be intuited insofar as the Front of incomplete immanence – which we are ourselves – self-regards itself in its own immediacy. An 'intuition of a possible better being' (Bloch, 1985a, p. 116) is co-posited in the disjunctive present of immanence. When we relate to ourselves without the contrivance of a mirror, without the past traces of mere material repeatability, mere material possibility, a *real* self-relatedness, then; we intuit the incompleteness of our own and the world's being. We intuit how that incompleteness that we ourselves are relates to itself: as a question devoid of a clear response yet anticipating one. At least, such is Bloch's line of argument as I have reconstructed it here.

How can design and anticipation studies more broadly be affected by this utopian intuition? For design, the intuition primarily clarifies that its practices and modes of ideation are never just a passive copying of or corresponding to something that already *exists*; design never simply bows to the repeatability of possibility, in Poli's sense of that term. Rather, design is always an anticipation of something that does *not* exist but *could* if design were truly anticipative of what anticipation might actualize. After all, when it salvages its mandate of living in potentiality – again, in Poli's sense of that term – design has the construction and not just the destruction of imagination on its conscience. Likewise, for anticipation studies more broadly, Bloch's utopian intuition, I think, signals a lighthouse warning, an exhortation against shipwrecking on the rocks of premature limits that foreclose the radically open reality it inhabits. The 'urge to correspond to oneself' (Bloch, 2000, p. 171), to realize a 'genuine possession' (Bloch, 2000, p. 168) – a tendency at the heart of all human designs – is for Bloch an expression of the anticipative ground of our existence, the full realization of which is a real possibility. At the beginning of this chapter, I quoted Poli and Miller (2010, p. 1) to the effect that analyzing anticipation 'is an effective way to unpack the future'; I also mentioned how Poli (2010, 2017) had led us to ask the question of what makes human anticipation ontologically possible. Bloch's utopian philosophy provides us with a response to this question: it is the ontological potentiality of fulfilment, the real 'capacity of arrival' (Bloch, 1962, p. 448) that renders anticipation possible. For Bloch, then, the future will be not appropriately unpacked if it is not seen as harbouring this capacity within itself as a possibility.

Acknowledgements

This chapter was written during a Visiting Fellowship at the Institute of Modern Languages Research (IMLR) at the School of Advanced Study, University of London. I thank the IMLR for their support.

Notes

1 For Aristotle (2004) the essence of a thing is what it *was* to be that thing. See also: Poli (2006, pp. 74–75).
2 The essence of a thing is what it *will be* to be that thing. See also: Poli (2006, p. 75).
3 All translations of Bloch's German works are my own.
4 For a contemporary discussion on capitalism and design, see Brassett and Marenko (2015, p. 4).
5 Aristotle's *Metaphysics* states that necessity's primary sense is '[w]hen something does not admit of being otherwise', making 'it impossible for something to be other than it is' (Aristotle, 2004, pp. 120–121). However, anticipation is coextensive with freedom and a lack of causal antecedence. To claim that anticipation is necessary is to paradoxically state that something is restricted in an openness to its own transformation, in an 'openness forwards' into a newness of itself (Bloch, 1962, p. 14).

6 '[T]rue being, as *ontōs on*, is not yet' (Bloch, 1970, p. 223).
7 A disposition is a fact 'with an anchor in the future' (Poli, 2011, p. 70).
8 Bloch was a materialist. For him, materialism equates to a philosophical perspective that attempts to explain the world *from out of the world*, without recourse to a transcendent principle (Hudson, 2019). In Bloch (1972, p. 17), matter is conceived of in a similar way: 'the word matter originated from *mater*, meaning a fruitful world-womb, experimenting with forms, figures' and 'shapes of existence'. Matter is conceived of as 'midwife' to its own autopoesis (see Bloch, 2019; Moir, 2020).
9 See Pelletier (2020) for critical remarks on the transcendence-free nature of Bloch's ontology: 'From [Max] Scheler's viewpoint [a contemporary of Bloch's], Bloch's rejection of every transcendental predetermination, with the exception of the utopian Good that is meant negatively in all our anticipative acts, may well have seemed to provide no roadsign [*sic*] for the human quest and praxis' (Pelletier, 2020, p. 44).
10 I cannot go into detail on these concepts of self-referentiality and impredicativity (see: Poli, 2017, pp. 193–206). My concern in the relevant section below will be with Poli's theorization of matter and its application to utopia.
11 For an excellent survey of the philosophical sources of Bloch's ontology, see: Pelletier (2018).
12 It seems here that 'the [...] categorical openness' of fulfilment itself is denied (Poli, 2006, p. 77).
13 '[P]otentiality makes sense only if the thesis is accepted that *reality is open*' (Poli, 2009, p. 20; original emphasis).
14 Thus, not only are there utopias of the left and right (Poli, 2017), but utopias internal to the left itself, reformist and revolutionary utopias, utopias of possibility and utopias of potentiality (cf. Poli, 2006).
15 '[T]hat we humans are, represents only an untrue form, to be considered only provisionally' (Bloch, 2000, p. 200).
16 Intuition assumes a fissured presence in the history of philosophy. It has tended to concern two incompatible types of immediate knowledge: sense-perception's empiricism and intelligible perception's rationalism (Caygill, 1995, pp. 262–264). The intuition meant here is akin to a direct perception which just sees without conceptual mediation; a seeing, not a thinking (see: Gram, 1981; Estes, 2010; Nassar, 2012). The transition from the psychology of the darkness of the just lived moment to its ontological register does not remain incontrovertible in Bloch; it is not philosophically demonstrated as such. This is perhaps due to its embeddedness in intuition. For an alternative reading of intuition and the future, see Jancsary (2019), who explores this in connection to Henri Bergson's process philosophy.
17 Bloch develops a concept of immanence opposed to Spinoza, for 'Spinozan [sic] immanence is devoid of distance' (Chepurin, 2017, p. 74).
18 In his *Jugendmanuskript* 'On Force and its Essence' (from 1902), the youthful Bloch interprets Kant's 'thing in itself' – the inaccessible correlate of our experience – as 'the objective phantasy' (Bloch, 1985a, p. 5; see also: Catalani, 2020).
19 Given Bloch's immersion in the German tradition, it is not fortuitous that art (however questionable its qualities) is the organon of this revelation. Indeed, in the earlier stages of Schelling's work, art is seen as 'the true organon of philosophy' (Schelling, 1993, p. 14). Art is 'a concrete analogue of intellectual intuition, the one place where producing and intuiting fully coincide' (Vater, 1993, p. xxiv; see also: Harris, 1983, p. 15; Inwood, 1993, pp. xii–xiii; Schelling, 1993, p. 12). The issue of the organon of German philosophy is intimately tied to the French Revolution. In its aftermath, the 'German road' to revolution found its way divided between, on the one hand, its organon lying in an aesthetic, 'spectorial position' of the 'French original'; or on the other, a political realization of philosophy – a tradition leading to Karl Marx (Kouvelakis, 2018, pp. 11–12).

References

Agar, J. (2014) *Post-Secularism, Realism and Utopia: Transcendence and Immanence from Hegel to Bloch*. Oxon and New York, NY: Routledge.

Aristotle (2004) *The Metaphysics*. Trans. H. Lawson-Tancred. London: Penguin Books.

Bloch, E. (2019) *Avicenna and the Aristotelian Left*. Trans. L. Goldman and P. Thompson. New York, NY: Columbia University Press.

Bloch, E. (2000) *The Spirit of Utopia*. Trans. A.A. Nassar. Stanford, CA: Stanford University Press.

Bloch, E. (1998) 'Self-Portrait Without Mirror', in Hemacher, W. and Wellbery, D.E. (eds.), *Literary Essays*. Trans. A. Joron et al. Stanford, CA: Stanford University Press, pp. 192–196.

Bloch, E. (1995) *The Principle of Hope*. Trans. N. Plaice et al. Cambridge, MA and London: The MIT Press.

Bloch, E. (1988) *The Utopian Function of Art and Literature: Selected Essays*. Trans. J. Zipes and F. Mecklenburg. Series: 'Studies in Contemporary German Social Thought'. Cambridge, MA and London: The MIT Press.

Bloch, E. (1985a) *Philosophische Aufsätze zur objektiven Phantasie*. Frankfurt am Main: Suhrkamp.

Bloch, E. (1985b) *Ernst Bloch Briefe, 1903–1975*: Erster Band. Eds. K. Bloch, J.R. Bloch, A. Frommann, H. Gekle, I. Jens, M. Korol, I. Mülder, A. Münster, U. Polka, and B. Schmidt. Frankfurt am Main: Suhrkamp.

Bloch, E. (1975) *Experimentum Mundi: Frage, Kategorien des Herausbringens, Praxis*. Frankfurt am Main: Suhrkamp.

Bloch, E. (1978) *Tendenz-Latenz-Utopie*. Frankfurt am Main: Suhrkamp.

Bloch, E. (1972) *Das Materialismusproblem, seine Geschichte und Substanz*. Frankfurt am Main: Suhrkamp.

Bloch, E. (1970) *Tübinger Einleitung in die Philosophie*. Frankfurt am Main: Suhrkamp.

Bloch, E. (1962) *Subjekt-Objekt: Erläuterungen zu Hegel. Erweitere Ausgabe*. Frankfurt am Main: Suhrkamp.

Brassett, J. (2015) 'Poised and Complex: The Becoming Each Other of Philosophy, Design and Innovation', in Marenko, B. and Brassett, J. (eds.), *Deleuze and Design*. 'Deleuze Connections' Series. Edinburgh: Edinburgh University Press, pp. 31–57.

Brassett, J. and Marenko, B. (2015) 'Introduction', in Marenko, B. and Brassett, J. (eds.), *Deleuze and Design*. 'Deleuze Connections' Series. Edinburgh: Edinburgh University Press, pp. 1–30.

Brentano, F. (1975) *On the Several Senses of Being in Aristotle*. Trans. R. George. London: University of California Press.

Burke, P. (2005) 'Creativity and the Unconscious in Merleau-Ponty and Schelling', in Wirth, J.M. (ed.), *Schelling Now: Contemporary Readings*. Bloomington, IN: Indiana University Press, pp. 184–206.

Catalani, F. (2020) 'Anticipation as Critique: Objective Phantasy from Ernst Bloch to Günther Anders', *Praktyka Teoretyczna*, 1(35), pp. 149–166.

Caygill, H. (1995) *A Kant Dictionary*. Oxford: Blackwell Publishers.

Chepurin, K. (2017) 'Beginning with Kant: Utopia, Immanence, and the Origin of German Idealism', *Russian Journal of Philosophy and Humanities*, 1(2), pp. 71–90.

Claussen, D. (2008) *Theodor W. Adorno: One Last Genius*. Trans. R. Livingstone. London: The Belknap Press.

Dunne, A. and Raby, F. (2013) *Speculative Everything. Design, Fiction, and Social Dreaming*. Cambridge, MA and London: The MIT Press.

Estes, Y. (2010) 'Intellectual Intuition: Reconsidering Continuity in Kant, Fichte, and Schelling', in Breazeale, D. and Rockmore, T. (eds.), *Fichte, German Idealism, and Early Romanticism*. Amsterdam and New York, NY: Rodopi, pp. 165–177.

Goudeli, K. (2002) *Challenges to German Idealism: Schelling, Fichte and Kant*. Basingstoke: Palgrave Macmillan.

Gram, M.S. (1981) 'Intellectual Intuition: The Continuity Thesis', *Journal of the History of Ideas*, 42, pp. 287–304.

Hales, D. (2015) 'Re-Designing the Objectile', in Marenko, B. and Brassett, J. (eds.), *Deleuze and Design*. 'Deleuze Connections' Series. Edinburgh: Edinburgh University Press, pp. 139–172.

Harris, H.S. (1983) *Hegel's Development: Night Thoughts (Jena 1801–1806)*. Oxford: Clarendon Press.

Hudson, W. (2019) 'Ernst Bloch and the Philosophy of Immanence', *Cosmos and History: The Journal of Natural and Social Philosophy*, 5(1), pp. 155–164.

Inwood, M. (1993) 'Introduction', in Hegel, G.W.F. (ed.), *Introductory Lectures on Aesthetics*. Trans. B. Bosanquet. London: Penguin Books, pp. ix–xxxvi.

Jancsary, J. (2019) 'The Future as an Undefined and Open Time: A Bergsonian Approach', *Axiomathes*, 29(1), pp. 61–80.

Jensen, K.M. (1978) *Beyond Marx and Mach: Aleksandr Bogdanov's Philosophy of Experience*. London: D. Reidel Publishing Company.

Kouvelakis, S. (2018) *Philosophy and Revolution: From Kant to Marx*. London and New York, NY: Verso.

Lenin, V.I. (1972) *Materialism and Empirio-Criticism*. Peking: Foreign Languages Press.

Löwy, M. (1976) 'Interview with Ernst Bloch', *New German Critique*, 9, pp. 35–45.

Mach, E. (1914) *The Analysis of Sensations and the Relation of the Physical to the Psychical*. Trans. C.M. Williams. Chicago, IL and London: Open Court Publishing.

Merleau-Ponty, M. (2004) *Maurice Merleau-Ponty: Basic Writings*. Ed. T. Baldwin. London: Routledge.

Moir, C. (2020) *Ernst Bloch's Speculative Materialism: Ontology, Epistemology, Politics*. Leiden and Boston, MA: Brill.

Nassar, D. (2012) 'Spinoza in Schelling's Early Conception of Intellectual Intuition', in Förster, E. and Melamed, Y.Y. (eds.), *Spinoza and German Idealism*. Cambridge: Cambridge University Press, pp. 136–155.

Pelletier, P. (2020) 'On Ernst Bloch's Moral Theory', *Praktyka Teoretyczna*, 1(35), pp. 35–50.

Pelletier, L. (2018) 'Ernst Bloch's Ontological Realism, Considered from its Sources', in Zimmermann, R.E. (ed.), *Ontologische Probleme der Grundlegung nach Schelling und Bloch*. Hamburg: Verlag Dr. Kovač, pp. 29–48.

Poli, R. (2019) 'Pragmatic Utopias', in Poli, R. and Valerio, M. (eds.), *Anticipation, Agency and Complexity*. Cham: Springer, pp. 1–10.

Poli, R. (2017) *Introduction to Anticipation Studies*. Cham: Springer.

Poli, R. (2011) 'Steps Towards an Explicit Ontology of the Future', *Journal of Future Studies*, 16(1), pp. 67–78.

Poli, R. (2010) 'The Many Aspects of Anticipation', *Foresight*, 12(3), pp. 7–17.

Poli, R. (2006) 'The Ontology of What is Not There', in Malinowski, J. and Pietruszczak, A. (eds.), *Essays in Logic and Ontology*. Amsterdam and New York, NY: Rodopi, pp. 73–80.

Poli, R. and Miller, R. (2010) 'Anticipatory Systems and the Philosophical Foundations of Future Studies', *Foresight*, 12(3), pp. 1–4.

Rehmann, J. (2020) 'Ernst Bloch as Philosopher of Praxis', *Praktyka Teoretyczna*, 1(35), pp. 75–94.

Rodgers, P.A. and Bremner, C. (2018) 'The Design of Nothing: A Working Philosophy', in Vermaas, P.E. and Vial, S. (eds.), *Advancements in the Philosophy of Design*. Cham: Springer, pp. 549–564.

Schelling, F.W.J. (1993) *System of Transcendental Idealism (1800)*. Trans. P. Heath. Charlottesville, VA: University of Virginia Press.

Scognamiglio, C. (2010) 'Anticipation and Future Vision in Nicolai Hartmann's Ontology', *Foresight*, 12(3), pp. 50–58.

Siebers, J. (2018) 'Ernst Bloch's *Geist der Utopie* after a Century: A Janus-Faced Reading on the Trail of Hope', in Silberman, M. (ed.), *Back to the Future: Tradition and Innovation in German Studies*. Oxford: Peter Lang, pp. 37–62.

Siebers, J. (2013) 'Ernst Bloch's Dialectical Anthropology', in Thompson, P. and Žižek, S. (eds.), *The Privatization of Hope: Ernst Bloch and the Future of Utopia*. Durham, NC and London: Duke University Press, pp. 61–81.

Truskolaski, S. (2020) '"*Etwas Fehlt*": Marxian Utopias in Bloch and Adorno', *Praktyka Teoretyczna*, 1(35), pp. 167–185.

Vater, M. (1993) 'Introduction: The Odyssey of Consciousness', in Schelling, F.W.J. (ed.), *System of Transcendental Idealism (1800)*. Trans. P. Heath. Charlottesville, VA: University of Virginia Press, pp. xi–xxxvi.

Vilmar, F. (1965) 'Welt als Laboratorium Salutis: Zur Aufhebung von Transzendenz und Immanenz in der Philosophie Ernst Blochs', in Unseld, S. (ed.), *Ernst Bloch zu ehren: Beträge seinem Werk*. Frankfurt am Main: Suhrkamp, pp. 121–134.

5 Are scenarios creative?

Questioning movement and innovation in anticipation practices

Fabrice Roubelat

Introduction: scenarios and innovation

Witnessing the many papers published over the last decade in scholarly journals such as *Futures*, *Technological Forecasting and Social Change*, *Foresight*, or the newcomer *Futures and Foresight Science*, we can see that futures work is burgeoning. As part of this development, scenario planning and scenario design are being positioned as core methodological frameworks of anticipation literature, as well as of anticipatory organizational practices. From a methodological and conceptual perspective, scenarios explicitly refer to the dichotomy between the exploratory and the anticipatory design of the future (Ducot and Lubben, 1980; Crawford, 2019). In exploratory mode, organizations use scenarios to design prospective futures from past and present processes; while anticipatory scenarios start from imagined or hypothetical views of futures considered as end-states. But is this anticipatory mode, necessarily, a creative way to design scenarios? And do scenarios still reach their promise of bringing novelty to introduce the 'unthinkable' (Kahn, 1962) in decision-making processes? However we see it, designing future-oriented scenarios is a sensemaking exercise, which can be creative or not (MacKay and McKiernan, 2010). So, how might we approach these questions about the relationship between anticipation and scenario design?

Management scholars Brad MacKay and Peter McKiernan (2010) note that, together with innovation, the concept of creativity has become a cornerstone of scenario literature (see also: Sarpong and Maclean, 2011). As Benoît Godin (2016) stresses, innovation had nothing to do with creativity before the twentieth century and the history of the concept of innovation encounters either positive or negative considerations. Prior to this, innovation was considered a 'vice', 'something explicitly forbidden by law and used as a linguistic weapon by the opponents of change' (Godin, 2016, p. 21) before it became a 'buzzword' cast with concepts of initiative, entrepreneurship and creativity. Anyway, can scenarios be considered that creative when creativity is not so much 'thinking outside the box' as finding new boxes in which to think (de Brabandere and Iny, 2010)? Innovation systems are challenged by the act of looking forward in the context of an ever-moving world (Berger, 1957), with organizations trying either to look for new ideas or 'the reapplication of old ideas in a new way' (De Smedt et al., 2013, p. 434). Creativity and innovation embrace a multiplicity of processes to be explored, uncovered or transformed in scenarios.

Like scenarios, innovation processes are defined less by fixed situations than by the different developments that stakeholders are able to create over time. For example, if the future of autonomous vehicles can be designed from standards that need to be negotiated in order to set up new industrial and societal paradigms, such vehicles, as well as related contracts and services, could also be continuously challenged by consumers' practices and behaviours; which will, themselves, be transformed over time. As work on innovation in relation to foresight practices has argued, '[a]lthough the innovation process is now much more open and receptive to social influences, further progress calls for a greater involvement of stakeholders' (De Smedt et al., 2013, p. 434), the concept and value of 'stakeholder' has also become increasingly important in scenario design (Wangel, 2011; Cairns et al., 2016a, b). Offering the possibility to engage a wider range of individuals and organizations, the stakeholder-based perspective invites us to design scenarios as 'moving action' processes (Marchais-Roubelat and Roubelat, 2016). Scenarios in motion necessarily involve creativity and innovation as they issue from stakeholders' capacities to act while simultaneously transforming the multiple branches of action processes. To think in motion is, therefore, creative. This is because scenarios – which are designed as a response to those actions that challenge stakeholders' capacities to act – need to cope with ever-changing rules, practices and contexts. Action thus appears as a source of creativity in scenario design, questioning the emergence of movement at work and suggesting designing scenarios in an iterative process of rule transformation. Focusing on stakeholders' acts, creative scenarios challenge the concept of 'paradigm' (as suggested by Kuhn, 1996), while rhythms of movement change our way to design scenarios.

This chapter will investigate the issue of creativity in anticipation practices, with special attention to scenario design. The first section stresses the difficulty of thinking scenarios in motion, beyond the seeming importance of end-states and the constraints of plausibility. It will then address the relationship between anticipation and stakeholders' own capacities to act within scenarios designed in this way. The second section examines, in an iterative and potentially never-ending process, how action may serve as an Ariadne's thread that allows us either to explore the ramifications of anticipatory action beyond weak signals, or to imagine novel futures from a *tabula rasa*. Questioning the end of paradigms, the concluding section explores the irreversibility of personas' and stakeholders' capacities to act and the rhythms of action over time.

Transforming scenarios as moving action processes

If scenarios are about novel futures, anticipation practices face the issue of the ability of the mind to think ever-changing action processes, which we develop in the first subsection, 'Mobility and the creative mind', named after concepts from philosophers Henri Bergson (1859–1941) and Gaston Berger (1896–1960). In the second subsection, 'Creativity, motion and capacities to act', we discuss the problem of stakeholders anticipating their own transformations and, beyond this transformation issue, creativity in their capacities and incapacities to act.

Mobility and the creative mind

Behind innovation and creativity, scenario literature stresses the need for uncovering surprises and disruption, as well as the difficulty in anticipating such novel futures (Burt, 2007). As a result, more than a field of knowledge, scenarios have been primarily conceived as a method for developing creativity and innovation. In his seminal works on scenarios, Herman Kahn defined scenario thinking with the apparent *aporia* 'thinking the unthinkable' (Kahn, 1962). Kahn has little interest investigating the frontiers of plausible futures (Ramírez and Selin, 2014), because the notion of the 'cone of plausibility' (after Taylor, 1993; Miller, 2011) reduces the creativity of scenarios by focusing on those that can operate with thinkable futures. On the contrary, considering scenarios as 'an aid to imagination' (Kahn, 1962, p. 143), Kahn invites us to look for innovation out of the cone and to be open to questioning successive branching points as well as encountering extreme, impossible or unfamiliar futures. In such creative scenarios, new ideas would provoke discomfort (Ramírez and Selin, 2014) rather than the security of probability or plausibility.

As a networking and a sensemaking process (Roubelat, 2000; Wright, 2005), scenario planning encounters many pitfalls such as 'groupthink', excessive fragmentation of or a lack of innovation (Schoemaker, 1998; Roubelat, 2000; Wright and Cairns, 2011) and has to face a lack of imaginative flexibility (Burt and Nair, 2020). Such rigidities are investigated by Bergson in order to explain the difficulty in thinking a world in motion. In *The Creative Mind* (1946; a translation from the French title *La pensée et le mouvant*, literally 'thinking and moving', originally published in 1938) he discusses the ability of the mind to think mobility, writing: 'it substitutes for the continuous the discontinuous, for mobility stability, for the tendency in process of change it substitutes fixed points which mark a direction of change and tendency' (Bergson, 1946, p. 221). But Bergson (1946, p. 223) adds that 'our mind is able to follow the reverse procedure. It can be installed in the mobile reality, adopt its ceaselessly changing direction, in short, grasp it intuitively'. For scenario thinking, the question rests in the choice between the intellectual comfort of scenarios designed as new end-states or new paradigms and the discomfort of scenarios that could never be captured because they are always in motion.

This issue of dynamic states is a keystone of early futures literature. In proposing the concept of 'prospective anthropology', futurist and philosopher Gaston Berger stresses first that the world is moving and that our actions transform it. But, he explains, the 'acceleration of change' does not affect 'all elements of the world' equally, which are transformed with different rhythms provoking 'time lags' difficult to manage (Berger, 1957, p. 421). Berger notes that, 'the consequence of our acts will take place in a world completely different from that in which we performed them', while 'in order to face new situations we are condemned to an effort of continual invention' (Berger, 1957, p. 420). The 'prospective attitude' proposed by Berger sums up this effort of continual invention and the underlying creativity it requires: looking forward, looking outward, looking in depth, taking chances, thinking sustainability – all dynamic perspectives in a moving frame. He explains further that, 'in the game we must play today, the rules are constantly modified while the number and properties of the pieces change during the same match' (Berger, 1957, p. 422). From such a viewpoint, scenarios can no longer be designed from the grounded certainties of exploratory planning or the teleological certainties of the end-states of anticipatory scenarios, but need to anticipate ever-changing rules, as well as stakeholders' moves and transformations. Both Bergson's and Berger's approaches invite us to design scenarios 'in motion', so that the concept of *state* fades in front of the action of stakeholders, constantly changing rules and the dynamic interaction between every act and everything. All of this continuously challenges scenarios.

If rules are constantly changing and if stakeholders are being transformed over time, the creation of scenarios should include such movements, as well as the time lags created by differences of pace and rhythm across a world with an uneven distribution of futures. Such innovations then serve as new branches for creativity in scenario design. Furthermore, scenarios undergoing such transformations need also to incorporate the tensions created when stakeholders continually shift their (emotional, commercial or other) positions and the breaks and delays that appear when the games they play in the scenario process diverge from that scenario's apparent agenda. Scenario branches (Kahn, 1962, p. 143) are more than disruptions leading to specific end-states (Van Notten et al., 2003). On the contrary, a moving scenario is a never-ending creative process that continually assesses the ramifications of itself over time. In this way, a scenario becomes the beginning of a moving history, rather than its end.

Creativity, motion and capacities to act

Over the last decade, scenario literature has been increasingly focused on the role of stakeholders in scenario design (Wright and Cairns, 2011). This position has evolved from the important notion that futures thinking is a field of power for developing capacities to act (Jouvenel, 1964). In order to understand the value and the operational capacities of stakeholders to act, to innovate, to challenge rules and to create new ones, it is important that we explore branches of scenarios in motion. To be of interest by creating novel ideas, a scenario needs to move, to explore tensions (rather than treat them as problems to be definitively solved). It needs to understand that stakeholders emerge and that others disappear and that organizations, businesses or territories undergo constant transformation. In business-as-usual scenarios, where for example energy consumption is reduced without changing value chains, stakeholders' moves are not that creative: they remain playing within standard, unchanging rules, in stagnant games that fail to evolve. In surprising or extreme scenarios, a stakeholder's capacities to act are radically transformed. This leads us to wonder whether such stakeholders are engaged in a paradigm shift (Kuhn, 1962); which does not mean the end of history as one paradigm moves into another but, rather, characterizes a competition between proliferating paradigms in which the rules for directing stakeholders' acts are in constant motion.

In philosopher of science Thomas Kuhn's (1962) approach to scientific paradigms, a paradigm shift emerges as a creative response to an anomaly; while action within a paradigm remains a fairly simplistic puzzle solving that involves little creativity. Within a paradigm, capacities for action are defined by that paradigm. When an anomaly occurs, these capacities are useless and the various stakeholders have to create both a new paradigm and the capacities for acting within that paradigm. This endless innovation process requires a continuous and mutual transformation of action-rule couples and a constant refoundation of stakeholders' capabilities. Such creativity provokes what was considered an anomaly in one paradigm to move from occurring at the end of a story to the beginning of a new one (Sardar, 2015). In scenario design, looking forward offers an ever-moving perspective to imagine new capacities to act. On the other hand, when a paradigm ends, through scenario building we can highlight the incapacity to act of untransformed stakeholders while showing, at the same time, how new ones create new rules. Scenarios of transformation such as these need to question the next rule and the upcoming move, as these may reveal an anomaly that leads to the next paradigm shift; scenarios thought this way are not mere puzzle solving activities. However, because the upcoming move is not that obvious to anticipate, scenario thinking needs to look outward; for innovation may come from an emergent stakeholder, or rule changes arrive from beyond the limits of thinkable futures. All of which does not necessarily include the innovation of new capacities to act but rather brings a new organization of these capacities.

Thinking the unthinkable, then, is a risky enterprise. Creativity cannot be guaranteed to generate new capacities to act, to make new rules that make sense for stakeholders and become for them an innovative scenario. To renew their capacities to act, stakeholders have to move and then be able to imagine their own transformation in scenarios to be designed as action processes. Therefore, in the following section, I will examine how action may be used as the source of creativity in scenario design.

Action and creativity in scenario design

This examination will resolve into two moments: in the first I will focus upon the ways in which designing scenarios from action challenges the different ramifications of action, weak signals and the role of the *tabula rasa* in future scenarios. In the second subsection, 'Scenario design and rule transformation', I will stress that anticipation practices may become a never-ending process; as creativity may either concern the emergence of new rules or their novel transformation over time.

Questioning the emergence of movement at work

If innovation introduces something new, adapts something old for a novel context or transforms something into something else, the emergence of the change process becomes an important issue in futures literature (Fuller, 2019). One wonders whether or not the origin of creativity across any of these emerging processes can be captured. Nevertheless, however creativity may be located in dynamic emergence, if change is underway scenarios will need to stress its role in a movement either anticipated or uncovered and explored. Scenario design then becomes an activity of time travel in the past, the present and the future: which aims, first, to imagine what is moving or what could move – for example, stakeholders, things, the flows of material or immaterial goods, capital (and so on) or nature. Second, scenario design searches for the seeds of innovation and change: for example, things that have happened and which could be anomalies within a paradigm and so are worth developing into a scenario. Third, looking forward invites scenario designers to create in anticipation something that does not yet exist or that stakeholders have rejected as not, yet, acceptable but which could be a seed for innovation.

Finding out the seeds of innovation does not appear that creative, as scenarios working in this way are based on ideas that have already emerged and have to be developed as drivers for the future. Such innovation is not completely new and relates to the concept of 'weak signals' (Schoemaker et al., 2013; see also: Rossel, 2010): a concept of Ansoff's from the 1970s – that continues to serve as a threshold for futures thinking (Holopainen and Toivonen, 2012; and the whole 'Weak Signals' special issue of *Futures* journal of which this article is a part: Rossel et al., 2012). In such an approach, the relevant environment or horizon are scanned to look for either new movements or changes in trajectories, which then can serve as new branches in designing specific scenarios. Through weak signals, the future is already at work. As a result, there is no creativity in the elicitation of the movement that will ground the scenario; here, creativity comes *after* the signal – with ramifications for action and the diffusion of innovation[1] – which is transformed and distorted over time to create unexpected, if not unthinkable, futures. A signal is not creative by itself, therefore, but from the multiple actions it opens for the future, different scenarios being creatively designed from a single signal, which can be considered as a movement uncovered in the present and the near past.

As a contrast with looking forward, another side of scenarios explores the anticipation of facts that have no connection with the present and the past, facts that cannot be uncovered through the exercise of scanning but from imagination. Here we are closer to a pure creative process. However, we may question whether, for stakeholders engaged in this type of scenario process, it is really possible to cut the ramifications of past and present actions. Is it possible to imagine, in anticipation, a fully disruptive future history? This further requires a reconsideration of the balance between the data and facts on the one hand and the flow of imagination that creates a new developmental process on the other. Consequently, the implications of any scenario may be inferred either from established facts and any transformation they might manifest, or from future conjectures that cannot be grounded in anything, except the creative imagination.

This is not new. In *The Obsolescence of Oracles* (1936) by the Platonist philosopher Plutarch (c. 45–120 CE) the character Lamprias, who seems to speak for Plutarch, states the importance of 'the prophetic current and breath' (Plutarch, 1936, p. 469). He makes the difference between 'the best of seers' who 'is the intelligent man, following the guidance of that in his soul which possesses sense and which, with the help of reasonable probability, leads him on his way'. As a contrast:

> that which foretells the future, like a *tabula rasa* is both irrational and indeterminate in itself, but receptive of impressions and presentiments through what may be done to it, and inconsequently grasps at the future when it is farthest withdrawn from the present. Its withdrawal is brought about by a temperament and disposition of the body as it is subjected to a change which we call inspiration.
>
> (Plutarch, 1936, p. 469)

Before the end of the dialogue, Lamprias stresses that 'the power comes from the gods and demigods, but, for all that, it is not unfailing nor imperishable nor ageless', while 'there are some who assert that the things above the moon also do not abide, but give out as they confront the everlasting and infinite, and undergo continual transmutations and rebirths' (Plutarch, 1936, p. 501). To situate this argument, it is worth noting that Plutarch, in addition to being a philosopher, had a lifelong position serving as one of the two priests of the Temple of Apollo at Delphi.[2] As a result, Plutarch's discussions of the Delphic oracle offer both philosophical and theological perspectives; and so, the theological dimension of these dialogues should not be neglected. By discussing the nature of the prophetic breath, then, Lamprias encourages us to anticipate the future as a *tabula rasa*, so that our imagination can be free to design it as a worldview separated from the present.

From this Plutarchian perspective, scenarios face a dilemma: to be creative from grounded facts or theories; or to create from a blank slate. Moving scenarios, as I have been arguing for here, can be designed from two different inferential processes (Jungermann, 1985): forward induction moving from the present to the future; and backward induction – like backcasting (Vergragt and Quist, 2011) and anticipatory scenarios (Ducot and Lubben, 1980; Crawford, 2019) – from the future to the present. In order to be methodologically creative, then, forward induction considers an imaginary future as the beginning of new ramifications both in and of a new paradigm, rather than as the end-state of a current paradigm. Backward induction, on the other hand, moves from this imaginary future to create consequences in the present. Such creative futures might be unthinkable, utopian or dystopian, though these characterizations would depend upon any stakeholders' assessments of these scenarios, which may vary over time.

If, as futurist, engineer and mathematician Pierre Massé (1967) emphasizes, future-creating facts may not always be those that come first to mind, how might 'moving scenarios' develop interest – both in themselves and as a response from their stakeholders – notwithstanding their drivers as either grounded or imaginary? Futurist Paul Schoemaker (1998, p. 435) cautions that designing too many scenarios is a trap for scenario planning ('Pitfall Number Seven' of twenty). Thus, though designing moving scenarios opens a Pandora's box of numerous signals, imaginary movements, implications and complications, choices must be made. Another difficulty to consider is generated by the concept of 'weak signals', mentioned already above. The concept of 'weak signals' of the future found in the present has been critically evaluated in recent times with the realization that most of them are not that weak (Alloing and Moinet, 2016) and neither are they that emergent. Challenging the concept of 'weak signals' echoes the debate on the value of scenarios (Inayatullah, 2009). These concerns force us to consider how we may value moving scenarios, given their production of complex multiples and their often difficult to discern characteristics.

In a moving scenario, a movement emerges from action, imagined creatively, whatever the processes (inferential or otherwise) at work in its design and development. A creative scenario is not of interest because it is supposed to be plausible (Walton et al., 2019), or extreme (Goodwin and Wright, 2010), as the way to assess a scenario will differ from one stakeholder to the next. A scenario is of interest because it points to new movements to be developed from action. To be creative, then, scenarios cannot be considered as fixed points – therefore, a scenario can only be interesting if it is in movement.

Designing scenarios in motion reveals functions and dysfunctions not only of stakeholders' acts in any scenario, but of their capacities and incapacities to act too; and while change is never-ending (Van de Ven and Sun, 2011) and creativity limitless, the rules for action need to be challenged in an iterative process over time. For it is through action that the thread allowing us to keep our way through the complex maze of moving scenarios will be spun. The next subsection will examine, a little further, the relation between such actions, the rules that generate them and moving scenario design.

Scenario design and rule transformation

We have seen moving scenarios influence facts, events and trends, as well as be implicated in the production of paradigmatic surprises. They also invite us to question those ever-changing rules that impact relations of power and dominance (Marchais-Roubelat and Roubelat, 2016). The concept of 'transformation' that such scenarios participate in relates to the dynamic production of 'pioneering new game rules and transgressing boundaries' (MacDonald, 2012, p. 284). As this characteristic of designing moving scenarios transforms the rules by which their action manifests itself, it also reframes (Ramírez and Wilkinson, 2016) and questions those strategies that articulate stakeholders' interest in them. Scenarios that undergo movement are implicated in the production of strategic action (Marchais-Roubelat and Roubelat, 2015). They, therefore, also drive an iterative process exploring movements that result from sets of rules challenged in action over time. This approach to scenario design states that a scenario can be defined by the rules of stakeholders' engagements. As long as they play by these rules, the action takes place in the related scenario. If one of the rules, or more, changes – even slightly – the action moves to a new scenario. As a result, scenario design encounters two main issues: first, relating to the creative design of the rules; and the second, concerning how to imagine the transformation of the rules over time.

Designing moving scenarios – and managing the strategic action in which this whole process is implicated – suggests dealing with different viewpoints from three interacting rules: a rule regarding action; an institutional rule; and a rule relating to operational matters. These three rules are supplemented with a steering guide, which serves to assess the transformations of the rule set. The first viewpoint is the one which focuses on stakeholders' acts: determining what can and cannot be done in the scenarios. Through the action rule, the objective is to infer or imagine who is acting, as well as their limitations. A scenario under this rule is thus defined by stakeholders' acts over time and by those that cannot be played. This tension between acting and non-acting pushes the limits of the action process at work in the scenario, as the action rule incorporates new acts as well as rejecting acts that would not be sustainable in the scenario. Within this rule, time and space are also of interest to locate when and where stakeholders' acts can be performed or not. As a result, creativity in the action rule deals not only with the acts to be played, but also with their possible performance in time and space.

The second viewpoint, the institutional rule, regards stakeholders' strategies and policies. This rule engages with either the justification for, or the opposition to, the scenario. This rule highlights tacit or explicit agreements between stakeholders, as well as the norms and values they express and the role played by the organization as another stakeholder entity. This stresses the tension between tradition and innovation and the issues that result from the transformation of norms and values over time; in this way, the institutional rule explores issues around regulation, engagement, ethics, politics and community in relation to strategy and scenario design. Here, relations of dominance and subordination, as well as the empowerment or disenfranchising of new or disaffected stakeholders, can all be examined (Cairns and Wright, 2019). It is in relation to the institutional rule, then, that transgressive strategies offer the possibility to design creative scenarios as well as question issues of power (Bourgeois et al., 2017). Here, we are outside any box or any cone, not only dealing with crisis and emergency management, but also correlating new strategic scenarios from the action of emerging stakeholders, whose actions radically change the underlying values of the institutional rule.

Third, the operations rule deals with the organization of stakeholders' interactions within the scenario. In these terms, stakeholders both need to facilitate the performance of the scenario and manage their relations according to the constraints that they face; thus, these can be regarded operationally. This rule specifies stakeholders' capabilities, partnerships and strategic moves as actions within the scenario. Organizational innovation thought from this perspective, then, implies a creativity in thinking the operational capabilities that develop the partnerships needed to make the scenario sustainable over time. Be they creative or not, such capabilities will make the action rule possible, thereby affording capacities to act; while creativity at the level of the organization is required to overcome those incapacities to act that are promoted by the concerted influence of operational constraints and oppositions to the movement of the scenario. While we have treated these rules separately, we can see here that they are interwoven, setting the conditions for and affecting each other.

Such a scenario rule set is not conceived to be an end-state. On the contrary, strategic action scenarios are designed to be challenged by new movements and transformations. In these scenarios, the steering rule overlooks the rule set. As long as the rules run together, stakeholders are engaged in the performance of the scenario. As a result, the next iteration of the exercise attempts to imagine, or discover, how the scenarios undergo various transformations and into what: such as transfers, oscillations, stalemates and phase lags (Marchais-Roubelat and Roubelat, 2015). While transformations suggest exploring shifts from one scenario to another, oscillations between scenarios suppose that any scenario can be played again; with creative interjections being focused upon combinations of scenarios. In a stalemate, stakeholders are trapped in a scenario, which means that creative acts do not succeed in making a transformation. Finally, in phase lags, some stakeholders do not play the game at all, ignoring the rules of engagement within a specific (albeit moving) scenario, preferring instead to perform parallel scenarios. Across all these modalities, scenarios are not designed to be fixed points but to be moving over time, in a never-ending process of change, purposefully and continuously challenging the development of the scenarios, as well as being challenged by them. As a consequence, we will discuss in a final section the issue of the end of paradigms, for this is where moving scenarios and creative anticipation make an impact.

Conclusion and further research: scenarios and the end of paradigms

Processes of never-ending change, as we have been discussing in relation to scenario design, rules of action and imaginative anticipation, invite us to challenge the concept of paradigm itself. For how can an ontological frame last long enough to be called a paradigm under these conditions? These processes also ask us to question the concept of irreversibility: a concept that is presupposed by both paradigm change, in Kuhn's (1996) sense,[3] and the linear, forward-pointing boxes or cones that try to capture the future. Beyond this concept of irreversibility, further research is proposed in the second concluding subsection 'Probing scenario rhythms over time' to question sustainability and ephemerality in scenario design.

Challenging irreversibility

In anticipating the transformations of the rules of the game in their very enactment, moving scenarios are able to incorporate innovations and stakeholders' acts that open new capacities to innovate. At the same time, they can close down other prospective futures, as some either existing or new capacities to act will not be available as creative, moving scenarios develop. This assumes a certain irreversibility to these scenarios, which need to be played out fully even if they ask us to 'think the unthinkable' (Kahn, 1962). The notion of irreversibility actually closes some branches of action from such a scenario while, nevertheless, opening new ones (Ramani and Richard, 1993). Insofar as they challenge paradigms (Roubelat, 2006; Wayland, 2015, 2019), moving scenarios can offer possibilities to investigate new consequences that arise from different anomalies. Other ramifications, however, can become closed, as some puzzle solving capacities are no longer possible as scenarios develop. The concept of irreversibility suggests that, while a transformation has occurred, oscillations are no longer possible; and, further, stakeholders who neither accept nor understand irreversibility are playing within or according to a phase lag. On one hand, in the moment of a paradigm shift, some acts are either no longer possible or are useless, because the action rule has changed. On the other, when new acts do not modify the set of rules that govern them and can be performed according to this rule set, the scenario is not challenged, and the underlying movement does not create any irreversibility.

If we focus upon stakeholders' acts within our moving scenarios, we can see that these scenarios propose an action-based perspective that makes it possible that a stakeholder, or a group of stakeholders, can create irreversibilities over time. Thought in this way, stakeholders' strategies are embedded in scenarios, with strategic options assessed in relation to either their capacities or incapacities to act within them. Stakeholders' empowerment – and the changes in relationships of dominance and power in which they are implicated – are thus occasions to change rules and create irreversibilities. Phase lags – generated when some stakeholders simultaneously play multiple roles on different stages, as we have seen – suggest that stakeholders' performance may question both the concept of irreversibility and the possibility for stakeholders to move from one scenario to another. From a methodological viewpoint, this questions the capacity for stakeholders to anticipate and to be engaged in different scenarios, as well as to manage branches between these scenarios, which could lead to oscillations.

Designing scenarios from stakeholders' viewpoints and taking into account their capacities to act, is an important challenge to the way scenarios are created: with each stakeholder engaged in their scenario, interacting with other stakeholders whose own scenarios branch in different directions. Giving stakeholders capacities to act and to play their own scenario, even insofar as generating new selves or personas (Fergnani, 2019; Vallet et al., 2020), not only offers the possibility for designing many creative scenarios, it also challenges the concept of paradigm: stakeholders playing their own stories, investing scenarios with personal and group interests, with personal irreversibilities that may not be of interest at a global, iterative, level but, nevertheless, may be the source of local developments.

Probing scenario rhythms over time

If scenario rules are ever-changing, as I have been arguing, and if stakeholders (as individuals, groups, communities and organizations) are being continuously transformed over time, the rhythms of scenario transformations can be brought into question, offering us the opportunity to 'probe ephemeral futures'. As applied to scenarios, the issues of acceleration raised by Berger (1957) suggest that scenarios would be moving always faster. As a consequence, each rule set would be ephemeral as stakeholders' capacities to act would need to be re-organized as rules change. Even if we were not to design endlessly moving scenarios (in the ways that we have been discussing throughout), simply thinking scenarios in motion would need a reflection on the rhythms and paces of action that permeate the various creative processes involved. This is apparent in the logic of fashion design's evolving and evanescing temporalities.[4] And so, creativity develops in relation to the variations of transformational temporalities, with each movement following its own evolving space and time.

Would such a fragmentation of time and of scenarios be endless? The end of regular temporalities and the end of long-range rules could become a new set of relationalities, with stakeholders navigating from one scenario to another. The fragmentation and ephemerality of scenarios puts into question the sustainability of their movements in scenario design. With transgressions and disruptions acceptable actions within moving scenarios, the inability to move from one scenario to the next risks stagnating the exercise, even challenging the existence of many stakeholders. Are such scenarios creative? Or would creativity relate only to the development of slow living and the emergence of sustainable communities (Botta, 2016)?[5]

Phase lags (see also Berger's (1957) 'time lags') also invite us to pay attention to stakeholders playing the rules at different rhythms, in scenarios with costless capacities to act that would be designed to be transformed according to circumstances, with stakeholders moving with agility from one scenario to another. Scenarios, albeit ephemeral, would remain paradigms with evolving rule sets. The end of paradigms and of scenarios appears to be designed from a position of 'fleeing forward': racing into the unknown, anticipation-free. Here, stakeholders are either unable to anticipate any scenario or refuse to engage with imaginatively driven anticipation practices in order to focus on action – a present action bereft of any modelled future. These stakeholders would be short-sighted and would be developing capacities only in the flow of action, without any awareness of the rhythms of their actions and their transformations over time.

Such a viewpoint suggests a critical research proposition that challenges anticipation practices designed from a long-range perspective, with multiple, moving scenarios developing along with mutable rule sets that govern action within them. We imagine that the 'creative mind' would also bring out the issues of mobility and ephemerality, irreversibility and iterability, complexity and criticality (Brassett, 2015). Anticipating anticipatory and moving scenarios promises much.

Notes

1 Editors' note. A key progenitor of 'diffusion theory', especially in innovation and invention, but also more widely, was Gabriel Tarde (1843–1904), famous nineteenth-century sociologist, criminologist and Professor of Modern Philosophy at the Collège de France. See: Tarde (1903); on Tarde in relation to sociology of innovation, see Kinnunen (1996). On Tarde and his relationship to many of the philosophies included in this book, see: Toscano (2007). The classic business innovation text on diffusions, which pays its debt to Tarde, is Rogers (1962).

2 Editors' note. This is also known as the Delphic Oracle, one of the most famous Panhellenic oracles; inscribed in the temple is the phrase 'know thyself' (γνῶθι σεαυτόν [*gnōthi seauton*]).

3 Editors' note. We also note that not only is Kuhn's irreversible linearity in question, but also his eurocentrism. Physicist and philosopher of science, Arun Bala (2006) provides a more civilizationally dialogical account of the development of 'modern science', encountering: Indian, Arabic, Chinese, Ancient Egyptian and African work. This recognition spreads the dynamism of intellectual, creative and innovative drivers of Roubelat's argument here even further.

4 Editors' note. See also Brassett and O'Reilly in this volume (Chapter 10), on fashion and being fashionable as acts of contemporaneity.

5 Editors' note. See also O'Reilly in this volume (Chapter 6) and his discussion of Nowotny (2016) and her critique of disruptive innovation.

References

Alloing, C. and Moinet, N. (2016) 'Les signaux faibles: Du mythe à la mystification', *Hermès*, 76, pp. 86–92.

Bala, A. (2006) *The Dialogue of Civilizations in the Birth of Modern Science*. New York, NY: Palgrave Macmillan.

Berger, G. (1957) 'Sciences humaines et prévision', *La Revue des deux Mondes*, 3, pp. 3–12.

Bergson, H. (1946) *The Creative Mind*. Trans. M.L. Andison. New York, NY: The Philosophical Library.

Botta, M. (2016) 'Evolution of the Slow Living Concept within the Models of Sustainable Communities', *Futures*, 80, pp. 3–16.

Bourgeois, R., Penunia, E., Bisht, S., and Boruk, D. (2017) 'Foresight for All: Co-Elaborative Scenario Building and Empowerment', *Technological Forecasting and Social Change*, 124, pp. 178–188.

de Brabandere, L. and Iny, A. (2010) 'Scenarios and Creativity: Thinking in New Boxes', *Technological Forecasting and Social Change*, 77(9), pp. 1506–1512.

Brassett, J. (2015) 'Poised and Complex. The Becoming Each Other of Philosophy, Design and Innovation', in Marenko, B. and Brassett, J. (eds.), *Deleuze and Design*. 'Deleuze Connections' Series. Edinburgh: Edinburgh University Press, pp. 31–57.

Burt, G. (2007) 'Why are we Surprised at Surprises? Integrating Disruption Theory and System Analysis with the Scenario Methodology to Help Identify Disruptions and Discontinuities', *Technological Forecasting and Social Change*, 74(6), pp. 731–749.

Burt, G. and Nair, A.K. (2020), 'Rigidities of Imagination in Scenario Planning: Strategic Foresight through "Unlearning"', *Technological Forecasting and Social Change*, 153, pp. 1–14.

Cairns, G. and Wright, G. (2019) 'Making Scenario Interventions Matter: Exploring Issues of Power and Rationality', *Futures and Foresight Science*, 1(e10). doi:10.1002/ffo2.10.

Cairns, G., Goodwin, P. and Wright, G. (2016a) 'A Decision-Analysis-Based Framework for Analysing Stakeholder Behaviour in Scenario Planning', *European Journal of Operational Research*, 249(3), pp. 1050–1062.

Cairns, G., Wright, G. and Fairbrother, P. (2016b) 'Promoting Articulated Action from Diverse Stakeholders in Response to Public Policy Scenarios: A Case Analysis of the Use of "Scenario Improvisation" Method', *Technological Forecasting and Social Change*, 103, pp. 97–108.

Crawford, M.M. (2019) 'A Comprehensive Scenario Intervention Typology', *Technological Forecasting and Social Change*, 149, pp. 1–27.

De Smedt, P., Borch, K. and Fuller, T. (2013) 'Future Scenarios to Inspire Innovation', *Technological Forecasting and Social Change*, 80(3), pp. 432–443.

Ducot, G. and Lubben, G.J. (1980) 'A Typology for Scenarios', *Futures*, 12(1), pp. 51–57.

Fergnani, A. (2019) 'The Future Persona: A Futures Method to Let your Scenarios Come to Life', *Foresight*, 21(4), pp. 445–466.

Fuller, T. (2019) 'Anticipation and the Normative Stance', in Poli, R. (ed.), *Handbook of Anticipation*. Cham, Switzerland: Springer, pp. 93–108.

Godin, B. (2016) 'Making Sense of Innovation: From Weapon to Instrument to Buzzword', *Quaderni*, 90, pp. 21–40.

Goodwin, P. and Wright, G. (2010) 'The Limits of Forecasting Methods in Anticipating Rare Events', *Technological Forecasting and Social Change*, 77, pp. 355–368.

Holopainen, M. and Toivonen, M. (2012) 'Weak Signals: Ansoff Today', *Futures*, 44(3), 'Weak Signals' Special Issue (April), pp. 198–205.

Inayatullah, S. (ed.) (2009) 'Special Issue: Scenario Symposium', *Journal of Futures Studies*, 13(3), pp. 75–156.

Jouvenel, B. de (1964) *L'Art de la conjecture*. Monaco: Editions du Rocher.

Jungermann, H. (1985) 'Inferential Processes in the Construction of Scenarios', *Journal of Forecasting*, 4(4), pp. 321–327.

Kahn, H. (1962) *Thinking about the Unthinkable*. New York, NY: Horizon Press.

Kinnunen, J. (1996) 'Gabriel Tarde as a Founding Father of Innovation Diffusion Research', *Acta Sociologica*, 39(4), pp. 431–442.

Kuhn, T. (1962) *The Structure of Scientific Revolutions*. Chicago, IL and London: The University of Chicago Press.

MacDonald, N. (2012) 'Futures and Culture', *Futures*, 44(4), pp. 277–291.

MacKay, B. and McKiernan, P. (2010) 'Creativity and Dysfunction in Strategic Processes: The Case of Scenario Planning', *Futures*, 42(4), pp. 271–281.

Marchais-Roubelat, A. and Roubelat, F. (2016) 'Dominance, Stakeholders' Moves and Leadership Shifts: New Directions for Transforming Futures', *Futures*, 80, pp. 45–53.

Marchais-Roubelat, A. and Roubelat, F. (2015) 'Designing a Moving Strategic Foresight Approach', *Foresight*, 17(6), pp. 545–555.

Massé, P. (1967) 'Les attitudes envers l'avenir et leur influence sur le présent', in *Étapes de la prospective*. Paris: PUF, pp. 335–344.

Miller, R. (2011) 'Being without Existing: The Futures Community at a Turning Point? A Comment on Jay Ogilvy's "Facing the Fold"', *Foresight*, 13(4), pp. 24–34.

Nowotny, H. (2016) *The Cunning of Uncertainty*. London: Polity Press.

Plutarch (1936) 'The Obsolescence of Oracles', in Plutarch (ed.), *Moralia*, vol. V. Trans. F.C. Babbit. Loeb Classical Library. Cambridge, MA: Harvard University Press, pp. 347–501.

Ramani, S.V. and Richard, A. (1993) 'Decision, Irreversibility and Flexibility: The Irreversibility Effect Re-Examined', *Theory and Decision*, 35(3), pp. 259–276.

Ramírez, R. and Selin, C. (2014) 'Plausibility and Probability in Scenario Planning', *Foresight*, 16(1), pp. 54–74.

Ramírez, R. and Wilkinson, A. (2016) *Strategic Reframing. The Oxford Scenario Planning Approach*. Oxford: Oxford University Press.

Rogers, E.M. (1962) *The Diffusion of Innovation*. New York, NY: The Free Press.

Roubelat, F. (2006) 'Scenarios to Challenge Strategic Paradigms: Lessons from 2025', *Futures*, 38(5), pp. 519–527.

Roubelat, F. (2000) 'Scenario Planning as a Networking Process'', *Technological forecasting and Social Change*, 65(1), pp. 99–112.

Rossel, P. (2010) 'Making Anticipatory Systems More Robust', *Foresight*, 12(3), pp. 73–86.

Rossel, P., Miller, R. and Jorgensen, U. (eds.) (2012) 'Weak Signals', *Futures*, 44(3), (April), pp. 218–228.

Sardar, Z. (2015) 'Postnormal Time Revisited', *Futures*, 67, pp. 26–39.

Sarpong, D. and Maclean, M. (2011) 'Scenario Thinking: A Practice-based Approach for the Identification of Opportunities for Innovation', *Futures*, 43(10), pp. 1154–1163.

Schoemaker, P.J.H. (1998) 'Twenty Common Pitfalls in Scenario Planning', in Fahey, L. and Randall, M. (eds.), *Learning from the Future*. Chichester: Wiley, pp. 422–432.

Schoemaker, P.J.H., Day, G.S. and Snyder, S.A. (2013) 'Integrating Organizational Networks, Weak Signals, Strategic Radars and Scenario Planning', *Technological Forecasting and Social Change*, 80, pp. 815–824.

Tarde, G. (1903) *The Laws of Imitation*. Trans. E. Clews Parsons. New York, NY: Henry Holt and Co.

Taylor, C.W. (1993) *Alternative World Scenarios for a New Order of Nations*. Strategic Studies Institute, U.S. Army War College. Collingdale, PA: Diane Publishing Co.

Toscano, A. (2007) 'Vital Strategies. Maurizio Lazzarato and the Metaphysics of Contemporary Capitalism', *Theory, Culture and Society*, 24(6), pp. 71–91.

Vallet, F., Puchinger, J., Millonig, A., Lamé, G. and Nicolaï, I. (2020) 'Tangible Futures: Combining Scenario Thinking and Personas – A Pilot Study on Urban Mobility', *Futures*, 117, pp. 1–16.

Van de Ven, A.H. and Sun, K. (2011) 'Breakdowns in Implementing Models of Organizational Change', *Academy of Management Perspectives*, 25(3), pp. 58–74.

Van Notten, P.W.F., Rotmans, J., van Asselt, M.B.A. and Rothman, D.S. (2003) 'An Updated Scenario Typology', *Futures*, 35(5), pp. 423–443.

Vergragt, P.J. and Quist, J. (2011) 'Backcasting for Sustainability: Introduction to the Special Issue', *Technological Forecasting and Social Change*, 78(5), pp. 747–755.

Wangel, J. (2011) 'Change by Whom? Four Ways of Adding Actors and Governance in Backcasting Studies', *Futures*, 43(8), pp. 880–889.

Walton, S., O'Kane, P. and Ruwhiu, D. (2019) 'Developing a Theory of Plausibility in Scenario Building: Designing Plausible Scenarios', *Futures*, 111, pp. 42–56.

Wayland, R. (2015) 'Strategic Foresight in a Changing World', *Foresight*, 17(5), pp. 444–459.

Wayland, R. (2019) 'Three Senses of Paradigm in Scenario Methodology: A Preliminary Framework and Systematic Approach for using Intuitive Logics Scenarios to Change Mental Models and Improve Strategic Decision-Making in Situations of Discontinuity', *Technological Forecasting and Social Change*, 146, pp. 504–516.

Wright, A. (2005) 'The Role of Scenarios as Prospective Sensemaking Devices', *Management Decision*, 43(1), pp. 86–101.

Wright, G. and Cairns, G. (2011) *Scenario Thinking: Practical Approaches to the Future.* Basingstoke, UK: Palgrave Macmillan.

6 Becoming other-wise as the practice of anticipation

John O'Reilly

> My thoughts are to show me *where I stand*, but they are not to tell me where I am going to – I love the ignorance about the future and I do not want to perish in light of impatience and the anticipation of *augured* things.
>
> Friedrich Nietzsche (quoted in Stegmaier, 2016. Original emphasis)

> Tomorrow, is there a possibility for me to become otherwise than what I am? This is not just a question of a private struggle with the self, but of the social terms by which identities are supported and articulated. In this sense, it is always in the context of a certain constellation of social power that I am able to pose the question of my own becoming differently.
>
> Judith Butler (interviewed, in Reddy and Butler, 2004)

Practice stitch_ preface: the anticipatory space-time of spacesuit practice

When US President John F. Kennedy gave his famous speech at Rice University on 12 September 1962, setting the target to go to the Moon within the decade, it was 'because that goal will serve to organize and measure the best of our energies and skills'. The rhetorical splendour of the President's speechwriters is structured by traditional management 'science'. The intangible values of purpose and leadership are auditable in terms of organization and measurement. Section 5.1 of ISO 9001 on Quality Management Systems is on 'Leadership and Commitment'. In the age of management systems, we no longer have to 'show our commitment' as it is now auditable and therefore transparent.

But perhaps there are other versions of anticipation and leadership quite unlike traditional heroic leadership models. From the perspective of Actor Network Theory (ANT), an anticipatory leadership is created from a practice that, according to Suzanne Jane Perillo in her study of leadership in education, is constructed from:

> non-human entities acting in relation with sociality as co-participants. Conceivably, such entities may include buildings, gardens, research papers, offices, pathways, seating, signage, doors, stairways and other examples of

> materiality that emerge as relevant to actual experiences of leadership in everyday practice. Such an approach is underpinned by the assumption that leadership is an open and fluid practice, possibly shaped, in large part, by factors that emerge and construct novel and unanticipated leadership effects.
>
> (Perillo, 2008, p. 194)

Furthermore, she argues, these unanticipated leadership effects emerge in unanticipated locations, through a relational knowledge structure that anticipates the new world in which this new entity emerges, as well as providing that new world with the tools to sustain it. It is perhaps also worth noting that JFK's vision of heroic leadership, both at an individual level and at a geopolitical level, is one that has emerged from a history of 'privilege' and performance of whiteness (Liu and Baker, 2016). Nicholas De Monchaux's (2011) account of the creation of a protective suit for spaceflight in *Spacesuit: Fashioning Apollo* is an ANT version of how a network's human, material and semiotic actors fashion (a discipline which is changed by its relationship to engineering which has also been swerved by its collision with engineering) the new and the world which it will inhabit. In this case it is the imagined world of spaceflight actualized by an emergent knowledge network which seeps out of an inhospitable cranny.

De Monchaux composes his book in 21 separate sections which reflect the 21 layers of the spacesuit worn by Neil Armstrong. In this way De Monchaux challenges or at least draws attention to narrative formation, its linear flow. In effect he fashions a method of construction modelled on clothing construction rather than traditional narrative. De Monchaux is also materializing what he sees as the flat ontology of the project, of human and non-human rather than the heroic organizational narratives of mainstream histories of spaceflight.

A suit, as de Monchaux points out, is not simply a matter of clothing. 'Suit' in English also means being suited to, being adapted to an environment. So, the notion of a spacesuit can be considered more generally as any suit that may anticipate the ecology of a particular environment. It protects and liberates. And if that sounds like a bad strapline for a 1960s underwear advertisement, it is because the success of putting a man on the Moon is largely due to a type of knowledge production, gender, materials and organizational practice excluded from creating the future. To travel to an alien planet with known and unknown dangers we do not need the military-industrial complex, we need seamstresses who make elegant underwear.

There were two different kinds of knowledge production and practice which fed into NASA's networked organizational ecosystem: a dominant knowledge with exclusionary forms of representation; and the Other-wise. There was Hamilton Standard of the United Aircraft conglomerate, one of the biggest US aerospace engineering companies. This organization suited NASA's need for hard documentation and record-keeping, which is key to actualizing it not just as an organization but as a science and technology organization with all the institutional, social and commercial value generated by this documentation. Bruno Latour highlights how industrial drawing not only anticipates what an object might look like but anticipates the many different systems and measurements required to anticipate the coming into being of the new:

> Realms of reality that seem far apart (mechanics, economics, marketing, scientific organization of work) are inches apart, once flattened out onto the same surface. The accumulation of drawings in an optically consistent space is, once again, the 'universal exchanger' that allows work to be planned, dispatched, realized, and responsibility to be attributed.
>
> (Latour, 1986, p. 25)

A traditional engineering company embedded in the atmosphere and systems of procurement of a government organization, Hamilton Standard understood the knowledge production and engineering media of charts and memos required in systems management. The design, management and circulation of these documents both mobilize resources and interests, but also in effect create the organization. Latour is known for his criticism of the traditional discipline of sociology which assumes the concept of the social as the 'backstop' idea, a tautology mobilized to explain the final context (Latour, 2005b). Latour argues that it is exactly the same with the notion of 'organization', whose actualization as an entity is an act of anticipation through documentation, recording and devices which materialize.

> we take for granted that there exist, somewhere in society, macro-actors that naturally dominate the scene: Corporation, State, Productive Forces, Cultures, Imperialisms, '*Mentalités*', etc. Once accepted, these large entities are then used to explain (or to not explain) 'cognitive' aspects of science and technology. The problem is that these entities could not exist at all without the construction of long networks in which numerous faithful records circulate in both directions, records which are, in turn, summarized and displayed to convince. A 'state', a 'corporation', a 'culture', an 'economy' are the result of a punctualization process that obtains a few indicators out of many traces. In order to exist these entities have to be summed up somewhere.
>
> (Latour, 1986, p. 26)

Then there was the International Latex Corporation (ILC) who were better known to consumers as Playtex, a brand embedded in American culture for making bras and girdles. Playtex's initial (1962) partnership working under the lead of Hamilton Standard for NASA did not work out, partly because of the different cultures in both companies. One was a company with a heritage in lingerie, the other in the military-industrial complex. One worked with the human body, flesh and skin, the other with machines. The military-future was hard and diagrammed in inflexible ways, but the future needed to be soft and Other-wise. As de Monchaux points out (2011, p. 182):

> ILC not only prepared the human body for an extreme environment, but also imbedded its own corporate functioning and identity in a new skin. The bureaucratic epidermis, as much as its layered spacesuit design, enabled the company to finally and fully, incorporate itself into the Apollo program.

One of the key frustrations that NASA had with ILC was the inability to provide diagrams for the suit production. For NASA, it was only the organizational performance of the act of documentation that rendered any matter real and authentic. Embedded in the operational function of a style of documentation, this documentation is how time itself comes into play in an organization, as 'history', as official history of organization in the making. De Monchaux cites a NASA procurement official, whose complaints to headquarters display the curious organizational norms operational in large companies for whom the glue of systems information is as important as the products developed. The official noted that prototype suits being delivered by ILC did not have the relevant paperwork: 'After 1 August 1964, all such material not properly shipped will be subject to rejection and will be considered not furnished at all' (de Monchaux, 2011, p. 214). The map is the territory. Yes, NASA loved ILC's work, which far outstripped the hard material alternatives that had been on offer: engineers simply were not suited to producing garments, even ones as technologically accomplished as spacesuits.

The managers at ILC would later admit that the paperwork they produced in order to keep the circulation of the work moving was a 'smokescreen'. If we stitch Whitehead's (1967, p. 191), 'Immediate existence requires the insertion of the future in the crannies of the present' into 'anticipatory creativity requires the squeezing of the future into the organizations of the present', we might see that sometimes the act of smoke screening . . . is an anticipatory practice. That a smokescreen is not necessarily an act of deception, but a construction that enables belief; a smokescreen is a non-human expression of leadership that enables belief in a future to emerge. Process Organization theorist, Tor Hernes (2014, p. 165) calls this 'event formation':

> Leadership may be seen as the power to bring to life and sustain belief in a certain event formation stretching into the past and into the future, but there may be times when the need arises to make a break, at least temporarily, with the past in order to create commitment towards a new future.

The smokescreen is in effect a kind of materialized sympathy, an emergent relational entity composed of misdirection, belief and an engagement with the major language, the dominant discourse and the forms of representation within an organization.

ILC's process emerged from the tailoring and it was the handcraft knowledge of each individual seamstress that was impossible to formalize in the organizational models that NASA required. As one of the seamstresses noted afterwards, 'no two people sew alike'. The suits were sewn to minute tolerances, fitted to the body of each individual astronaut. It was not the only part of the project where the innovation techniques of engineering were no match for older craft skills. The computer backup system of binary code was 'hand-crafted in fabric like "ropes" of thin wire and magnetic washers' (de Monchaux, 2011, p. 284). This practice is similar to that identified by Sadie Plant in *Zeroes and Ones* (1997), not least in the re-insertion of bodily practice into the previously industrial production process of space technology (see also: Plant, 1995).

The precision of the seamstresses in making the spacesuit extended their skills beyond the current forms of their bodies and tools. The fabric for the suit 'was sewn with an accuracy smaller than the sewing needle's eye' (de Monchaux, 2011, p. 209). Each suit comprised hundreds of feet of seams that, stitch-by-stich, had to be exact, as any minute deviation meant the entire fabric would be discarded. The anticipation of this suit emerged not from representational diagrams but from a practice of human and non-human that created new pathways for organizational structuring and encounter. De Monchaux (2011, p. 211) writes:

> some of ILCs most effective engineers, such as Robert Wise, took weeks of sewing lessons from the seamstresses to better understand how fingers, fabric, and thread interacted to build up the suit's complex assemblies. In practice, the craftswomen were allowed, and even encouraged, to suggest improvements in procedures and assemblies as they were continually developed.

In order for ILC to help NASA anticipate the ILC spacesuit being worn on the Moon, the ILC spacesuit used the skin of 16mm film to successfully enter NASA's ecology. An ILC team went to Dover High School, in Delaware near the Playtex factory with a 16mm film camera. As an alternative to more conventional methods of motion studies, and without generating any of the data normally required, they shot footage of an 'astronaut' in a sports-field playing football. At the semiotic level, where better for the future of America to arrive and come home than the high-school football pitch? For three hours, the man in the spacesuit could move around as easily as the engineer without the suit who was throwing the football (de Monchaux, 2011, p. 199). This performance of anticipation and creativity produces the immediate existence of both ILC and NASA in relation to a future in which success has been modelled – in the scientific sense and the fashion sense. Modelling as trying-on-the-future, seeing what it looks like, turns out to be more effective as an anticipatory practice than the conventional measurements. That visual documentation persuaded NASA that the suit had the required mobility that the harder, alternative suits lacked. If, as de Monchaux says, Playtex's design – of the suit and of itself – enabled ILC to connect with NASA, NASA was also transformed, redesigned. In de Monchaux's narrative this 16mm film is organizationally coded as documentation. His narrative moves away too quickly, but one might also see this film – of a man in a pressurized spacesuit tethered to a tube, playing American football with an engineer in a short sleeve shirt – as a piece of avant-garde cinema. The validation of the material for giving sufficient and superior mobility to other suits for walking on the Moon could only be anticipated as an aesthetic, through a different kind of sensing and recording. This organizationally emergent form of conceptual art is an anticipatory expression of the Other-wise.

Film in the mid-twentieth century, the recording matter of film for the military-industrial complex, functioned as what Deleuze and Guattari (1986) call a 'minor language' within the bureaucratic documentation that constructed the industrial and corporate organization as much as the bricks and mortar spaces in which rubber-stamped forms circulated. Film narrative was an environmental idiom that emerged from the linear text systems of bureaucracy, that also served as a medium to compose an extended public for new government apparatus. (For the sake of demonstrating US technological and military power, the 1947 Bikini Atoll nuclear bomb tests were filmed in order to create publicity. The military used over 1.5 million feet of film stock and one million stills, so that they created a scarcity of film worldwide (de Monchaux, 2011, p. 40). How the past of 1947 appears to us compared to other years of the period is a special effect of the lack of film stock.)

This practice of the Other-wise was already folded into the company. Because Hamilton was a traditional aerospace company, with a pipeline of career-paths and knowledge systems, managers in the company found the knowledge background of ILC's designated Apollo engineers indigestible. ILC's Apollo manager was originally hired at ILC because he had repaired TVs for the company's Chief Executive. Son of an auto-dealer, manager George Durney was drafted into the air force as a mechanic in the Second World War, then became a pilot. Post-war he was a sewing-machine salesman at Sears and Roebuck (de Monchaux, 2011, p. 191). Durney's practice combined what would traditionally be called hard and soft skills.

De Monchaux constructs his story as layers, echoing Bruno Latour's isolation of 'eleven distinct layers' (1994, p. 54) in the journal *Common Knowledge* in order to show the complex co-creation of human and non-human actors. Perhaps we need to consider the practices of the Other-wise as the 'uncommoning' of common knowledge; this 'uncommoning' being the practice of seeing how excluded common knowledges may function as useful methods for anticipating the future in the crannies of the present. The Other-wise as practices of anticipation as the stitching that creates the practice of stitching the 'possibility for me to become otherwise than what I am'.

Stitch_ introduction

Recently, I was reading sociologist Helga Nowotny's *Cunning of Uncertainty* (2016), a work that pulls together the conceptual history and discourses of the increasingly pervasive idea of uncertainty. In a twenty-first century world defined by the disruptions of technology, financial crises and climate change, she argues, the concept of uncertainty has become an essential and potentially benevolent disposition to help us deal with the future. The irony is that with uncertainty held as the answer to disruption, it has progressively come to be defined and anticipated in terms of disruption (as we will see).

In effect/affect, in an age whose time and development has been characterized, framed and defined through disruption, the future – as filtered through the experience and category of disruption – is something that happens to us. Organizations, companies and governments can develop scenarios and plan the future through a whole set of foresight and storytelling tools that have been available to us since the work of Herman Kahn at least since the 1960s. But while disruption is the future arriving to change and perhaps destroy an organization and its business, disruption has also arrived as a set of businesses ready to guide other businesses through disruption. Nowotny (2016, p. 23) calls disruption 'one of the most aggressive slogans' of recent times, accusing Clayton Christensen's classic *Innovator's Dilemma* (1997), as an important originator of the disruptive attitude and discourse. Here Christensen describes the ongoing, relentless process of disruption – largely in terms of technology and its impact – in such a way that previously informed organizational decisions are no guide to how we might anticipate the future. Nowotny couches her description of Christensen's concept of disruption in the intense meteorological language of excess/destruction: 'storms', 'hurricane', a 'trail of destruction' (Nowotny, 2016, p. 23): echoing the pre-Enlightenment experience of the Gods. Moreover, unlike Joseph Schumpeter's (1934) vision of innovation – as a process with multiple modalities and vectors, from new kinds of financing, processes, products and organization – disruption, argues Nowotny (2016, p. 23), only promises its own future: 'disruption only preaches the hope of salvation against the very damnation it describes: disrupt and you will be saved'. Uncertainty and the future are constantly being resolved in a process rendered by Nowotny in theological language, echoing Agamben's (2007) discussions of 'profanation': the making secular of the once divine. Nowotny's language is perhaps directed at the power of the discourse of disruption in the business world, a commercial totem that offers certainty and belonging among fellow businesspeople.

Nowotny highlights journalist Jill Lepore's (2014) feature for *The New Yorker*, 'What the gospel of innovation gets wrong', where Lepore diagnoses the shift in what Michel Foucault (1994) would call 'epistemes', regimes of truth.

> The eighteenth century, [Lepore] writes, embraced the idea of progress; the nineteenth featured evolution; the twentieth touted economic growth;

> and now we have innovation. Disruption is an atavistic concept. It is 'a theory of history founded on a profound anxiety about financial collapse, an apocalyptic fear of global devastation, and shaky evidence'.
>
> (Nowotny, 2016, pp. 23–24)

Lepore's identification of the conceptual reduction of innovation and disruption is a move similar to Schumpeter's (1983) connection of economics and innovation. This reduction and equivalence is part of the popular business idiom, expressed in headlines as though it is an equivalence expressed in the business discourse: *Forbes* (2020), 'Timeless Truths About Disruptive Innovation'; *Management Today* (2020), 'Change, we are told, is the only constant in a fast-changing world, making ongoing innovation and the so-called "disruptive leadership"'.

And because, according to Nowotny, disruption only begets disruption, there is no future, only the certainty – in the language of design thinking – of disruption's iteration.[1] For Nowotny there is no anticipation without uncertainty. She cites the example of Brazilian author Cristovão Tezza's (2007) autobiographical fiction about his relationship with his mentally disabled son, whose story concludes with their relationship being forged more deeply by the pair's shared passion and uncertainty around their football team. This, she says, is 'the joy of anticipation' (Nowotny, 2016, p. 145): the experience of not knowing but the anticipation of shared discovery. Though Nowotny elides her insight, anticipation is a joyful activity. To borrow Nietzsche's phrase, we should consider anticipation to be a 'joyous science' (2018). Anticipation is affective and, as we will see, ethical. Furthermore, anticipation is more than a feeling even while being animated by feeling. And we might keep the word 'feeling' open to the outside by considering it an activity of exploring and connecting with a space's ecology and its affective connections: a specific kind of sensing. Anticipation as the activity of 'feeling out' something, of 'sending out feelers', a reconnoitre of the future to make a new present.[2]

Anticipation as the practice of research. This anticipatory practice is uncertain. But if the finding out is uncertain, so is the 'found-out'. The found-out, your team's fortune at the weekend re-arranges your present and past. It is that orientation in starting the week, in a dark cloud or on a high, due to how the team we support fared over the weekend.

Nowotny highlights the affective power of anticipation, but stops at teasing out its implications, how the affective re-arranges our ordering and sensing of time. She speculates, regarding anticipation that:

> What is it in our adult lives that makes these moments so rare when we remember them as abundant in childhood? The titillation of hardly being able to wait: the intense anticipation which outweighs the joy over what is anticipated [. . .] Of having reached some kind of saturation in enjoying the frivolous tension that uncertainty can induce?
>
> (Nowotny, 2016, p. 145)

Anticipation is experienced as a saturation of the self, something which exceeds the self, to produce new perception of the future in the present. If we think of it in educational terms, and anticipation is a kind of scaffold for inhabiting the future in pedagogical terms it is a 'threshold moment': a moment at which we change from one state to another through learning. This event produces what educational theorists Meyer and Land (2003) call, 'troublesome knowledge'. 'Troublesome' because the old ways-of-knowing of the pre-learning state are challenged by what is learned and the new state is uncomfortable. There is an ontological aspect of this that exceeds the epistemological. The (super-)saturation of knowledge through learning is always ontological, in which case the *becoming* of learning requires an excessive saturation of the self.

We may, therefore, consider the ways in which anticipation never remains in a simple and static saturated state. That it is incapable of being and staying only full. That it is the discipline that expands capacity through the excess of uncertainty: a capacity that is, itself, never completed and is always surpassed. The promise of anticipation, as *some kind of* saturation, is its joy in uncertainty, it exceeds. This is what psychologist of creativity Mihaly Csikszentmihalyi (1996) calls flow: where inherited boundaries of self and environment become fluid, generating something new; where the volume of any activity is exceeded by creative practices. Thought this way, anticipation becomes an iteration of creativity.

So, when we 'consider the ways in which anticipation is never saturated', as I offered at the opening of the preceding paragraph, the anticipation of us overflows the 'me' typing this and 'you' making different kinds of sense of it. In truth, if we are following each other in the research of which this chapter is an event, we are already overflowing these positions, practices and temporalities of reading and writing: with 'us' being whatever we are now, in the future, positioned as writer and reader; 'you' who I anticipate reading this and 'me' who you anticipate having written, or in the act of writing these words. Anticipation is a practice which cannot be saturated (definitely). It cannot saturate because, by its very nature as a practice, it is always open to being changed by the future that develops in relation to the present anticipated (Rosen, 2012; Poli, 2017). The practice of anticipation opens frames, environments and even words to the futures that are already here but occluded; it is also affective. It is partly the anticipation of an experience of overflow, of the overflow of experience – the anticipation of going to a concert, a movie, a fashion show, the anticipation of penalties at the end of a football match. It is the affective anticipation of the future changing the present.

This gesture towards opening up of temporality is what I anticipated when I was reading Nowotny's (2016) *Cunning of Uncertainty* with my MA Innovation Management students. Nowotny's preface, entitled 'It could be otherwise', poses the question of the future as difference. This preface is an example of what Gérard Genette (1991) calls 'paratext': a set of practices and discourses around and within the text of a book which positions the text in some way. The paratext surrounding books (or films, Boxsets, music, shopping) may include anticipations such as previews or reviews and, as such, is not always necessarily included in a text's acts of deliberate contextualization. But the overflow I would like us to attend to, now, is the overflow of text and meaning provided by the design of the pages. This is how Nowotny's text appears in the layout in my 2016 Polity Press edition, typeset in 10.5 on 12pt Sabon.

> What does the cunning of uncertainty do to predictions? Probabilities are a forceful reminder that reliance on past extrapolations has limited value because things can be other-wise. The cunning of uncertainty suspends routine. It deflects the linearity of direct cause-effect links, reassuring and indis- pensable as they may be for the routines of daily life.
>
> (Nowotny, 2016, p. 35)

Who might have anticipated the line breaking where it did? With the hyphen sent from a typesetting and printing future to collide with the title of Nowotny's preface, to send it hurtling into our classroom in King's Cross, London and then into this chapter. If anticipation is about finding out as a creative process in itself, not an outcome-oriented problem-solving, but the practice of research, it is also about things anticipating us from the future. To consider another anticipation in this paragraph, the Other-wise deflects the *pensable*, the thinkable of the routine.

Philosophically speaking, this hyphen can be thought Other-wise as Latour's *Ding* (2005a), as the thing that gathers people round a concern that also divides a people. This prompts us to consider anticipation, as the hyphen, as the immediate experience of the future inserted in the crannies of the present. The straight lines of tools such as PESTEL used in foresight work mistake the outcomes generated by assumed contextual categories, for recognizable futures rather than the product of epistemologies closed to their own construction, to the regimes of truth that made them and to the Other-wise that would change them. Foucault would describe such epistemologies and their tools as regimes of practice which:

> possess up to a point their own specific regularities, logic, strategy, self-evidence and 'reason'. It is a question of analyzing a 'regime of practices' – practices being understood here as places where what is said and what is done, rules imposed and reasons given, the planned and the taken for granted meet and interconnect.
>
> (Foucault, 1991, p. 75)

Nowotny tells us with the other hyphen in the passage quoted above, if uncertainty deflects the linearity of 'cause-effect', the Other-wise will be a creative practice eliciting the future to change a present that will never have happened. This deflection,[3] we will see, is also one that acts upon knowledge hierarchies, on official knowledges and practices, as well as unofficial, excluded ones. The cunning of uncertainty, part of the Other-wise practice of taking us into the unknown-unknown, saturates anticipation, and separates the affective re-subjectivation of anticipation from the entitlement to the future of expectation. If deflection and deviation are practices of anticipation, it has a heritage in a mythology of innovation going back to the ancient Greeks. Latour (1994, p. 29) reflects on the different kinds of practices and knowledges required by the ancient Greeks:

> In the myth of Daedalus, all things deviate from the straight line. The direct path of reason and scientific knowledge – episteme – is not the path of every Greek. The clever technical know-how of Daedalus is an instance of metis, of strategy, of the sort of intelligence for which Odysseus (of whom the *Iliad* says that he is polymetis, a bag of tricks) is most famed.

Anticipation is ontological swerving and deviation, of the present to open a future that does not emerge from epistemological straight lines. The present will have expanded its capacity (overflowed itself) to create a future in which it can sustain itself. Anthropologist Arjun Appadurai (2014) argues that this is what research does; and, for him, this is an everyday notion of research, expanded beyond the agendas, interests and knowledge production of universities and corporations. In short, for Appadurai (2014, p. 167), research is linked to what he calls the 'capacity to aspire', a capacity which gives the power of orientation. He says: 'the capacity to aspire is thus a navigational capacity' (Appadurai, 2014, p. 69), and those without this capacity (he argues the poorer members of society) have fewer opportunities to experiment with this capacity, to learn from failure. In the UK for example the aspiration of getting your child into a 'good school' requires a degree of research, experimentation, learning and practice. Anticipation in this way is an actualization of the capacity to aspire to a future not already prefabricated for us. In Appadurai's language foresight research (like much academic research) is parochial, its concerns are limited. Research, says Appadurai, is political and while it is required to be systematic its formats and spaces are not proscribed:

> taking part in democratic society requires one to be informed. One can hardly be informed unless one has some ability to conduct research, however humble the question or however quotidian its inspiration.
>
> (Appadurai, 2006, p. 176)

In the terms of this chapter, we might say that anticipation – as a form of creative researching – is also the capacity to anticipate beyond the given discourses of needs. Anticipation as research is always excessive: exceeding and overflowing the given epistemologies. The hyphen of the Other-wise is an artefact created by our research; and this hyphen which connects and separates, begins to anticipate the kind of discipline anticipation might need to engage with if it is to stay open to the Otherness of the future.

In his paper, 'Nietzsche's Orientation toward the Future' (2016), philosopher Werner Stegmaier explores how Nietzsche's philosophy of interpretation was also, in fact, a philosophy of anticipation: with Nietzsche concerned about how his work would be read and misread by future humans. For Stegmaier, the faultline of this reading/misreading is the divide created by his conceptualization of the future itself:

> When working on his *Zarathustra*, he notes that we have to 'guess the conditions under which future humans live – because such *guessing* and *anticipating* has the **power** of *a motive: the future* as that what we *want affects* our now' [. . .] Orientation is always about exploring a situation with respect to how one can act in it in order to master it, instead of being mastered by it. Orientation is much more oriented toward the future than toward norms. 'Guiding thought: we have to take the future as *binding and decisive* [*maaßgebend* {*sic*}] for all our value judgements – and not look for the laws of our actions **behind** us!' [. . .] Norms restrain the future, instead of opening it.
>
> (Stegmaier, 2016, p. 388; emphases, italics and bold texts are Nietzsche's)

As Nietzsche argues we can orientate ourselves in multiple futures and, most importantly, create the selves each of them generates. Like all research, the challenge of anticipation is not the requirement to know more – which in terms of foresight is different to sensemaking, or wayfinding, or PESTEL – but to create the orientations whereby the future affects us now (see, for example: Rosen, 2012; Poli, 2017). Embedded in these more conventional creative futures' processes are norms of cognition and perception. This has been a problem for traditional futurology, futures thinking and foresight, which are underpinned strongly by the *norms* and values that uphold them. I will examine these in the next section. This direction is most obvious in the justly celebrated work of Pierre Wack for Shell in the 1970s, whose scenario work envisioned the oil crises of the time and enabled the company to be prepared, to a degree, for what happened (Wilkinson and Kupers, 2014). Scenario work is about Wack envisioning risk, threat and opportunity for a company, providing a corporate orientation for the future that clearly worked for Shell. But instead of inserting the future in the crannies of the present, it future-polluted those crannies. The values and ontological orientations of this scenario work, whose context was the age of the birth of environmentalism, meant that it failed to recognize and could not register the planetary havoc and destruction which fossil fuel production was already causing and which would pose the ultimate threat to corporate survival.

In order to generate a creative orientation for anticipation we will need to become Other-wise.

Stitch_ _ _ mapping the corporate imaginaries of futures

> The illusion that Western Civilization could create the problem and solve it is facing its limits.
>
> Walter Mignolo, *The Darker Side of Western Modernity*

The recent history of foresight and trend forecasting has been well-documented (for example: McClanahan, 2009; Millett, 2003; Beck, 2016; Beck and Bishop, 2016; Powers, 2018). Nevertheless, I will outline some of the ways in which it has been represented and constructed; this is to show what is at stake in anticipatory practices that might be able to create a sufficiently liminal space of knowledge/power, to make a future not already policed by pre-formed knowledge boundaries. Though the institutional, political and social genealogy of anticipation from the worlds of futurology, foresight and forecasting is not the focus of this chapter, in order to mobilize a concept and practice of anticipation, I will note that this area of 'futures' is a function of a particular 'episteme' (Foucault, 1972). That is, such a 'futures' produces, and is the product of, knowledge that emerges at a particular historical intersection. Alongside culturally and economically reproduced concepts of the future, the organizations involved in futures research also produce history for business, governments and universities that validates the tools used to produce the future.

Historians and practitioners of futures studies, such as Edward Cornish (2004), situate its emergence as an idea and a cultural and social entity during the industrialization of the eighteenth and nineteenth centuries. It is suggested that, culturally, the growth of scientific methodologies began to open up a predominantly religious worldview whereby people began integrating the idea that life's divine purpose, direction and providence could include ideas of 'progress' that would comprise ways of improving the lives of people (Cornish, 2004, p. 174). In effect, it is the story of 'enlightenment'. German historian Reinhart Koselleck's *Futures Past: On the Semantics of Historical Time* (2004), highlights the shift from the prophetic future of European Christianity, to a predictive future of early modernity with its embedded concept of progress. Before the eighteenth century, any change in the everyday could be understood as a sign of the divine vision, folded into an iteration of God's will, the 'new' being inconceivable.

> Until the eighteenth century it was an almost universally accepted doctrine that one could, from the history of the past, learn lessons for the future. Knowledge of what had been and foreknowledge of what was yet to come remained connected through a quasi-natural horizon of experience, within which nothing essentially new could occur. This was as true of a believing Christian awaiting the End as of a Machiavellian man of politics.
>
> (Koselleck, 2004, p. 58)

In *Futuring: The Exploration of the Future* (2004), Cornish claims that ideas of the future highlight the birth and disillusionment of the idea of progress. In the twentieth century, and especially after the Second World War (though not exclusively), the future as knowledge production became professionalized, socialized and commercialized. If, in the Global North of the twentieth century, the future as a discourse, discipline and object of study was produced alongside economic growth and as an economic commodity itself, the future as a discourse also produced own celebrities.

Cornish, one of the founders of the World Future Society in 1966, highlights former members as: designer and systems theorist Buckminster Fuller; military strategist Herman Kahn, famous for his work *On Thermonuclear War* (1960), as well as the concept of Megadeath which he coined to describe over one million casualties; Gene Roddenberry, producer of TV series *Star Trek*; and anthropologist Margaret Mead. At this moment, the future itself was made famous with Herman Kahn as the model for Stanley Kubrick's character Dr Strangelove, Buckminster Fuller and his Geodesic Dome modelling future isolated city living, and Gene Rodenberry narrativizing a multinational, multialien techno-future for TV living rooms. The dominant discourse of the future of the 1960s is progress through technological supremacy: from missions to the Moon to dinners in a pill.[4]

On the death, in 2016, of Alvin Toffler – popular futurist and author of *Future Shock* (1970) – *The New York Times* (Manjoo, 2016) quoted Cornish from 1978, thus:

> Congressmen and their staffs are searching for ways to make government more anticipatory [. . .] They're beginning to realize that legislation will remain on the books for 25 or 50 years before it is reviewed, and they want to be sure that what they do now won't have an adverse impact years from today.

The piece positions the study of the future as a matter of good governance; and good governance means taking account of bad governance. The more fundamental problem is not the inability to review legislation, but the belief that the future can be legislated. Furthermore, this legislation is specified as the outcome of knowledge production from a political class: elected politicians; the group of educated individuals who aspire to and have been trained in passing through sanctioned institutions of knowledge production; and the policymakers and legislators who know the formats and protocols of how plans for the future are enacted and consequently how the future is, first, performed and, second, how particular performances of the future are legitimated. The 'future' is performed in the present. Sociologist and Science and Technology Studies scholar John Law describes that, in the world outside/beyond its own production as knowledge, 'methods are practices that tend to *enact* realities as well as describing them' (Law, 2009, p. 240; original emphasis). In this way, knowledge practices are 'more or less *performative*' (Law, 2009, p. 240; original emphasis). What is at stake for Law in his unpacking of the *relationality* of Actor Network Theory, is that unless we notice the specificity of actors (and actants) in the production of networks and in their specific organization of knowledge, the social outcomes will already have been prescribed by the use of unexamined epistemologies.

This form of expert anticipatory government that is sub-contracting the future to private organizations and think tanks, enacts a legitimate and legitimizing future through the normalization and classification of regimes of practices (Foucault, 1991).[5] The futures produced in corporate settings (involving and creating certain kinds of experts, processes, spaces, co-creations and their interrelations) are coded using history-specific but atemporal epistemic algorithms. In doing this, these futures bodies not only code but also create the future. Foucault (1991, p. 75) explains:

> It is a question of analyzing a 'regime of practices' – practices being understood here as places where what is said and what is done, rules imposed and reasons given, the planned and the taken for granted meet and interconnect.

Regimes of practices codify spaces of knowledge, such as institutes and think tanks: with objects such as datasets; rules such as PESTEL, backcasting, scenarios, trend analysis, gaming, scanning; and the 'taken for granted' professionals and experts. Farhad Manjoo (2016) concludes his piece for *The New York Times* with a nod to the affective danger and threat of a world where the future is not produced by those professional bodies; wondering whether 'without soliciting advice from a class of professionals charged with thinking systematically about the future, we risk rushing into tomorrow headlong, without a plan'. Imagine what it might be like to go somewhere without a plan, without a goal and result already known in advance? Imagine experimenting instead of strategizing? Imagine taking social science epistemologies out of their consultancy vacuum-seal. The future without a plan offers a dangerous experience of speed, where our upright, vertical bodies tilt on their axis, accelerating like a brainless superman. Planning is insurance against the future, protection against crashing into it. The use of professional futures and foresight practices and the ruling and coordinating of our lives by a plan, are a kind of strategic-hedging: an exercise that creates the belief that at best, engaging in the activity is in itself an indemnity against the future.

Cornish's (2004) book highlights a history of futures built around similar groups of experts. He begins with Plato's vision of utopia in *The Republic*, runs through Enlightenment thinkers, to writers such as Jules Verne and H.G. Wells in the nineteenth and early twentieth centuries and fiction writers of dystopias such as Aldous Huxley, George Orwell, Kurt Vonnegut and Anthony Burgess (whose stories I will return to in the next section). Cornish summarizes the various approaches, methods and tools – the 'regimes of practices' – for creating futures. Like most synopses of futures research, the pivotal moments in Cornish's future telling are the Enlightenment and the two World Wars of the twentieth century, and the different geopolitical, cultural and disciplinary futures that emerged in the post=war period. In France, for example, philosopher and businessman Gaston Berger established in Paris the *Centre International de Prospective* (International Centre for Foresight) and the following year set up the journal *Prospective*. The interdisciplinary strategy of *Prospective* was complemented by journalist and economist Bertrand de Jouvenel (1967) and his Ford Foundation-supported 'Futuribles' project. ('Futuribles' is a neologism mashing 'future' with 'possible'.) The title itself declares a methodological preference in moving futures research and prediction out of the domain of the scientific towards something more creative.

Arguably, the most prominent public image of the futurist by the 1960s was Herman Kahn, who worked for the RAND corporation (RAND being an acronym for 'research and development'). Project RAND emerged from the US Air Force, who gave a contract to the Douglas Aircraft Company (they helped produce the B-17 Flying Fortress) in 1946 to explore inter-continental warfare (Cornish, 2004, p. 195). RAND split from Douglas Aircraft and was initially funded by the Ford Foundation becoming in effect the first modern 'think tank', developing among other things the Delphi method of foresight. Herman Kahn joined in 1947, published *On Thermonuclear War* in 1960 and left to set up the Hudson Institute in 1961, which included some heavyweight institutional thinkers such as French philosopher Raymond Aron, sociologist Daniel Bell and novelist and scholar Ralph Ellison whose 1952 novel *Invisible Man* documented the experience of a black man in 1920s American south and 1930s Harlem (I will return to this). Kahn's notorious work on military strategy (along with Kubrick's reconstruction), gave him social power. The mythology of Kahn and his 'unthinkable futures', was captured by journalists at the time as, '"a roly-poly, second-strike Santa Claus" and "a thermonuclear Zero Mostel." He is supposed to have had the highest IQ on record' (Menand, 2005). Kahn also developed a commercial offer built around aspects of his futures-thinking which he sold to organizations. He called it: 'Corporate Scenarios'.

Elke Seefried, Professor of Modern History at the University of Augsburg, breaks down the development of post-war research into the future into three types of knowledge production by the 1960s: the empirical and positivistic characterized by Herman Kahn in the US (with techniques such as gaming, scenarios and computer simulations); the normative ontological represented by Bernard de Jouvenel (which included notions of the 'good life' in its consideration of the future); and a 'critical and emancipatory' approach reflected in the work of historian Robert Jungk, political scientist Ossip K. Flechtheim and philosopher and pioneer of peace research Johan Galtung (Seefried, 2014, p. 4). Seefried argues that in the 1950s and 1960s futures research was driven by the notion of feasibility and planning, intellectually orbiting around theories of cybernetics and 'systems thinking'. By the 1970s she sees a shift from, broadly, an East-West orientation in the research of futures (a function of the Cold War) to Globalized futures characterized, for example, by Pierre Wack's scenario work for Shell, a methodology he brought from Kahn's Hudson Institute.

Equally, Jenny Andersson, CNRS fellow (*Centre National de la Recherche Scientifique*) and researcher at the *Centre d'études européennes* at Sciences Po, examines the direction of futures in the 1970s. She maps two contrasting approaches: that of OECD's Interfutures group; and The Club of Rome with its report *The Limits to Growth* (1972) which examined the computer simulations of economic and population growth that challenged the political consensus. Andersson shows that the circulation of frameworks, concepts and tools among planners and future researchers of the inter- and post-war periods was well-documented. But, she argues, this field of research and its focus on the long term since the 1970s has had less examination:

> this second mode in planning implied a notable expansion of planning rationalities, both with reference to the 'long term' as temporal horizon, and to the global scale. Future research, for instance the scenario method experimented with by Interfutures, originated in technological forecasting and Cold War strategy. Both focused on the idea of the 'long range,' a category produced by nuclear strategy and ballistic research.
>
> (Andersson, 2017, p. 324)

Andersson draws attention to the military framing of the horizons of time that future research is guided by and, in military parlance, the means by which it targets the future and sees the future as a target. To extend this martial thinking more explicitly, the introduction and awareness of 'complexity' – the conceptual innovation of interwar and post-war futures research – can be seen in this 'gaming'; and 'scenarios' are in effect the working out and measurement of time as a variation of 'collateral damage', no matter how benign the scenario. In any case, Andersson highlights the extent to which the funding and organizational drivers for research into the future were embedded in the Cold War, and in the disciplinary knowledges of colonialism and empire that emerged in the West alongside Enlightenment and modernity. These categories have immense discursive power too, which I will turn to in a moment. Obviously, RAND was already embedded in that culture. It was not just personified in Herman Kahn but in researchers such as Olaf Helmer who believed that their work at RAND enabled a 'social technology' for prediction and decision-making equivalent to that of biology and chemistry (Andersson, 2012, p. 1411). Andersson argues that sociologist Daniel Bell, who founded the Institute for the Future, saw forecasting as a tool 'that would enable the modern state to foresee coming developments requiring a shift in political strategy, a new form of collective rationality in a stage-driven process of modernization' (Andersson, 2012, p. 1417). De Jouvenel, who had moved in fascist circles in the interwar period and joined Hayek's Mont Pelerin Society; their work (according *Guardian* journalist and climate activist George Monbiot (2007)) was in effect the Petri dish for what would become known as neoliberalism) initially received funding from the Ford Foundation. The Ford Foundation was seeking a planning and strategy expert to champion the thinking and values of Western democracy against the threat of the Soviet Union, and saw de Jouvenel's *Futuribles* project, 'as a liberal alternative to the long-term planning emerging in Third World countries under the influence of Marxism' (Andersson, 2012, p. 1419).

There was another tradition, represented by Ukrainian-born Marxist Ossip Flechtheim who coined the word 'futurology' in the 1940s (Kessler, 2011; Andersson, 2012; Seefried, 2017). Flechtheim fled Nazi Germany in the 1930s and eventually ended up in Switzerland where he made contact with the Institute for Social Research in Geneva and left for New York when Max Horkheimer moved the Institute to Columbia University – Herbert Marcuse and Erich Fromm taught there. He never got a position there, ending up teaching political science at Atlanta University, a college largely attended by black students. Flechtheim's vision of futurology was inflected by his left-wing politics. These political views he believed were the reason he never got a permanent post in a US university despite the fact that he never held the anti-Americanism of his intellectual peers (Kessler, 2011, p. 189). He had been a long-standing proponent of a left-wing critique of communism but believed that the most powerful tool in its defeat was the tractor rather than the tank – the promise of food and a better life changes the future, war destroys it (Kessler, 2011, p. 186). Flechtheim created the term futurology in a paper entitled 'Teaching the Future' published in the *Journal for Higher Education* (1945), where he put forward the case for a curriculum that would provide the knowledge resource in order to highlight trends and future problems. The sub-heading of the paper – 'A contribution to the intellectual and moral growth of the participants' – gestures towards Flechtheim's vision of futurology as a discipline as more than a technocratic or business tool.

> Instead of consulting the stars, the 'futurologist' of 1945 can get his clews [sic] from historians and sociologists, from philosophers and psychologists, from political scientists and economists. He can make intelligent use of a tremendous reservoir of knowledge, though some of the theories found may differ as widely as those of Arnold Toynbee and Ralph Turner, of Pitirim Sorokin and R. M. MacIver, of Sigmund Freud and Erich Fromm. The analyst of the future would have to find his way between the optimism of Trotzky [sic] or Croce and the pessimism of Henry Adams or Berdyae.
>
> (Flechtheim, 1945, p. 461)

The curriculum of the student of futurology would also include Italian fascist, philosopher, artist and occultist Julius Evola and utopian fiction such as Huxley's *Brave New World*. Flechtheim returned to post-war West Germany and, by the 1960s, his socially engaged vision of futurology found itself in tune with the politically and ecologically minded student activism of the times (Andersson, 2012, p. 1412). Andersson highlights Flechtheim's (1971, p. 8) selection and citing of two definitions of futurology from two different encyclopaedias on each side of the Iron Curtain:

> Futurology: Systematic and critical analysis of future issues.
>
> (From the West German *Brockhaus Encyclopädie*, vol. 6, 1968.)

> Futurology: A term denoting the attempts of bourgeois scientists and planners in imperialist states to prognosticate future developments in capitalism.
> (From the East German *Kleines Politisches Wörterbuch*, 1968.)

When discourses of professional futures research of any type in the post-Cold War world (especially if the client is government, NGOs or the private sector) presents itself as a neutral methodology (see *The New York Times* above), it is striking to see futures research as fundamentally and politically charged. Flechtheim's futurology was coming from a more obviously self-aware political perspective, it belongs to the family of academicized knowledge production.

Andersson's considered, critically reflective history is limited by the disciplinary blind spot that it is working in and with. Andrew Sanchez's (2019) review of Andersson's comprehensive study of knowledge production, in the journal *Social History*, highlights the 'episteme' (Foucault, 1994) that governs the restrictive framing of a problem. Sanchez (2019, p. 274) writes:

> One leaves the book wondering how visions of the future within, for example, the Black Power Movement or the Women's Liberation movement diverged from the almost-exclusively white and male practitioners of expert future prediction. Unfortunately, Andersson's otherwise insightful text gives the impression that the only significant conceptions of the future were those espoused by the technocrats.

This has two corollaries. First, as the nascent discipline of anticipation grows from this technocratic vision of futures (even with its up-to-date epistemology from biology), in the absence of an ethology driving its politics, it risks reproducing biopolitical practices such as profiling and what is called 'Prevent' in the UK. It is about 'chronopolitics' (which I will address in the concluding section). In his paper on nanotechnology – via *Back to the Future* and *The Terminator*: two pop culture explorations of knowledge production (the first a kind of citizen science and the other the result of a lack of corporate anticipation) – Mario Kaiser (2015, p. 166) writes that, 'at the core of prevention, we find the desire for a normalization and conservation of the present in relation to dangerous futures'. For the UK government anticipation is folded into the technocratic pre-criminology of Prevent and the technocratic behaviourism of 'Nudge' in government policy. The Behavioural Insights Unit (BIU) was established in the UK Cabinet Office in 2010, based around the work of Richard Thaler and Cass Sunstein, articulated in their best-selling book *Nudge: Improving Decisions about Health, Wealth, and Happiness* (2009). The work of the BIU is highly regarded; and *The Guardian* reported (2018) that in 2018 they earned £14 million with 40 per cent of that coming from overseas. However, it is a model whose vision of the future is organized around constraint and the limited pathways of algorithms. Pat Kane, who is lead curator of Nesta's[6] *FutureFest* and author of *The Play Ethic* (2005) – and its innovation consultancy offshoot – is critical of Nudge. Its horizon, he argues, misses alternative socio-biological theories around play as a creative model for ways of being and learning, play as an activity that gives people the capacity to engage creatively both within and outside the given rules and whose pleasure is partly derived from unexpected outcomes. In an interview with *Varoom* magazine (2015, p. 59) Kane says:

> What this socio-biological emphasis does is two things – one, to make their interest in play become an interest in full human flourishing, rather than as another lubrication of a route to market; and two, to get them to attend to the full spectrum of current mind-science around play, innovation and creativity – which is much bigger than the current trends of 'behavioural economics' and 'choice architecture' (known as the Nudge school), dominant in quite a lot of policy and business. They think we're all Homer Simpsons at heart, needing to be shown the error of our over-optimistic, misperceiving ways. I think we're all as equally Lisa Simpsons, constantly searching for the right contexts and conditions for our self-expression and creative urges.

I would push this further and suggest that Lisa Simpson anticipates the conditions for creativity by actively creating them.

Second, the future in wider, popular culture is expressed in many different idiolects and forms: from the sociopathic – or perhaps cosmopathic – excess and sci-fi revisionism of *Rick and Morty* to the cyborg extravagances of Japanese Anime. While such cultural forms may be increasingly mainstream, they are outsider forms. Through policy, models and formats the governmental and corporate imagination perform the future as a boundary of insider knowledge and practice defined against and validated by the exclusion of outsider knowledge and practice.

Stitch_ _ _ _ anticipating when knowledge is made for cutting

The restricted vision that comes from the dominant epistemologies and disciplinary genealogies of futures studies, means that anticipation needs to anticipate not only a different future for futures studies, but towards a different past as well as from a different past. It needs to anticipate that futures research was always inherently political in its range of key actors (military, corporate, government departments), its embedded geopolitical standards of 'modernity' and consequently its Enlightenment perspectives.

One way to conceive of this mobilization and formation of interests, knowledges and perceived needs, is through the concept of what Foucault (1980) calls *dispositif*, and which is most commonly translated into English as 'apparatus', albeit a not entirely adequate translation. In an article on *dispositif*, Jeffrey Bussolini (2010, p. 86) cites one of Foucault's translators, Graham Burchill, as saying that

English has no equivalent to describe what Foucault's use captures; that is, to 'designate a configuration or arrangement of elements and forces, practices and discourses, power and knowledge, that is both strategic and technical'. Italian philosopher Giorgio Agamben (2009, p. 14) gives a more direct characterization:

> I shall call an apparatus literally anything that has in some way the capacity to capture, orient, determine, intercept, model, control, or secure the gestures, behaviors, opinions, or discourses of living beings. Not only, therefore, prisons, madhouses, the panopticon, schools, confession, factories, disciplines, juridical measures, and so forth (whose connection with power is in a certain sense evident), but also the pen, writing, literature, philosophy, agriculture, cigarettes, navigation, computers, cellular telephones and – why not – language itself, which is perhaps the most ancient of apparatuses [. . .].

This concept formed part of Foucault's wider project to make power, and its various iterations, central to our understanding of how different disciplines and knowledges have emerged since the renaissance. These *dispositifs* have shaped: individuals and governments; commonly received ideas of history, continuity and progress; the practices and behaviours of the newly professional classes and the populations they constituted and classified with the knowledge systems required in the development of capitalism, the state and empire. Most of all, the *dispositif* is a history of the present (Garland, 2014, p. 373).

Foucault's approach is not about having a better understanding of the past. It is about re-thinking the past as the product of a *dispositif*: an arrangement of forces and political struggle to engage with those forces as they activate the present – our subjectivities, practices, cognition, official bodies of knowledge, norms – and therefore circumscribe, but do not limit, a future. In this way, Foucault's method is informed by both Nietzsche's method of genealogy and the concept of 'chronopolitics'[7] (which I address in the next section). Foucault's approach, which he outlines in an analysis of Nietzsche (Foucault, 1977), is similar to what the German philosopher called '*wirkliche Historie*', translated as 'effective history' (Nietzsche, 2007): a history that does not attribute a post-match, retroactive continuity that reconciles the many different directions, moments, movements, displacements into an inevitable development, the past as something we re-trace from the present (Foucault, 1977, p. 152). Foucault, rather, highlights the affective dimension of practising effective history. This is crucial to any practice of anticipation. And though he is attracted to 'liberation' Foucault is wary of its discursive power, the cultural and social authority it acquires and delivers. Which, furthermore, sets limits on what he calls 'practices of freedom': ethopoetic practices that enable the ethical/aesthetic creation of self:

> History becomes 'effective' to the degree that it introduces discontinuity into our very being – as it divides our emotions, dramatizes our instincts,

> multiplies our body and sets it against itself. 'Effective' history deprives the self of the reassuring stability of life and nature, and it will not permit itself to be transported by a voiceless obstinacy toward a millennial ending. It will uproot its traditional foundations and relentlessly disrupt its pretended continuity. This is because knowledge is not made for understanding; it is made for cutting.
>
> (Foucault, 1977, p. 154)

Unlike more conventional theories of power, for Foucault the power of 'power' lay in the fact that it was not simply hierarchical and negative, but was diffuse and productive. Rather than power being 'omnipotent', a capacity sometimes attributed to it, Foucault talks about the 'omnipresence of power'. Though he would not put it in these terms, it is a folding into each other of the chronopolitical and the geopolitical. Power is omnipresent not because it displays the hand of God gathering everything towards a divine destiny, 'but because it is produced from one moment to the next, at every point, or rather in every relation from one point to another. Power is everywhere; not because it embraces everything, but because it comes from everywhere'. Here Foucault's analysis of power echoes that of Deleuze and Guattari in *What is Philosophy?* (1994) in their choice to shift the analysis of power from history to geography. Being produced from one moment to the next, in a fold that forms its exteriors and interiors from, on one side: historical disciplines, practices and knowledges; and on the other, subjectivities, bodies and organizations. Power is emergent and its specific emergent formation, its temporary stability, presupposes the resistance to power from which it differs – power is a relation of forces. Being emergent means that to understand how power moves, mobilizes and gathers, to identify its specific formations and patterns – and, therefore, to generate insight and map where it may emerge next in terms of the effects and affects in the world which it is producing – requires more than simple data gathering. It requires understanding the measurements, knowledges, institutional bodies that visualize the 'normalizing' of power, the power of 'normalizing' and the wider production of practices which follow from the ongoing configuration of norms.

Foucault shows how power can be tracked through its affects and effects, by attending to the relation created in its specific formations, organizations and technologies. In this way, his theory of power is, simultaneously: a theory of 'how' rather than 'what'; a theory of movement and visibility, rather than a theory of substance or instrument. In his book on Foucault, Gilles Deleuze tracks the allusively effective and affective nature of Foucault's power. Deleuze (1988, p. 60) writes:

> An exercise of power shows up as an affect, since force defines itself by its very power to affect other forces (to which it is related) and to be affected by other forces. To incite, provoke and produce, (or any term drawn from analogous lists) constitute active affects, while to be incited or provoked,

> to be induced to produce, to have a 'useful' effect, constitute reactive affects.

Neither active nor re-active are primary but emerge from their encounter, as Deleuze points out; what is being affected has the capacity for resistance. This is why Foucault looks like a historian. His different analyses of power require an empirical approach alighting on the data of specific concrete practices, in particular instances, that articulate historically singular rules which establish limits and norms for how we think and act in the world. These rules are in practice: immanent. They are in foresight practice, in tools and methods and methodologies, as we have seen; they confirm and universalize their own validity in their knowledge production, in the very fact of being actualized.

This is a strategically emergent network of power, composed of heterogenous elements, a 'formation which has as its major function at a given historical moment that of responding to an *urgent need*' (Foucault, 1980, p. 195; original emphasis). The heterogenous elements around futures include: the military; business; government from policy-makers to departments planning investments in infrastructure, healthcare; a variety of academic disciplines from economics to sociology and design; computer software and statisticians; charitable bodies; newspapers and magazines and publishers; and the 'public' imagined and formed by the *dispositif*. The immediate 'urgent need' in future studies is to dispel uncertainty in order to make decisions by capturing the future as a product of knowledge that can then be de-risked. Sociologist Kai Eriksson points out that as the *dispositif* creates an object of knowledge it also creates futures research as 'experience' (Eriksson, 2005). I will give myself the disciplinary licence to use this word 'experience' with the power of the double-meaning it has in French which Gilles Deleuze highlights – 'experience' and 'experiment'. This reading of experience as also experiment highlights the temporariness of a particular constellation of power and how, according to Foucault, a discursive formation of power also generates unexpected 'resistances':

> This thematical core [of a *dispositif*] defines an area of experience that manifests itself in and through the mutually constitutive interrelationships among theoretical discourses, social power relationships, and self-relationships of the self, as well as in the whole range of different practices, institutions, and systems involved. Dispositif is to be understood as a network of relationships that, in a given historical period, organizes the field of power and knowledge as both an object of speech and a field of experience. Thus, it does not only indicate the historical institutions involved but also opens an ontological dimension to be examined.
>
> (Erikson, 2005, pp. 600–601)

Eriksson highlights some key modes through which we can understand how people engage with a *dispositif* such as futures research. While the knowledge produced is valued because it is information produced by a group of experts, this production also sets epistemological perspectives and limits (epistemes). In turn, these also set the ontological boundaries of futures research, limiting its knowledge production and its pool of actors, which then become understood as the conditions of experience. Timothy O'Leary (2009, p. 83) explains Foucault's changing concept of experience as follows:

> Foucault is now working with a multi-layered notion of experience; and it is one which is not accessed through individual awareness, but through an analysis of what he now calls 'practices.' We can study the forms of experience, he says, through an analysis of practices – as long as we understand practices as 'systems of action . . . inhabited by forms of thought'.

Foucault's analyses of power, knowledge production and the objects and subjectivities created was always about mapping and following practices. As an antecedent of Actor Network Theory (ANT) and Science and Technology Studies (STS), Foucault's discourse analysis maps the relations between different 'actors' (human and non-human) in the *dispositif*. It is useful to understand this genealogical relationship between Foucault, ANT and Latour, because: first, ANT re-situates Foucault's project in the organizational systems of industrial and post-industrial capitalism to show how specific futures (the products and projects of innovation) are constructed. Second, ANT's micro-mapping of practices and the objects and materials which enable knowledge production, its formats and inscriptions as well as the expertise, relationships and circulations that assemble into what Latour calls an 'association' (2005b). Though Latour has been surprisingly dismissive of Foucault (see: Latour and Crawford, 1993), both these thinkers show how power is emergent. Consequently, both of them are intrigued by scale (through the examination of technologies of measurement and classification) and the analysis of apparently local practices. Their work rethinks conventional sociological perspectives of macro, meso and micro, folding each around and within the others. For Foucault, subjectivity located here, now, is produced via institutions, disciplines, practices that also produce notions of space itself. Indeed, by paying attention to what Foucault calls the 'regimes of truth' of normative knowledge production, Latour and Foucault show how new forms of knowledge and experience emerge.

From Foucault's perspective, it is not just power that is productive of new objects, subjects, temporalities and spatialities. It is also what is excluded by power and the norms set to uphold it; what 'resists' normative forms of knowledge, with its own strategies, subjectivities and expressions of knowledge. In an interview with schoolchildren (admittedly highly philosophically aware French schoolchildren) Foucault (1977) speaks of a conventional analytical approach that focuses upon particular moments when governing systems have been threatened. The genesis of such threats, he argues, is normally attributed to famine, the raising of taxes or lack of work. Foucault argues that official knowledge, in this case the discipline of history in 1970s France, is only capable of framing change and the rise of popular movements antagonistic to the establishments as reactive: for example, to the levying of ever-increasing taxes or an empty stomach. No one imagines it might be because people want to exercise power.

There is a common articulation of change between the critical thinking he proposes and what he calls 'subjugated knowledges' (Foucault, 1980). This intersection of criticality and subjugation occurs in the struggle against normative knowledges whose power rests in not even recognizing, in not being capable of recognizing, the voice or idiom in which new thought is expressed. There is a section from a 1976 lecture by Foucault that is worth quoting in full as it begins to connect his early work focusing on the power/knowledge couple, with his later work that explores care of the self as a creative practice. Subjugated knowledges, says Foucault, do not seek approval from what he calls 'the established regimes of thought' (Foucault, 1980, p. 81). Subjugated knowledges, insofar as they are recognized as knowledge at all,

> have been disqualified as inadequate to their task or insufficiently elaborated: naive knowledges, located low down on the hierarchy, beneath the required level of cognition or scientificity. I also believe that it is through the re-emergence of these low-ranking knowledges, these unqualified, even directly disqualified knowledges (such as that of the psychiatric patient, of the ill person, of the nurse, of the doctor – parallel and marginal as they are to the knowledge of medicine – that of the delinquent etc.), and which involve what I would call a popular knowledge (*le savoir des gens*) though it is far from being a general common sense knowledge, but is on the contrary a particular, local, regional knowledge, a differential knowledge incapable of unanimity and which owes its force only to the harshness with which it is opposed by everything surrounding it – that it is through the re-appearance of this knowledge, of these local popular knowledges, these disqualified knowledges, that criticism performs its work.
>
> (Foucault, 1980, p. 82)

Criticism, or creative thought, opens up a genealogy of thinking hidden by what Foucault calls 'globalizing discourse' revealing (using a word which would have more affective relevance 40 years later) their *privilege* and their status as 'avant-garde'. This term captures both the orientation and the military origin of futures research. To continue the genealogy of the military in futures research, we might call traditional futures the 'avant-garde of the corporate imagination'.

As futures/futurology/foresight wraps its scarf around its face and slips out the back door as the alternating genders of hurricanes, trolling nationalisms and environmental migrants burst into the futures-workshop-room spilling over with post-it notes, how will anticipation avoid becoming another Davos fad, another name-tagged and power-pointed avant-garde, giving government, universities and business a taste of something a little risqué? You might already have anticipated that Foucault's work offers critical possibilities for a world in which the political and the commercial are now visibly and publicly entangled; a world where individual commitment to 'making a difference' in the boardrooms, in the workplace, in the normative visibility of bring-your-own coffee cups.

Though it is not often noted in research in the rich fields of Foucault, his work is ultimately a study of change. It is therefore of crucial importance to any kind of futures work, otherwise such work is doomed to reproduce existing value and knowledge horizons in its outputs unless it engages in the critical thinking around self and world which will then enable the creation of new and just futures. Foucault (1984, p. 50) explains:

> The critical ontology of ourselves has to be considered not, certainly, as a theory, a doctrine, nor even as a permanent body of knowledge that is accumulating; it has to be conceived as a attitude, an ethos, a philosophical life in which the critique of what we are is at one and the same time the historical analysis of the limits that are imposed on us and an experiment with the possibility of going beyond them.

The possibility of going beyond them means examining the embedded values and forces that colonize the future.

Stitch_ _ _ _ _ chronopolitics and the anticipation of the Other-wise

The future as a practice of the Other-wise happens in ways that the discursive practices organizations often hide and occlude. Organizations according to Tsoukas and Chia (2002) are both products of change and the tools which then seek to stabilize change. Organizations are constantly changing through internal and external dynamics – think of the future, rather than 'seeping in' but as something 'seeping out' from the 'crannies of the present', using Whitehead's description. The future as living, generative excess. Musician Brian Eno gives an example of how the excess of an aesthetic body (such as an instrument or music technology or a throat), expressed as a technological mistake or glitch, is in fact the anticipation of the technology's future present and its future past. The given social and technical codings, its semiotic organization, cracks and ripples futures.

> Whatever you now find weird, ugly, uncomfortable and nasty about a new medium will surely become its signature. CD distortion, the jitteriness of digital video, the crap sound of 8-bit – all of these will be cherished and emulated as soon as they can be avoided. It is the sound of failure: so much modern art is the sound of things going out of control, of a medium pushing to its limits and breaking apart. The distorted guitar sound is the sound of something too loud for the medium supposed to carry it. The

> blues singer with the cracked voice is the sound of an emotional cry too powerful for the throat that releases it. The excitement of grainy film, of bleached-out black and white, is the excitement of witnessing events too momentous for the medium assigned to record them.
>
> (Eno, 1996, p. 283)

Experiences such as 'distortion', 'jitteriness', 'crap sound', are the failures of the current technologies which define the norm. This emergent failure in music is the anticipation of a new language that requires the practice of listening for the future. The future isn't a matter of seeing (though the language of foresight, from seers to envisioning to scenarios locates it there), it is also a matter of ears, mouths, noses and skins. Eno is also describing how the past and the future emerge simultaneously beyond the semiotic codings of nostalgia. The chronological emergences and shifts involved in Eno's 'cracked' voice and CD distortion are echoed in Mark Fisher's exploration of 'hauntology' in his paper, 'The Metaphysics of Crackle' on the foregrounding of the sound-sample of crackle from vinyl records on musicians such as Tricky, Burial and artists on the Ghost Box label. Fisher (2013, p. 48) writes that

> If the metaphysics of presence rests on the privileging of speech and the here-and-now, then the metaphysics of crackle is about dyschronia and disembodiment. Crackle unsettles the very distinction between surface and depth, between background and foreground. In sonic hauntology, we *hear* that time is out of joint. The joins are audible in the crackles, the hiss . . .

Dyschronia is a concept invented by music critic Simon Reynolds to describe temporal displacement, disturbing the pastiche of time construction of postmodernity (Fisher, 2013, p. 47). The crackle is an anomaly that is sending a signal (in the semiotic and social senses) and a biotechnological aesthetic creating affect. In this way it resonates with Foucault's commentary on the work of philosopher of science George Canguilhem (Foucault, 1998) where he argues that what is characteristic of life is that it is capable of 'error', of 'mistake'. This is not a question of deficient epistemologies, or only that question but also a question of ontology, ethics and aesthetics – how do we learn to listen for contingency? What technology of the self might I need to anticipate to be transformed by anomaly before the 'deficient' becomes a disease of the system, before it becomes pathological? The pathology is always a pathology of time, the arrival of the too-soon in the unready-now.

Fisher's argument is rooted in the excess bubbling up of the lost futures from the past – seeping out, let's say from the crannies of the present. The sonic 'hauntology' he is describing is inspired by Derrida's *Spectres of Marx* (1994). It was a response to the fall of the Soviet Union and its empire and how, according to Derrida, this moment generated a psychological-state-of-mind among policy-makers that resembled the 'triumphant' phase in Freud's theory of mourning – the death of communism and the celebration of market capitalism.

But Fisher (who sadly could not envision a future and committed suicide in 2017) creatively deviated Derrida's concept of hauntology to open up a more complex notion of chronopolitics – though he doesn't call it that, that's effectively what it is. Fisher locates his bifurcation of hauntology in philosopher Martin Hägglund's reading of Derrida's deconstruction which Hägglund argues can be seen as an exploration of 'broken time'. Fisher writes

> Referring back to Hägglund's distinction between the no longer and the not yet, we can provisionally distinguish two directions in hauntology. The first refers to that which is (in actuality is) no longer, but which remains effective as a virtuality (the traumatic 'compulsion to repeat', a fatal pattern). The second sense of hauntology refers to that which (in actuality) has not yet happened, but which is already effective in the virtual (an attractor, an anticipation shaping current behaviour). The 'spectre of communism' that Marx and Engels had warned of in the first lines of the Communist Manifesto was just this kind of ghost: a virtuality whose threatened coming was already playing a part in undermining the present state of things.
>
> (Fisher, 2014, p. 19)

Chronopolitics is affective. First, it emerges as affective trauma from the past-present which creates a future that can only sustain the trauma, albeit in formations specific to a culture and society, deactivating the future as different and new as opposed to the same. Nevertheless, a hauntological analysis of trauma, of broken time, is required to enable the moves of the chronopolitical. Second, hauntology as a disciplinary resource of chronopolitics also enables the activation of the future as a creative anticipatory attractor – a future-present that can then intervene in the past-present like alien ghost-activists redistributing futures. Both of these directions have already been activated in the theory-art of afrofuturism, and I will briefly highlight some of its chronopolitical consequences in the work of Black Quantum Futurism.

Black Quantum Futurism (BQF) is a project based in Philadelphia, created by lawyer Rasheedah Philips and poet and musician, Camae Ayewa. They have created a practice of anticipation located partly in Afrofuturism (Eshun, 1998; Womack, 2013; Anderson and Jones, 2016), quantum physics and Black/African thinking around time and space. The signifier 'Black' is not just mobilizing 'skin pigmentation, race, lineage, and cultural identity [. . .] but also refers to the Blackness that permeates deep space, what is commonly known as "dark matter". It encompasses the Blackness or darkness that permeates mental space and inner space' (BQF, 2015, p. 13).

This issue of 'blackness' and 'dark matter' – the invisible matter that the visible depends on – is that there are consequences for the anticipatory researcher: of not being aware that there is dark matter configuring realities; in not being Other-wise. This plays out methodologically and consequentially in the delivery of the outcomes of research for which dark/black matter is occluded: rich creative futures closed off by the unanalysed political and social discourses generating the subjectivities and horizons of the futures researcher. Connected to this are the unrecognized, unacknowledged pasts whose unacknowledgement means that the futures of corporate imaginations can perpetuate endemic imperialism. Tobias Van Veen, Research Fellow in Media and Cultural Studies at University of California Riverside writes in his paper 'The Armageddon Effect' that:

> Afrofuturism is as much a recovery project of a revisionist past as it is an imaginary of a future otherwise. Afrofuturism seeks to displace temporality from its whitewashed visions.[8]
>
> (Van Veen, 2016, p. 83)

So as much as chronopolitics is a kind of emergency response to broken time, to time broken from injustice and racism, it is also spatial. The chronopolitical activist deviates time in order to save future-space from being already occupied and colonized by the corporate imagination and their imperial avant-garde. The work of Black Quantum Futurism works on real estate threatened by gentrification and the unreal estate of NASA history, recovering an alternative time of US space travel, on Earth and to the Moon. The 2018 event, 'Community Futurisms: Time and Memory in North Philly 002 – Black Space Agency', intersected an art exhibition with community projects that traversed issues of housing, technology and business. BQF's Afrofuturism is a technology that emerges when interconnections and intersection between pasts and futures are anticipated in events such as performance, renting and rights workshops, open mic sessions where voice is not framed by the participatory design models, activist panels and visioning sessions that often code how 'the public' should interact with professionals working for them.

BQF's 'Community Futurisms' is an active creation both of community and time as a practice of concept generation and chronopolitical intervention. For example, the Community Futures Lab took place from May 2016 to April 2017 in the Sharswood neighbourhood of Philadelphia. This area had 'a name' as a poor neighbourhood, parts of which were now being demolished, parts undergoing redevelopment, but also with a history of successful business and a thriving arts scene (Black Quantum Futurism, 2020, p. 13). The methodologies of BQF enable them to see this as a space-time collapse of a community; for developers – and perhaps for a city already desperately underfunded – it is through the extraction of future time and conversion into rents, assets and taxes that capital occupies the future and the past.

For BQF chronopolitics involves enabling orientation, for people who have been displaced – Afrofuturism is about displacement and diaspora. Hence a crucial technology for BQF is the map and helping others map, with different systems and methodologies: 'quantum event maps, housing journey maps, sonic maps and communal memory maps' (Black Quantum Futurism, 2020, p. 17). This mapping is not located in calendar time. Time in modernity is irrevocably connected to colonialism and the British Empire – time was Britain's most significant export to the rest of the world. The International Prime Meridian conference of 1884 established Greenwich Mean Time as the reference point for the organization and circulation of time in modernity. Following Adam Barrows's analysis of time in *The Cosmic Time of Empire: Modern Britain and World Literature* (2011), I would argue that the imperial technology of GMT dissolved and liquified local space-times and associated practices and roles, in order to reconfigure and code local space for the time-space of Empire. Furthermore:

> struggles over standard time in the early twentieth century reveal the radical instability of that globally synchronized modernity, dependent on a continual colonization of social time requiring extensive capital investment, technological modification, legislative sanction, and cultural saturation.
>
> (Barrows, 2011, pp. 20–21)

BQF's anticipatory practice liberates the many different given codings of time, and in the Community Futures Lab project invited in self-selecting oral historians and oral futurists to express multiple temporalities. With their various forms of mapping:

> Time becomes something remembered, not something that defines and predates the creation of memory. In our workshops, we have groups create communal quantum event maps that allow them to struggle through some of the ways in which a community constructs communal time around a past, future or present event, composed of diverse and intersecting temporal rhythms and other event textures and features.
>
> (Black Quantum Futurism, 2020, p. 17)

In this way for BQF the anticipatory quality of the event is constructed through the immanent composition of methodologies, peoples, time and art. I'm using the term art according to the function Deleuze accords it in his essay 'What is the Creative Act?' (2007) as something which contains not information, nor even counter-information, but counter-information as an act of resistance. Deleuze points out that counter-information in itself, as a better epistemological framing of the world does nothing: 'In Hitler's time, the Jews arriving from Germany who were the first to tell us about the concentration camps were performing counter-information [. . .] No counter-information ever bothered Hitler' (Deleuze, 2007, p. 322). Traditional foresight simply with better, more expensive, more beautifully designed counter-information about a future, changes nothing. What makes an event anticipatory is creativity and Deleuze argues in this essay that disciplines which understand and define themselves as creative activities share the ability to invent 'space-time'. He elaborates on this activity in an interview with Italian Marxist Toni Negri, reflecting on how in creating new ways of being in the world requires moving beyond current forms of knowledge:

> if you believe in the world you precipitate events, however inconspicuous, that elude control, you engender new space-times, however small their surface or volume. It's what you call pietas. Our ability to resist control, or our submission to it, has to be assessed at the level of our every move. We need both creativity and a people.
>
> (Deleuze, 1995, p. 176)

'Community Futurisms: Time and Memory in North Philly 002 – Black Space Agency' is the space-time BQF created to connect the space and politics of 1960s technology with the space of housing and community. They located The Black Space Agency in the work of the Rev Leon H. Sullivan, who was a civil rights activist and Minister in Philadelphia in the 1960s. We could say Sullivan was an innovator in community action. In the late 1950s as pastor of Philadelphia's Zion Baptist church he challenged local businesses to interview and employ young black people and when only two responded, he responded by organizing a boycott. His activist work was so successful that Martin Luther King would later seek his advice on the organization of effective challenges to discrimination.

But Sullivan was most interested in developing the skills and knowledge that were needed in order for the black community to grow and attract investment. He set up and founded, in 1964, the first Opportunities Industrialization Centers of America. It was based in an abandoned jail. In 1962 he founded Zion Investments Associates, a co-op which began to develop the landscape in Philadelphia, in housing with Progress Aerospace Enterprises, the first black-owned aerospace company. Sullivan would later sit on the board of General Motors. Progress Aerospace Enterprises was set up after Martin Luther King died. Sullivan said, 'when the first landing on the Moon came, I wanted something there that a black man had made'.

The first Moon landing came about at exactly the same moment as people, not least those in the black community, began to question the amount of money being spent on the investment in this project rather than on deeply impoverished communities. There was a Poor People's March to Cape Canaveral. What is more, as BQF point out, there was a serious lack of diversity in NASA and black communities were dispersed and fragmented in order to build housing for NASA employees. According to the BQF website:

> Partly in response, NASA became involved in the design and applicability of spaceship materials in 'urban' housing, and created campaigns to increase diversity in employment. Much of this resistance and engagement with the space race from the Black community has been largely erased in popular memory.

To 'erased' we can add 'occluded'. This counter-information as resistance means that anticipating futures will go hand in-hand with anticipating the erased and occluded pasts of those futures too. As critic and artist, Kodwo Eshun (2003, p. 289) writes in 'Further Considerations on Afrofuturism':

> The field of Afrofuturism does not seek to deny the tradition of countermemory. Rather, it aims to extend that tradition by reorienting the intercultural vectors of Black Atlantic temporality towards the proleptic as much as the retrospective. It is clear that power now operates predictively as much as retrospectively. Capital continues to function through the dissimulation of the imperial archive, as it has done throughout the last century. Today, however, power also functions through the envisioning, management, and delivery of reliable futures [. . .] The powerful employ futurists and draw power from the futures they endorse, thereby condemning the disempowered to live in the past. The present moment is stretching, slipping for some into yesterday, reaching for others into tomorrow.

Foucault sees his version of History – genealogy – as counter-memory, where the hyphen is the counter that cuts memory from History. Genealogy, as an anticipatory act, doesn't travel back in time to restore History or to map the authenticity of a people, but to trace the deviations that have constructed our current space-time. Rather than framing foresight in the sealed logic of the possible, the plausible and the probable, it is analysing the deviation that helps us anticipate a future, that enables us to recognize the 'exteriority of accidents' (Foucault, 1977, p. 143), the matter that is forming our chronopolitics, that pre-distributes unjust futures. Eshun removes the cut, accelerating the act of counter-memory.

Anticipation must not only recognize what is at stake in the fact of chronopolitics, but like BQF, act to open up and restructure previously closed pathways of time blocking the future seeping out. If we are to commit to an anticipatory logic of emergent futures, we must address the issue of how time is fabricated in a variety of media (Parikka, 2018 p. 44), in pop culture, in art, but clearly also in the foresight and futures of the corporate imagination. Anticipation requires a creative chronopolitical imagination whose future is not already narrated and formed by the colonial archive. In his paper, 'Playing a Minority Forecaster in Search of Afrofuturism: Where Am I in This Future, Stewart Brand?' (2016) Communications scholar Lonny Avi Brooks, relates the story of how the actor Nichelle Nichols told Gene Rodenberry, producer of Star Trek, that she wanted to leave the series. Rodenberry asked her to reconsider and while reflecting on her decision on one evening in 1966 Nichols went to an NAACP dinner where she was introduced to a secret Star Trek fan, Martin Luther King. When she thanked him and told him it would be her final season, King asked her not to do so, as any trace of African American women in the future would have been removed. 'And that erasure,' writes Avi Brooks (2016, p. 152), 'persists in the scenarios created by professional forecasters and other producers of science fiction capital'. Anticipating the future as Martin

Luther King envisioned, requires a diversity of activisms, venues, platforms, crowds and research. Research of the corporate imagination has the veneer of successful futures in its cosmopolitanism, embodied in the expensive education and training of professionals from a variety of fields, and tools. Anthropologist Arjun Appadurai sees such conventional research as, by definition, parochial in its exclusivity to businesses and universities. For Appadurai, this is a matter of research, as he argues in his paper 'The Right to Research' (2006). I see it as the issue of anticipation of the Other-wise. Appadurai (2006, p. 176) writes:

> The uneven distribution of this capacity is both a symptom and a measure of poverty, and it is a form of maldistribution that can be changed by policy and politics. In the current context, I can only suggest that the capacity to aspire and the right to research are necessarily and intimately connected.

When time has already been unevenly distributed,[9] the past as well as the future, anticipation requires the crannies of the Other-wise. As Sverre Raffnsøe and Dorthe Staunæs write in 'Learning to stay ahead of time':

> To imagine what leading ahead involves demands picturing an ultimately uncontrollable and ever-changing Other, which may amplify yourself. What is needed is radical otherness, which may fuel and tune the processes of transformation in both body and mind to evoke the capacities for anticipating and thinking ahead.
>
> (Raffnsøe and Staunæs, 2014, p. 193)

Without the anticipation of the Other-wise the future is subject to occupation by its imperial and colonial past and its corporate present. Anticipation is the creative practice of cutting out the embedded, insular assumptions of the corporate imagination. It is the practice of stitching together a new space-time from occluded practices.

Notes

1 This notion of the iteration through certainty, uncertainty and disruption can also be seen in some of the key texts on Design Thinking: a trend in design and business that developed in the mid-2000s, with the main premise that in times of uncertainty and complexity, businesses would do well to engage the abductive thinking of designers rather than the reductive or even inductive logic of businesspeople. See: Brown (2009) and Martin (2009) as canonical in this field.

2 As will unfold over the following sections, anticipation as sensing and sensation is mobilized as an aesthetic practice that exceeds the representational communication of traditional scenario work. As Brassett and O'Reilly (2018, p. 75) suggest in the context of Roman poet and philosopher, Lucretius's Clinamen: 'Habit anaesthetizes and aesthetics terrifies with its newness. And so even in the inexorable flow of all things to habitual, probable, numbness there are opportunities for "islands" of creativity to emerge, even if they are terrible. Lucretius's entreaty to "let go of any terror of the new" (*desine qua propter novitate exterritus ipsa*) we will read not as requiring terror to be blunted, but that

the terror of the new should not be feared. To be open to the new, even if it causes such strong sensations as to be feared, is to remove the constraints of habit, of closed-minded dogma, and to deliver a system up to the possibilities that there are ways out of entropic fall'. Anticipation as aesthetic, which we will see in the work of Black Quantum Futurism later in this chapter, is thus a style of avoiding the anaesthetic of disruption, of avoiding being beaten into submission. (Editors' note. See also Brassett and Hales in this volume.)

3 There is a concept from Lucretius inspired by his adoption of Epicurean atomism: the swerve, which is useful to note here. I have written about this in terms of design and the value of serendipitous collisions to its practice in Brassett and O'Reilly (2018).

4 It is interesting to note that a more contemporary future is not only darker, but angry at this past future, shown in the slogan: 'Dude, where's my jetpack?' This is daubed on the urban fabric of Warren Ellis's acclaimed *Transmetropolitan* (2007–2009) comics, as well as discussed in other magazines and journals (see: Wilson, 2007). See also: Shaviro (2009, 2007).

5 This is also known as 'governmentality' in Foucault's later work (2004, 2008), as well as in Giorgio Agamben's (1998, 2011) following him. It is used as part of a critique of the neo-liberalization of politics, economy and the becoming-business of (State level) governing.

6 Nesta is a UK charity that started in 1998 as the National Endowment for Science Technology and the Arts, which funded all things from commercial invention to social innovation. Now it calls itself 'an innovation foundation' (Nesta, 2020).

7 Perhaps the most influential thinker of chronopolitics is French urbanist and philosopher Paul Virilio; see: Virilio (1989, 2006). Sharma (2013, p. 313) lists the variety of terms which chronopolitics covers: '"time-space compression"; "fast capitalism"; "the 24/7 world"; "liquid times"; "hypermodern times"; "the coming of immediacy"; "the dromocratic condition"; "the terrain of biopolitical production"; "the chronoscopic society"; and "chronodystopia"'. I will return to this concept in a later section in relation to theories of Afrofuturism.

8 Van Veen continues with a reference to Dery (1994) and his vision of futurologists as developers: 'the unreal estate of the future already owned by the technocrats, futurologists, streamliners, and set-designers – white to a man – who have engineered our collective fantasies'. This quote is from the dialogue in which he is said to have coined the term 'afrofuturism'. What we might see here as a metaphor is materialized in Black Quantum Futurism's response to gentrification through their community futurisms.

9 As science fiction writer William Gibson is, famously, supposed to have said, the future was 'already here it's just not evenly distributed'; see Gibson (2012).

References

Agamben, G. (2011) *The Kingdom and the Glory: For a Theological Genealogy of Economy and Government*. Trans. L. Chiesa (with M. Mandarini). Stanford, CA: Stanford University Press.

Agamben, G. [2006] (2009) 'What is an Apparatus?' in Agamben, G. (ed.), *What is an Apparatus? and Other Essays*. Trans. D. Kishik and S. Pedatella. Stanford, CA: Stanford University Press, pp. 1–24.

Agamben, G. (2007) *Profanations*. Trans. J. Fort. New York, NY: Zone Books.

Agamben, G. [1995] (1998) *Homo Sacer: Sovereign Power and Bare Life*. Trans. D. Heller-Roazen. Stanford, CA: Stanford University Press.

Andersson, J. (2017) 'Planning the Future of World Markets: the OECD's Interfuturs Project', in Christian, M., Kott, S. and Matějka, O. (eds.), *Planning in Cold War Europe. Competition, Cooperation, Circulations (1950s–1970s)*. Berlin and Boston, MA: de Gruyter GmbH, pp. 315–343.

Andersson, J. (2012) 'The Great Future Debate and the Struggle for the World', *American Historical Review*, , 5(117), pp. 1411–1430.

Anderson, R. and Jones, C.E. (eds.) (2016) *Afrofuturism 2.0: The Rise of Astro-Blackness*. Washington, DC: Lexington Books.

Appadurai, A. (2014) 'Interview: Arjun Appadurai', *Globalizations*, 11(4), 'The career of a Concept' Issue, pp. 481–490.

Appadurai, A. (2006) 'The Right to Research', *Globalisation, Societies and Education*, 4(2), pp. 167–177.

Avi Brooks, L. (2016) 'Playing a Minority Forecaster in Search of Afrofuturism: Where Am I in This Future, Stewart Brand?', in Anderson, R. and Jones, C.E. (eds.), *Afrofuturism 2.0: The Rise of Astro-Blackness*. Lanham, MD: Lexington Books, pp. 149–166.

Barrows, A. (2011) *The Cosmic Time of Empire: Modern Britain and World Literature*. Berkeley, CA: University of California Press.

Beck, J. (2016) 'The Future: RAND, Brand and Dangerous to Know', in Beck, J. and Bishop, R. (eds.), *Cold War Legacies. Legacy, Theory, Aesthetics*. Edinburgh: Edinburgh University Press, pp. 35–49.

Beck, J. and Bishop, R. (eds.) (2016) *Cold War Legacies. Legacy, Theory, Aesthetics*. Edinburgh: Edinburgh University Press.

Black Quantum Futurism (2020) *Space-Time Collapse II: Community Futurisms*. Ed. R. Phillips. Philadelphia, PA: House of Future Sciences Books.

Black Quantum Futurism (2015) *Theory and Practice*. Ed. R. Phillips. Philadelphia, PA: House of Future Sciences Books.

Brown, T. (2009) *Change by Design: How Design Thinking Transforms Organizations and Inspires Innovation*. New York, NY: HarperCollins Publishers.

Bussolini, J. (2010) 'What is a Dispositive?', *Foucault Studies*, 10, (November), pp. 85–107.

Christensen, C. (1997) *The Innovator's Dilemma*. Harvard, MA: Harvard Business School Press.

Cornish, E. (2004) *Futuring: The Exploration of the Future*. Bethesda, MD: World Future Society.

Csikszentmihalyi, M. (1996) *Creativity: Flow and the Psychology of Discovery and Invention*. New York, NY: HarperCollins.

Deleuze, G. (2007) 'What is the Creative Act', in Deleuze, G. (ed.), *Two Regimes of Madness. Texts and Interviews 1975–1995*. Trans. A. Hodges and M. Taorima. Ed. D. Lapoujade. New York, NY: Semiotext(e), pp. 317–329.

Deleuze, G. (1995) *Negotiations 1972–1990*. Trans. M. Joughin. New York, NY: Columbia University Press.

Deleuze, G. and Guattari, F. (1994) *What is Philosophy?* Trans. H. Tomlinson and G. Burchell. London and New York, NY: Verso.

Deleuze, G. (1988) *Foucault*. Trans. S. Hand. London: The Athlone Press.

Deleuze, G. and Guattari, F. (1986) *Kafka: Toward a Minor Literature*. Minneapolis, MN: University of Minnesota Press.

De Monchaux, N. (2011) *Spacesuit*. Cambridge, MA and London: The MIT Press.

Derrida, J. (1994) *Spectres of Marx: The State of the Debt, the Work of Mourning, and the New International*. New York, NY: Routledge.

Dery, M. (1994) 'Black to the Future: Interviews with Samuel R. Delany, Greg Tate, and Tricia Rose', in Dery, M. (ed.), *Flame Wars. The Discourse of Cyberculture*. Durham, NJ and London: Duke University Press, pp. 179–222.

Eno, B. (1996) *A Year with Swollen Appendices*. London: Faber.

Eriksson, K. (2005) 'Foucault, Deleuze, and the Ontology of Networks', *The European Legacy*, 'Toward New Paradigms' Special Issue, 10(6), pp. 595–610.

Eshun, K. (2003) 'Further Considerations on Afrofuturism', *CR: The New Centennial Review*, 3(2), (Summer), pp. 287–302.

Eshun, K. (1998) *More Brilliant than the Sun. Adventures in Sonic Fiction*. London: Quartet Books.

Fisher, M. (2014) *Ghosts of My Life. Writings on Depression, Hauntology and Lost Futures*. Winchester: Zero Books.

Fisher, M. (2013) 'The Metaphysics of Crackle: Afrofuturism and Hauntology', *Dancecult*, 5(2), pp. 42–55.

Flechtheim, O. K. (1971) *Futurologie: Der Kampf um die Zukunft*. Berlin: Verlag Wissenschaft und Politik.

Flechtheim, O. (1945) 'Teaching the Future', *The Journal of Higher Education*, 16(9), p. 460.

Foucault, M. (2008) *The Birth of Biopolitics. Lectures at the Collège de France 1978–1979*. Trans. G. Birchell. London and New York, NY: Palgrave Macmillan.

Foucault, M. (1991) 'Questions of Method', in Burchell, G., Gordon, C. and Miller, P. (eds.), *The Foucault Effect: Studies in Governmentality*. London: Harvester Wheatsheaf, pp. 73–86.

Foucault, M. (1988) *The Use of Pleasure: The History of Sexuality*, vol. 2. Trans. R. Hurley. Harmondsworth, Middx.: Penguin Books.

Foucault, M. (1984) 'What is Enlightenment', Trans. C. Porter, in Rabinow, P. (ed.), *The Foucault Reader. An Introduction to Foucault's Thought*. Harmondsworth, Middx.: Penguin Books.

Foucault, M. (1980) 'Two Lectures', in Foucault, M., *Power/Knowledge: Selected Interviews and other Writings 1972–1977*. New York, NY: Pantheon Books.

Foucault, M. (1977) 'Nietzsche, Genealogy, History', in Foucault, M., *Language, Counter-Memory, Practice: Selected Essays and Interviews*. Ed. D.F. Bouchard. Ithaca, NY: Cornell University Press.

Foucault, M. (1972) *The Archaeology of Knowledge*. London: Tavistock Publications.

Garland, D. (2014) 'What is a "History of the Present"? On Foucault's Genealogies and Their Critical Preconditions', *Punishment and Society*, 16(4), pp. 365–384.

Gibson, W. (2012) *Distrust that Particular Flavor. Encounters with a Future that's Already Here*. New York, NY and London: Viking.

Genette, G. (1991) 'Introduction to the Paratext', Trans. M. Maclean, *New Literary History*, 22(2), pp. 261–272.

de Jouvenel, B. (1967) *The Art of Conjecture*. London: Weidenfield and Nickolson.

Kahn, H. (1960) *On Thermonuclear War*. Princeton, NJ: Princeton University Press.

Kane, P. (2015) 'Pat Kane: The Play Ethic', *Varoom*, 30, pp. 56–61.

Kane, P. (2005) *The Play Ethic*. London: Pan.

Kaiser, M. (2015) 'Reactions to the Future: The Chronopolitics of Prevention and Preemption', *Nanoethics*, 9, pp. 165–177.

Kessler, M. (2011) 'Between History and Futurology: Ossip K. Flechtheim', in Fair-Schulz, A. and Kessler, M. (eds.), *German Scholars in Exile: New Studies in Intellectual History*. Lanham, MD: Lexington Books, pp. 173–211.

Koselleck, R. (2004) *Futures Past. On the Semantics of Historical Time*. Trans. K. Tribe. New York, NY: Columbia University Press.

Latour, B. (2005a) 'From Realpolitik to Dingpolitik or How to Make Things Public', in Latour, B. and Weibel, P. (eds.), *Making Things Public: Atmospheres of Democracy*. Cambridge, MA: The MIT Press, pp. 14–41.

Latour, B. (2005b) *Reassembling the Social: An Introduction to Actor-Network-Theory*. Oxford: Oxford University Press.

Latour, B. (1994) 'On Technical Mediation', *Common Knowledge*, 3(2), pp. 29–64.
Latour, B. (1986) 'Visualization and Cognition: Thinking with Eyes', in Kuklick, H. (ed.), *Knowledge and Society: Studies in the Sociology of Culture Past and Present*, vol. 6. Greenwich, CT: Jai Press, pp. 1–40.
Latour, B. and Crawford, T.H.H. (1993) 'An Interview with Bruno Latour', *Configurations*, 1(2), pp. 247–268.
Law, J. (2009) 'Seeing Like a Survey', *Cultural Sociology*, 3(2), pp. 239–256.
Liu, H. and Baker, C. (2016) 'White Knights: Leadership as the Heroicisation of Whiteness', *Leadership*, 12(4), pp. 420–448.
Manjoo, F. (2016) 'Why We Need to Pick Up Alvin Toffler's Torch', *The New York Times*. Available at: https://www.nytimes.com/2016/07/07/technology/why-we-need-to-pick-up-alvin-tofflers-torch.html (Accessed: 16 June 2020).
Martin, R. (2009) *The Design of Business: Why Design Thinking is the Next Competitive Advantage*. Boston, MA: Harvard Business Press.
Meadows, D.H., Meadows, D.L., Randers, J. and Behrens, W.W. (1972) *The Limits to Growth*. New York, NY: Universe Books.
Menand, L. (2005) 'Fat Man', *The New Yorker*. Available at: https://www.newyorker.com/magazine/2005/06/27/fat-man (Accessed: 16 June 2020).
Meyer, J.H.F. and Land, R. (2003) 'Threshold Concepts and Troublesome Knowledge: Linkages to Ways of Thinking and Practising', in Rust, C. (ed.), *Improving Student Learning – Theory and Practice Ten Years On*. Oxford: Oxford Centre for Staff and Learning Development (OCSLD), pp. 412–424.
McClanahan, A. (2009) 'Future's Shock: Plausibility, Preemption, and the Fiction of 9/11', *Symplokē*, 17(1–2), pp. 41–62.
Mignolo, W.D. (2012) *The Darker Side of Western Modernity: Global Futures, Decolonial Options*. Durham, NC: Duke University Press.
Millett, S.M. (2003) 'The future of scenarios: challenges and opportunities', *Strategy and Leadership*, 31(2), pp. 16–24.
Monbiot, G. (2007) 'How the neoliberals stitched up the wealth of nations for themselves', *The Guardian*, 28 August. Available at: https://www.theguardian.com/commentisfree/2007/aug/28/comment.businesscomment (accessed: 29 December 2020).
Nesta (2020) Home Page. Available at: https://www.nesta.org.uk/ (accessed: 29 December 2020).
Nietzsche, F. (2018) *The Joyous Science*. Trans. R. Kevin Hill. London: Penguin Books.
Nietzsche, F.W. (2007) *On the Genealogy of Morality*. Trans. C. Diethe and Ed. K. Ansell-Pearson. Cambridge: Cambridge University Press.
Nowotny, H. (2016) *The Cunning of Uncertainty*. London: Polity Press.
O'Leary, T. (2009) *Foucault and Fiction. The Experience Book*. London and New York: Continuum Books.
Parikka, J. (2018) 'Middle East and Other Futurisms: Imaginary Temporalities in Contemporary Art and Visual Culture', *Culture, Theory and Critique*, 59(1), pp. 40–58.
Perillo, S.J. (2008) 'Fashioning Leadership in Schools: An ANT Account of Leadership as Networked Practice', *School Leadership and Management*, 28(2), pp. 189–203.
Plant, S. (1997) *Zeros + Ones. Digital Women + the New Technoculture*. London: Fourth Estate.
Plant, S. (1995) 'The Future Looms: Weaving Woman and Cybernetics', *Body and Society*, 1(3–4), (November), pp. 45–64.
Poli, R. (2017) *Introduction to Anticipation Studies*. Anticipation Science Series, No. 1. Cham, Switzerland: Springer AG.

Powers, D. (2018) 'Thinking in trends: the rise of trend forecasting in the United States', *Journal of Historical Research in Marketing*, 10 (1), pp. 2–20.

Raffnsøe, S. and Staunæs, D. (2014) 'Learning to Stay Ahead of Time: Moving Leadership Experiences Experimentally', *Management and Organizational History*, 9(2), pp. 184–201.

Reddy, V. and Butler, J. (2004) 'Troubling genders, subverting identities: interview with Judith Butler', *Agenda*, 18(62), pp. 115–123

Rosen, R. (2012) *Anticipatory Systems: Philosophical, Mathematical and Methodological Foundations*. 2nd edn. New York, NY: Springer.

Sanchez, A.V. (2019) 'The Future of the World: Futurology, Futurists and the Struggle for the Post-Cold War Imagination', *Social History*, 44(2), pp. 273–274.

Schumpeter, J.A. (1983) *The Theory of Economic Development. An Inquiry into Profits, Capital, Credit, Interest, and the Business Cycle*. Trans. R. Opie. New Brunswick, NJ and London: Transaction Publishers.

Seefried, E. (2017) 'Globalized Science. The 1970s Futures Field', *Centaurus*, 59, pp. 40–57.

Seefried, E. (2014) 'Steering the future. The emergence of "Western" futures research and its production of expertise, 1950s to early 1970s', *European Journal of Futures Research*, 15(29), DOI 10.1007/s40309-013-0029-y.

Sharma, S. (2013) 'Critical Time', *Communication and Critical/Cultural Studies*, 10(2–3), pp. 312–318.

Shaviro, S. (2009) *Without Criteria. Kant, Whitehead, Deleuze and Aesthetics*. 'Technologies of Lived Abstraction' Series, No. 2. Cambridge, MA and London: The MIT Press.

Shaviro, S. (2007) 'You Will Never Own a Jetpack. Warren Ellis' *Doktor Sleepless*', in Ellis, W. (ed.), *Doktor Sleepless*, vol. 7. Rantoul, IL: Avatar Press.

Stegmaier, W. (2016) 'Nietzsche's Orientation toward the Future', *Journal of Nietzsche Studies*, 47(3), pp. 384–401.

Tezza, C. (2007) *O Filho Eterno*. Rio de Janeiro: Editora Record.

Thaler, R. and Sunstein, C. (2009) *Nudge: Improving Decisions about Health, Wealth, and Happiness*. London: Penguin Books.

Toffler, A. (1970) *Future Shock*. London: Bodley Head.

Tsoukas, H. and Chia, R. (2002) 'On Organizational Becoming: Rethinking Organizational Change', *Organization Science*, 13(2), pp. 567–582.

Van Veen, T.C. (2016) 'The Armageddon Effect', in Anderson, R. and Jones, C.E. (eds.), *Afrofuturism 2.0: The Rise of Astro-Blackness*. Lanham, MD: Lexington Books, pp. 63–90

Virilio, P. (2006) *Speed and Politics*. Trans. M. Polizzotti, New edn. New York, NY: Semiotext(e).

Virilio, P. (1989) *War and Cinema: The Logistics of Perception*. Trans. P. Camiller. London and New York, NY: Verso.

Wilkinson, A. and Kupers, R. (2014) *The Essence of Scenarios: Learning from the Shell Experience*. Amsterdam: Amsterdam University Press.

Wilson, D.H. (2007) *Where's My Jetpack?: A Guide to the Amazing Science Fiction Future That Never Arrived*. New York, NY: Bloomsbury USA.

Womack, Y.L. (2013) *Afrofuturism. The World of Black Sci-Fi and Fantasy Culture*. Chicago, IL: Lawrence Hill Books.

7 For a creative ontology of the future

An ode to love

Jamie Brassett

> The universe therefore is not wide enough for the range of human speculation and intellect.
>
> Longinus, *On the Sublime*, 35.3

> But in order to seize the possible and to actualize it, we need friendship, solidarity, happiness and the ability to take pleasure in physical relations.
>
> Franco 'Bifo' Berardi, *Futurability*, p. 99

Opening up to friendship

Philosophers have written often on friendship (for example: Aristotle, 1976; Derrida, 1997); maybe this is because we are engaged in a practice that includes the word 'love' in its name: *phílos* – beloved, love; *philosophíā* – the love of knowledge, wisdom, speculation and the pursuit thereof (Liddell and Scott, 1889). In 2018 John O'Reilly (co-editor of this volume) and I published a chapter in a book on the philosophy of design, discussing the creative value of collisions to designing (Brassett and O'Reilly, 2018a). From an account of the *clinamen*, The Swerve, in the atomist work of Lucretius (2005, 2007) we developed four possible directions designers, design agencies, studios or in-house design departments could take in order to avoid creative stagnation. These are: to map the collisions that may already be occurring in our practices; to produce diagrams of these colliding practices; to search out where randomness may be blocked in our practices; and to become fans of collisions (Brassett and O'Reilly, 2018a, p. 93). While we have known each other for some decades and have co-written other pieces in recent years (Brassett and O'Reilly, 2015, 2018b), that particular essay on collisions (Brassett and O'Reilly, 2018a) is of note both for what it argues and the actions it precipitates.

For Lucretius (writing in Rome, 1st century BCE), following the materialism and atomism of Epicurus (the Greek philosopher, 341–271 BCE), everything comes from atoms falling in a void. These follow their own paths on an incessant drop in laminar flows, which would lead to nothing were it not for the atoms being knocked slightly off course by the Swerve. They then collide, ricochet or clump to provide an otherwise sterile universe with things.

Creativity is both a consequence of and for the Swerve of all things. Here, already, is one of this chapter's important issues; indeed, one for the book as a whole: ontogenesis, or the creation and ongoing change of being. But more of this later.

For each of us, John and I, the particular configurations of the journeys we have taken as a result of our colliding (Brassett and O'Reilly, 2018a) manifest in our chapters for this book (the one you are holding now): the individual ones and those co-written. This meta-level evaluation highlights one particular manifestation of what our 2018 essay posited: that is, in terms of a fairly simplistic mapping of the trajectories of our lives as they have vectored towards collision points at particular spaces and times (for example, University of Warwick, 1988; London, 2013) and then rebounded in new directions. There is something more to be said of the swerve, collision and an emergent togetherness, however, in relation both to such issues of ontogenesis and to friendship too, for atomist Epicurus – important influencer of Lucretius – was well-known for his commitments to his friends and to friendship. According to the testimony of Cicero (a contemporary of Lucretius and a critic of Epicureanism), Epicurus says that 'of all the things which wisdom has contrived which contribute to a blessed life none is more important, more fruitful, or more pleasing than friendship' (Inwood and Gerson, 1994, p. 63). We are in good company, then, as this book coalesces around a number of material, intellectual and friendly collisions. In effect, friendship is the ethical and creative form produced from the relationality of collisions.

Though institutional philosophy, especially since the 1960s, has been given purpose in many different disciplines and contributed to the unfolding of many different debates (in sociology, politics, literature, history, architecture, design, organization and business studies, science technology and science studies, to name a few), this issue of friendship remains a little developed cog in the multiply connected engine of philosophical thinking. Nevertheless, despite, or because of, the pervasiveness of philosophy across the humanities, social sciences and design, philosophy from more recent times (for example, Derrida, 1997) has returned to thoughts of friendship. A love of wisdom that contrives friendship is developed in the work of Gilles Deleuze and Félix Guattari, notably in their (1994) *What is Philosophy?* 'These are not two friends who engage in thought', they write (1994, p. 69), 'rather, it is thought itself that requires the thinker to be a friend so that thought is divided up within itself and can be exercised' (translation modified). These philosopher friends, these lovers of wisdom and the creation of concepts, are not the originators of thoughts, not dowsers for the original springs of thinking. Rather, they are accelerators and decelerators of those thoughts and concepts that are already in motion, they are agents of the Swerve. Thinking happens and thinkers, friends, are in the middle of this happening. They can facilitate collisions or avoidances, blockages or liquefactions; or even, through some specific technique, impart some spin to generate a different type of swerve. That is, they create with concepts, even if the materials they work with are not original.

It is in this context that Deleuze and Guattari (1994) imagine friendship as a sympathetic emergence from a relational space, rather than something that is intentionally formed by any one person giving energy to inert matter in an empty void. Put this way, such an account of friendship harmonizes with philosopher and technologist Gilbert Simondon's (1924–1989) critique of hylomorphism (1989, 2009, 2013, pp. 39–51).

Matter and form

Hylomorphism is the concept suggested by Aristotle across much of his work whereby matter is inert until given form by some energetic, outside force. Aristotle's account of the relationship of form to matter is seen as one of the most important contributions of his work titled (in Latin translation) *De Anima*: on the soul, or *psuchê* in Greek (Duch, 2018).[1] What drives Simondon to critique, is the requirement in this account of relegating matter to being nothing more than an impotent stuff, which can only be given meaningful form by an outside force. For him, matter is able to form itself. Under particular conditions – notably, supersaturation and metastability – there is a mutual *modulation* of matter and form. This allows us to consider that processes of matter form themselves through a positive feedback between that which strives to fulfil itself and that which acts as resistance (Combes, 2013, pp. 3–6; Sauvagnargues, 2016, p. 69). So that clay placed into a mould modulates its boundaries through a formative relation between its movement and the edges of the mould. (This is one of Aristotle's and Simondon's favourite examples, respectively, of the definitive characteristic of hylomorphism and its undoing). Thus, we have information in both the matter and the form of the mould that work together (modulate) to make a brick. Miguel de Beistegui (2012, p. 171), in an essay on science, ontology and the relation between Simondon and Merleau-Ponty, articulates well Simondon's position on hylomorphism: 'There is no form that presides over the organization of matter; there is simply a series of processes of in-formation through which matter organizes itself'.[2] But it is just as important to recognize the materiality of the mould whose own form is set in relation to the clay-and-brick assemblage. The mould modulates and is modulated by the clay.

Just as privileging energetic form over matter can be undone in this account of mutual modulation and the multiple, complex relationships between matter, information and formation, so too can other creative constructions: not least, friendship. Deleuze and Guattari (1994, p. 3) write: 'Does [friend] designate a type of competent intimacy, a sort of material taste and potentiality, like that of the joiner with wood – the potential of the wood latent in the good joiner; is he the friend of the wood?' In designating the relationship between joiner and wood as 'friendship', there is a sense in which both the joiner and the wood can modulate and inform each other, as the potential of the wood is latent in the joiner and the potential of the joiner latent in the wood. We may, with this reading, extend the act of the realization of creative potential in this relationship the other way, so that the joiner can become joiner only by having their potential worked out by the wood. This is an important point and can be developed in relation to design practice.

Modulating sympathy and friendship

In an essay on Deleuze and design, theorist Hugh Crawford (2015) accesses this distributed, entangled and cooperative version of the creative bringing into being, via the concept of 'sympathy', which he develops using the work of architect and theorist Lars Spuybroek (2011). While a discourse of *empathy* has become a cornerstone of design thinking for almost two decades,[3] designing in a *sympathetic* mode is for Crawford both an ontological process and a living, mutually respectful friendship. This is what it means to create, to be creative: the production of a capacity that opens oneself up to an originality that encourages possibilities for others to be creative too.[4] It is not located in one, special type of person, or material, or even methodology. There are no tools that can work like magic to intensify creativity, even though it can be hard work. At most, sympathy is an attitude to friendship and a mutual modulation of potential for being.[5] We come back to the future in and as potential, a latent power to become that is not always recognized but when it is, allows for an array of developmental possibilities to be welcomed into existence (Souriau, 2009a, b).

There is anticipation here. Anticipation that pulls potential into being, that creates a situation in which this potential is actualized, thereby opening the anticipator to actualizing previously un-self-realized potential. I should qualify 'potential' here a little because it is different to the teleological Aristotelian idea of something dormant that gradually flourishes on a journey to actualization, thus placing this actual as the goal of the potential's growth. The way I am using the notion of 'actualizing … un-self-realized potential' here is slightly different. It involves the anticipation of a self whose elements are exterior to it, distributed across a plane of future adaptations that theoretical biologist Stuart Kauffman (across all his work) calls 'adjacent possible' (of which more below) and, most importantly, these possibles are not goals, but waypoints. The sympathy of friendship and the welcoming of the future coalesce around anticipation and its attendant ontological creativity.

This initiates a pair of thoughts, both of which are given a particular spin by Deleuze and Guattari, and which take their own trajectories while remaining relevant to this chapter. First: 'The creation of concepts in itself [i.e. philosophy] calls for a future form, for a new earth and people that do not yet exist' (Deleuze and Guattari, 1994, p. 108). Second: 'The *diagnosis* of becomings in every passing present is what Nietzsche assigned to the philosopher as physician, "physician of civilization," or inventor of new immanent modes of existence' (Deleuze and Guattari, 1994, p. 113. Original emphasis). The first provides a specifically future orientation for philosophy, philosophical creativity and ontological genesis. The second emphasizes that the philosopher-medic, as both experimenter and explorer, maps becomings in order to create new ontological possibilities.

Philosophy as conceptual creativity anticipates possible becomings of the people of the future and, at the same time, invents and is invented by these becomings. Importantly, though tangentially for the moment, the invention of modes of existence provides also a simultaneous account of their typologies: I am here deliberately echoing Deleuze's (1988, p. 23) definition of ethics from his reading of Spinoza: 'a typology of immanent modes of existence'.[6] Thus, future-focused creativity requires an immanent ethics – also known as sympathy – to ensure its system remains open to new possibilities for creative acts: where becoming something allows for something's becoming to take its own voyages.[7] Thinking, creating, experimenting and becoming are all actively coming into existence, are works in the process of being made, are unfolding and flourishing; and any ethics built into such a process can be either energized or closed down. 'To think is to experiment,' Deleuze and Guattari write (1994, p. 111), echoing a favourite concept of Deleuze's,[8] 'but experimentation is always that which is in the process of coming about – the new, remarkable, and interesting that replace the appearance of truth and are more demanding than it is'. Their anticipation of future existence is at once a mapping of future modes and a welcoming of these into being: a being that is never complete and is always in the process of becoming.[9]

To allow the future in, in order to recreate the present – as occurs in anticipation – is a trusting, sympathetic, friendly act. Friendship enjoys and anticipates the creation of itself through a number of relationships; and the anticipation of the modes of existence of future friends at our future funeral allows us a remodulation of our present existences.[10] Indeed, a deep and moving friendship is relayed in Derrida's (1995) eulogy for Deleuze, titled 'I must wander all alone', in the journal *Libération*. Here, as elsewhere (see Colombat, 1996 for a thorough evaluation), Deleuze's friendship is highlighted as one of his, many, traits; and, like Epicurus millennia before him, his friendship is also a philosophical event. After noting some aspects of Deleuze's work with which he never really agreed, Derrida (1995, online) writes:

> One day, I would like to try to explain how such an agreement on philosophical 'content' never excludes all these variances; that I don't know how to name or situate, even today. (Deleuze had accepted the idea of publishing, one day, a long, improvised conversation between us on this subject, and then we had to wait, to wait too long.) I know only that these differences, between us, never left room for anything other than friendship.

Even while both philosophers' works were concerned with similar issues, similar 'content', their points of divergence do not get in the way of friendship. Derrida is not sure that he can grasp the value of these differences but his wait to work them through, in friendly conversation with Deleuze, will have to last, now, for ever. We wait and can only take our own steps in friendship, and in anticipation, of the philosophical content to come.

Genealogies of anticipation

While anticipation is a concept with some pedigree, its promotion as a discipline within futures studies is fairly recent. Riel Miller, Roberto Poli and Pierre Rossel's (2013) *Working Paper No. 1. The Discipline of Anticipation: Exploring Key Issues* – delivered as part of a UNESCO project called 'Networking to Improve Global/Local Anticipatory Capacities' and supported by the Rockefeller Foundation – is explicit in determining the power and value of anticipation as a discipline in futures studies. For them, every act of engaging with the future involves anticipation in some way, but its specific concretization as a discipline is important as anticipation is best able to cope with the production of novelty within the universe. Creativity is fundamental to anticipation, both as a contextual validator and an intensifier of its capacities. While Miller et al.'s (2013) drive to make a discipline based upon 'scientific' activities within epistemological frameworks is questionable,[11] their insistence upon anticipation as a valuable activity is important not to ignore. In addition, Poli (2010, 2011, 2017; see also: Rossel, 2010) has been crucial in locating the relevance of anticipation to life in the twenty-first century. Encompassing a range of disciplinary references – for example, from biology to philosophy, physics to engineering, social sciences to linguistics and, of course, futures studies – Poli's work is astounding and his (2010) essay, 'The Many Aspects of Anticipation' is a comprehensive overview, carried further in his encyclopædic *Introduction to Anticipation Studies* (2017). In the essay, Poli (2010, p. 8) writes: 'Behaving in an anticipatory way means adjusting present behavior in order to address future problems', adding that an 'anticipatory entity (system or whatever) takes its decisions in the present according to forecasts about something that may eventually happen'. With this, Poli is faithful to Rosen's (2012, p. 313) definition: 'An anticipatory system is a system containing a predictive model of itself and/or its environment, which allows it to change state at an instant in accord with the model's predictions pertaining to a later instant'. These are the key elements of the concept of anticipation that we need: that future possibilities create present states. Here I use 'create'. Rosen most often uses 'cause'. As do other philosophers of the future; for example, Raimundas Vidunas (2019), who in an essay on 'delegated' causality, complex systems and Chinese philosophy, notes emergent phenomena as one of many instances of the future having an impact upon the present. Vidunas (2019, p. 82) writes: 'This concept of *delegated causality* should clarify much about emergence of whole new phenomena (Clayton and Davies, 2006), spontaneous order (Kauffman, 1993), synergy (Corning, 2005), functionality (Ariew et al., 2002), purpose and intention (Dennett, 1987)' (original emphasis). We will come across the influential theoretical biologist, Stuart Kauffman, later in this chapter; but note for now that Kauffman is not so enamoured of causality as Rosen or Vidunas. What is clear from Vidunas's discussion is that his causality operates 'downward' from higher states (that is, future ones) to lower (present) states. The antipathy to a reductionist paradigm within anticipatory systems, that is so evident in Rosen's work (Rosen, 1991), appears also in Vidunas's 'downward causation'. This anti-reductionism is one of the important aspects of the theory of anticipatory systems and places it firmly within other philosophical discourses about complexity. Vidunas lists many key protagonists in his paper: for example, Prigogine and Nicolis (1977), Maturana and Varela (1980), Kauffman (1993). However, Vidunas (2019, p. 93) repeats the unproductive cliché about the 'relativism' supposedly evinced by 'post-modernist' philosophers, referencing the agendas peddled against recent francophone philosophy so profitably by Allan Bloom (1987). Some francophone philosophers *are* important however: most notably the works of Isabelle Stengers (1997; see also: Prigogine and Stengers, 1984 and Stengers and Prigogine, 1997) and Michel Serres (1977). I will turn to Serres later in this chapter but will look now to Stengers's (1997) text on emergence, because the ontological concerns that it deals with – regarding the relations between wholes and parts – have become important, with Vidunas's introduction of different levels or layers of complexity and simplicity in accounting for the relative ontologies of emergence and anticipation.

Emergence

In *La Vie et l'artifice: Visages de l'émergence* [Life and Artifice: Faces of Emergence] (1997), Stengers critiques both the reductionist and holistic approaches that seek to explain how we should respond to emergence. Following materialist philosopher J.K. Feibleman's 'theory of integrative levels' (1954), Stengers argues that emergence itself emerges from and into the point of view of a third position that prioritizes neither the parts (atoms or smaller) to which a whole can be reduced, nor the ends which holistically determine how a group of elements can be thought. If any system's organization can only be explained with reference to a more complex level above or a less complex level below, then both a simplistic reduction (below) and a dominating holism (above) are in themselves inadequate. This leads Stengers (1997, p. 30) to write, with a strong Feiblemanian inflection, that:

> the question of emergence never poses *itself*. It is always actively asked. Indeed, it is only from the perspective of *a third term – establishing* the practice of articulating the relations between ends and means, whole and parts – that we can call forth this question. (Original emphases; my translation.)

Feibleman's first law of organization (1954, p. 59), states that: 'Each level organises the level or levels below it plus one emergent quality'. Furthermore, any level *depends* upon a lower level (Law Three: Feibleman, 1954, p. 60) without being reducible to it (Law Nine: Feibleman, 1954, p. 62) and its organization is *directed* at a higher level (Law Four: Feibleman, 1954, p. 61). Interestingly, when discussing this fourth law, Feibleman notes that while science should eschew the 'transcendental' and so avoid discussing levels or organizations 'teleologically', they can be characterized according to 'purpose'. This allows him (Law Five: 1954, p. 61) to state that 'for an organization at any given level, its mechanism lies at the level below and its purpose at the level above'; thus, providing a temporal qualification within a spatially articulated model.

'Purpose' is a curious concept and one that Feibleman feels less than comfortable using, as he feels it has an air of religiosity that belongs in a different register to science. A *driven* life is fairly common in philosophy and psychology: with notable contributions from Friedrich Schiller (1982) and Sigmund Freud (2003). Here the drives are located in the present fuelled by past experiences. A *purposeful* life, however, has touchpoints with philosophies of hope.[12] In a similar context, Whitehead has not been so squeamish about using terms recognized as religious. Across his work he makes reference to the Quaker concept of 'concern' – felt, for this religious group also known as 'Friends', as a divinely given imperative to act (Hall, 2013) – which, for Whitehead, and for Latour (2008) following him, is an important speculative philosophical concept. Purpose and concern, even without their religious tonalities, approach us from the futures in which they are articulated to the present where our responses to them can make change.

While Stengers mobilizes Feibleman's work in order to disrupt the easy qualifications of emergence as either reductive or holistic, the theory of levels plays an important part in other philosophical discussions. For example, in an essay titled 'Levels of Reality and Levels of Representation' (2004), mathematician Claudio Gnoli and futures philosopher Roberto Poli write of the relations between an ontology of levels (as we have been discussing) and epistemologies of representation, especially those dealing with the types of classificatory system used in library management. Gnoli and Poli (2004, p. 156) state that in Feibleman's analysis, levels can be regarded

> either as whole or as part. However, it is worth mentioning that the theory of levels has been intended by most of the scholars who have elaborated its details as a way to improve both the (traditional) theory of being and the theory of wholes.

Gnoli and Poli show that the relation between 'wholes' and 'parts' in the theory of levels is by no means unambiguous, indeed the ontological shift from one to the other is a key component of Feibleman's concept. For Stengers, that any level can be regarded as both a whole and a set of parts for another level, reinforces the processual relationality of the enterprise rather than thinking of either origins or goals as ideal forms. Stengers thus smooths any points of foundation or aim by smearing them through complex entanglements and multiple singularities, with the consequence that any ontological characterization of a level can fold back into itself as well as develop in relation to a 'third term'. Similarly, Gnoli and Poli find that many demands for the mathematical organization of bibliographic classification flounder on too rigid a construction of ontological position; Feibleman's fluidity seems to stand out.

There is an obvious alignment between this discussion of wholes and parts and Rosen's attitude to reductionism: with his advocacy of life exceeding its parts, chiming with Stengers's mistrust of a reductive explanation of emergent behaviour. Furthermore, when Rosen writes of a being producing a model of the future to incorporate into its present, which creatively constructs that present, we may recognize the creation of a more complex level of organization, brought to the present, living organism in such a way that the emergent life is different to what it would have been. This gives a purposeful recreation of the present according to pressures coming from the future; with a repurposed present providing new singularities according to which futures may emerge.

Stengers returns to this topic in *Thinking with Whitehead* (2011, p. 174), writing:

> For Whitehead, the parts do not constitute the whole without the whole infecting the parts. [. . .] If the body exists for its parts, it is because its parts are infected by such-and-such an obstinate aspect of what we call the body, but which, for them, is a portion of their environment; if the parts exist for each other and for the body, it is because the respective patterns of each are highly sensitive to any modification of the environment they constitute for one another.

Stengers presents an ontology that aligns Feibleman's levels with Whitehead's speculative philosophy. We may also see here a version of Rosen's 'downward causation' of a future construct 'infecting' the present through anticipation. A passage from Whitehead's *Modes of Thought* (1966, p. 21) – which Stengers uses (2011, p. 175) – takes this further:

> For example, our bodies lie beyond our own individual existence. And yet they are part of it. We think of ourselves as so intimately entwined in bodily life that a man is a complex unity – body and mind. But the body is part of the external world, continuous with it. In fact, it is just as much part of nature as anything else there – a river, or a mountain or a cloud. Also, if we are fussily exact, we cannot define where a body begins and where external nature ends.

As Whitehead delivers it here, neither wholes nor parts dominate, neither provide any final, totalizing, ontological principle. Moreover, each has an implication and entanglement with the other in their mutual characterizations.[13] It is worth labouring this point a little, for Rosen bases his adherence to a logic of anticipation as a counterweight to the reductionist approach so prevalent in contemporary science (especially physics).

The figure of Rosen

In *Life Itself* (1991), and its follow-up *Essays on Life Itself* (2000), Rosen poses that the question of life cannot be answered through reduction: life *qua* life, belongs to a higher order, a more complex system than that of the networked atoms (or smaller) that a reductionist account would explain. Such an approach appears obvious for someone schooled in 'relational biology'; which A.H. Louie describes as: 'the study of biology from the standpoint of "organization of relations". It was founded by Nicolas Rashevsky in the 1950s, thence continued and flourished under his student Robert Rosen. And I was Rosen's student' (Louie, 2016, p. 165; see also: Tuomi, 2019). Rosen, an accomplished mathematician and biologist, develops both through relational biology to infuse a doctrine as apparently absurd (to both everyday and scientific thought) as 'the future causes the present'.

An example of his position runs as follows: 'The physicist perceives that most things in the universe are not organisms, not alive in any conventional sense. Therefore, the physicist reasons, organisms are *negligible*: they are to be ignored in the quest for universality' (Rosen, 1991, p. 12; original emphasis). In *Anticipatory Systems* (second edition 2012; originally published 1985) Rosen is more conciliatory: a future-caused anticipatory present 'extends (but does not replace) the "reactive paradigm" which has hitherto dominated the study of natural systems, and allows us a glimpse of new and important aspects of system behaviour' (Rosen, 2012, p. viii). Nonetheless, the commitment to anticipation led Rosen to experience isolation from the scientific community. His foreword to the first edition of *Anticipatory Systems* (2012, p. v) tells that this:

> manuscript was written during the months between January and June in 1979. The writing was done under difficult circumstances. At that time, I had some reason to fear that I would no longer be able to personally pursue a program of active scientific research.

These 'difficult circumstances' anticipation scholar Mihai Nadin (2012a, pp. xviii–xxii; see also: Nadin, 2015a) outlines in more detail. They appear to revolve around the scientific community's disregard of the controversial aspects of anticipation – notably, the future entailing (or even causing) the present (Poli, 2010, p. 8; Louie, 2011, p. 25; Rosen, 2012) – and Rosen's adherence to his position.

The issues of friendship, even love, become foregrounded in this example, with Rosen's anxiety at losing his place within the community of scientists having a negative material impact upon his scientific creativity. Yet it is striking that a key group of friends, ex-students and family members – his daughter, Judith, participates in delivering the second edition of Rosen's book on anticipatory systems, as well as writing the Preface (Rosen, 2012, pp. xi–xiv) – work together in representing his work, as well as in taking it further: both within biology (for example: Louie, 2011) and outside (many of the researchers already listed in terms of futures studies, but also Nadin, 2010a, b, 2012a, b, 2015a, b and 2016). While we may not all have been friends with Robert Rosen, we may be (or are on the road to becoming) 'friends' of the concept of anticipation and of the potential it has for creating new relations, between future and present of course, as well as between different disciplines.

When discussing sympathy and friendship above, we encountered Deleuze and Guattari, from their final work together *What is Philosophy?* (1994, p. 3), writing:

> It is even more difficult to know what *friend* signifies, even and especially among the Greeks. Does it designate a type of competent intimacy, a sort of material taste and potentiality, like that of the joiner with wood – the potential of the wood latent in the good joiner; is he the friend of the wood? (Original emphasis)

Our friendship with anticipation, as a concept (the *material* of philosophy according to Deleuze and Guattari), can allow us insight into the potentials it has to develop new creations. But this also asks us to wonder about the role of anticipation in friendship, as the recognition of the sets of relations according to which it (friendship) can be constituted (that is, in anticipation). Maybe friendship, even love, needs to be pulled into being by a future construction of all the material relations that its organization requires? For to anticipate isolation and loneliness as a philosopher, biologist or simply as someone alive, is no future at all. The section that follows will look into these aggregates of love, anticipation and creative ontologies further, with special – but not exclusive – focus on the works of Lucretius and Michel Serres.

Love and war

The inexorable fall to death of atoms in the atomist ontology is the promise of identity as a fatal uniformity. This is one of the conclusions from Serres's (1977) examination of Lucretius: I will go into this in more detail soon. It is important to recognize here that Lucretius's creative ontology – and Serres's arrangement of this alongside more contemporary theories of nonlinear dynamics and complexity theory – requires both love and turbulence. These are indeed troubling yet productive and joyful collisions. Without love and turbulence there is no life, only atoms falling, falling, staying the same. There is nothing in these atoms that speaks of life. Or, thought another way around: as life becomes reduced to atoms it is gradually extinguished. This is very important for Lucretius in his quirky, deep and turbulent poem of creative ontology; just as we have seen with Rosen and Stengers. It is important also for Serres who castigates Western science's close association with war. Here, in this section we will encounter both of these positions in order to map an anticipatory account of a lovely, lively future/present.

There is much going on here. The issues of affective (generating collisions) and affected (collided) ontology are reminiscent of Spinoza (1996) and the ethics he draws from such relational processes. Here, we could also access the work of Deleuze, on his own and in collaboration with Guattari, in which relational ontological ethics are used to develop not only a critique of identity, representation and Being (Deleuze, 1994, 2001; Deleuze and Guattari, 1984, 1988; Guattari, 1989), but also to advocate a processual, politics of becoming (Deleuze and Guattari, 1988).[14] Still, important as these are, we can only take a little bite of them here. To take measure of creative ontology in Lucretius is to be interested in the ways in which love, nature and existence interrelate; and this necessitates looking at the opening of the first book of Lucretius's poem, 'Matter and Void':

> Life-stirring Venus, Mother of Aeneas and of Rome
> Pleasure of men and gods, you make all things beneath the dome
> Of sliding constellations teem, you throng the fruited earth
> And the ship-freighted sea – for every species comes to birth
> Conceived through you, and rises forth and gazes on the light.
> [. . .]
> Then beasts, the wild and tame alike, go romping over the lush
> Pastureland and swim across the rivers' headlong rush
> So eagerly does each pant after you, so do they heed,
> Caught in the chains of love, and follow you wherever you lead.
> All through the seas and mountains, torrents, leafy-rooted abodes
> Of birds, and greening meadows, your delicious yearning goads
> The breast of every creature, and you urge all things you find
> Lustily to get new generations of their kind.
> Because alone you steer the nature of things upon its course,
> And nothing can arise without you on light's shining shores,
> And nothing glad or lovely can be fashioned, I invite
> You Goddess, stand beside me, be my partner as I write
> *The Nature of Things* [. . .]
>
> (Lucretius, 2007, I.1–25)[15]

Some Lucretius scholars have wondered why such a philosopher – atomist and Epicurean – should have such an opening. Especially as he writes, just over 75 lines later of Agamemnon's sacrifice of his daughter Iphigenia, 'So potent was Religion [*religio*] in persuading to do wrong' (Lucretius, 2007, I.101).[16] That an anti-religious philosopher such as Lucretius opens with a call to a religious figure, speaks neither of a hypocritical attitude to religion nor of a simplistic adherence to poetic convention. The symbolism of Venus as love and procreation allows Lucretius a poetic intervention into the atomist and (Serres will say, 'complex') philosophic position. The ontological creativity encompassed by the term 'Venus' is projected through turbulence as, we know, she is born in the breaking waves of the sea. The collision of the poetic and the philosophical is thus a singular expression and materialization of them both, in their emergent entanglement. 'Because alone you steer the nature of things upon its course' Lucretius emphasizes (2007, I. 21). The nature of things *is* (because of) love. But if this deals with a loving, fecund present, what of the future? For this we must turn to the concept of strategy and Serres's critique.

In the multiplicitous world of designing (encompassing many practices and theoretical standpoints), strategy has become an increasingly important term (see, for example: Lockwood and Walton, 2008; Calabratta et al., 2016). This has coincided with the dominance of design thinking discourses since early in the twenty-first century (Brown, 2009; Martin, 2009. With comprehensive critical evaluations by Kimbell, 2011, 2012 and Johansson-Sköldberg et al., 2013) and its structural and evolutionary values axiomatized in the Danish Design Ladder (for example, Doherty et al., 2014; Whicher et al., 2015). This ladder seeks to show that for design to be used in the most mature ways possible, requires it to progress from lower ranking operations such as styling and move towards the more advanced forms of strategy and policy. Design comes of age when its practitioners and their services enter into C-suite boardrooms. To be strategic, the argument continues, is one way of displaying one's maturity. A fuller critique of design strategy, design thinking and related concerns deserves more than one chapter – indeed, John and I have embarked on such in previous work together (Brassett and O'Reilly, 2015) – however, I shall move to locate a loving critique of strategy in another way. And Serres will lead.

Loving complexity

To engage with strategy in this chapter is to question its easy acceptance by creative disciplines unused to, or not bothered by, its military values.[17] Crucial in achieving this is Serres's (1977) examination of the qualities of Lucretius's science that align with complexity theory and nonlinear dynamics, that he develops in a book titled *La Naissance de la physique dans le texte de Lucrèce. Fleuves et turbulences* (literally, 'the birth of physics in the text of Lucretius. Flows and turbulences').[18] For Serres, the dismissal of Lucretius's work as either quaint or wrong by generations of Western scientists, overlooks some of Lucretius's most important contributions; most notably, its presaging of complexity in discussing flows, turbulences and vortices (Clucas, 2005, pp. 72–73). But this ignorance of the truly creative properties of Lucretius's work is not the only problem here: not only has he been overlooked by Western science but shunned. Serres writes (1977, p. 135), with a sense of shame[19] that is palpable:

> It turns out that Western science has never ceased to choose otherwise than Lucretius; and there is nothing I can do because I am its slave. To opt for war, the plague. Blood, the fight and the bodies[20] on the bonfire. From Heraclitus to Hiroshima, Western science has never known anything other than martial nature. (My translation)

Like Lucretius, the collision Serres initiates of the poetic and the philosophical is the opening for creation. It is the opening and creation of 'Michel Serres' as the researcher and thinker of 'science' and 'literature'. Serres is already too late to remove the bloodstains of the battlefield from his hands, complicit, as he is, as a participant in the lists of science: as he has moved from mathematics to science to philosophy (and literature and culture) throughout his career (Serres and Latour, 1995, pp. 11–12). But while he is too late for the past, maybe Serres can be in time for the future? The 'gun-shy', ex-member of the Naval Academy – who says that he 'resigned from the navy because I didn't want to serve cannons and torpedoes. Violence was already a major problem – has remained so, all my life' (Serres and Latour, 1995, p. 7) – dismisses the principles of war and the generals and admirals, and all beholden to them, from the future; and thereby, we may add, from the present through anticipation.

It is here, at this moment where future and present involve one another, that Serres and Lucretius and Rosen converge: with creativity and love, future and present becoming one another in anticipation, in life. This is a life that is wrenched from the issues of war to be wrapped in the arms of loving creativity, where Mars slumbers and strategists are rendered impotent. For Serres, the disarming of Mars also shows a movement away from a repetitive reductionism that seeks only to find order and identity. He writes:

> The order of reasons is repetitive. Knowledge sequenced in this way, infinitely iterative, is only the science of death. A science of dead things and a strategy for putting to death. The order of reasons is martial. [. . .] The laws are the same throughout, they are 'thanatocratic'. There is nothing to know, to discover, to invent in repetition. It falls in the parallel of identity. Nothing new under the reign of the same. It is informational zero, redundancy.
>
> (Serres, 1977, p. 136. My translation)

Serres here bemoans the dominant rational scientific tendencies to shackle their systems to the dead weight of warlike futures past, driven to entropic stagnation by the dull motor of history either through a stifling path dependency or the reduction of everything to well-understood, thoroughly determined systems. 'The order of reasons is [. . .] "thanatocratic" [. . .]. Nothing new under the reign of the same': otherwise known, in design, business and organizations, as 'best practice'. His earlier, more plaintive cry – 'there is nothing I can do because I am [Western science's] slave' – giving way to something more insistent, more intense. Why should the 'thanatocratic' drive to death be an option? Why is a martially organized future-present even on the same page as one driven by love, creativity and friendship? Lucretius's opening ode to Venus is neither simply formalistic nor trivial: it is calling out to us – his future – as if we were there with Lucretius himself. *We* are his hope alive in the present (Nietzsche, 1997, p. 95[21]); we are with him to clear the ground for a different, more creative possibility. Lucretius and Serres produce the reader who is ready to anticipate, research and experiment, and the friends and lovers that will collide.

Law and treaty, physics and ethics

However, the relationship between Venus and Mars is not as simple as of her overcoming him with love. Indeed, Classics scholars are in dispute over whether Lucretius's Venus is triumphant (for example, Asmis, 1982) or whether they rule together (for example, Moorman, 2009). Venus has always found him attractive, enticing. She is one of the few deities in the pantheon not to regard Mars as objectionable with their affairs and sexual encounters causing strife among both their fellow gods and humans (Fairbanks, 1907; Graves, 2011). The mythology tells us that from the union of Venus and Mars issues Concordia, whose own mythos links her with order, peace and harmony.[22] Yet such a *Pax Concordia* is not entirely positive. Order may well happen by force, imposed by threats of violence if not respected; it looks as though Mars won after all and we were never vigilant enough to live without conquest. Maybe there is hope for us yet, in anticipation.

This brings us to an aspect of Serres's discussion that is more common in Lucretius scholarship: an examination of the roles of *fœdus fati* and *fœdera naturae*, fateful or natural treaties. (My use of 'treaty' undermines in particular the authority of what might commonly be called a '*law*' and the range of people and actors who act in accordance with it. I will return to this in more detail below.) Serres writes:

> Then Mars governs the world. He cuts bodies into atomic pieces and makes them fall. It is the *fœdus fati*: it is the law in the sense of physics – this is the case. It is also the law in the sense of dominant legislation – this is how they want it to be. Mars chose this physics: the science of the fall and of silence. This is the plague.
>
> (Serres, 1977, p. 137. My translation)

This fateful law sends atoms to their death in laminar flows, with no disturbance, presided over by the law of identity, of the same, without birth only death: the fateful urge of thanotocratic rationality. Serres continues:

> Nothing new under the reign of the same and under the same reign, conserved. Nothing new and nothing born: not of nature. It is death: eternally. The putting to death of nature, its birth repulsed. The science of this is nothing; it is calculably nothing. Stable, immutable, redundant. It recopies the same writings, with the same atom-letters. The law is the plague. Reason is the fall. The cause is iterated: death. Repetition is redundancy and identity is death. Everything falls towards zero: the nullity

> of information, the nothingness of knowledge, non-existence. *The Same is Non-Being.*
>
> (Serres, 1977, p. 137; original emphasis. My translation)

This scientific rationality of violence, of reductionism, of order and identity gives birth to nothing, certainly to nothing natural. Luckily, we have pockets of creative generation that erupt from these constraints and provide possibilities of escape, joyous flight into the arms of Venus. For Serres, from Lucretius, we have the *fœdera naturae*, the natural alliances that bring together those once trapped in identity to form the universe anew: vital and vibrant.

It is worth staying with the Latin term *fœdus* for a moment, for the directions it takes us in move away from Serres's use of 'law' (*la loi, le statut*). Classics scholar Elizabeth Asmis (2008, p. 141) chooses to focus upon this term in Lucretius's work as a way to bring together what had been two different systems in Epicurus's teaching: ethics and physics. She argues that, for Lucretius, *fœdus* indicated not a law but rather a treaty, a pact or compact. To say 'natural *law*' in the sense that we, today, have of 'the *laws* of physics' is somewhat wide of the mark when thinking of *fœdus*. To show this, Asmis (2008, pp. 142–143) provides four key characteristics of a treaty as the Romans would have understood the term: 'an agreement between self-governing states'; a pact 'designed to end conflict or promote cooperation'; as setting out 'conditions according to which different parties should act', often in terms of setting boundaries or limits within which actions should be constrained; and, finally, 'a treaty must be respected'. The second of these – ending conflict – is clear within Serres's work. The third – working within limits – is also apparent in Lucretius, especially when he discusses the immutability of individual atoms and their inability to change their character in themselves (see, for example, Lucretius, 2007, I. 584–598 and II. 299–302; discussed by Asmis, 2008, pp. 144–146). The fourth point – that a treaty must be respected – Asmis uses to highlight an important difference for the Romans between a treaty and a law. She writes: 'The term "law" suggests the necessity to submit to constraints; the term "treaty" suggests a flourishing of powers within limits' (Asmis, 2008, p. 146). Treaties are respected but laws must be obeyed (Asmis, 2008, p. 143); treaties deal in trust and good will and give each party making the pact, agency.

This is an important intervention in the argument I have been developing with Serres. For the ways in which natural treaties allow for the possibilities of multiplicitous, creative diversity, albeit within specific boundaries, are shared with *fœdera naturae* from Lucretius to Serres. It is as if, notwithstanding two millennia of distance, a pact can be made in the name of a more creative science or philosophy; or, indeed, a more creative life itself.

Asmis's outlining of the ability of the natural pact to promote flourishing within limits is apparent in Gilles Deleuze's essay 'Lucretius and the Simulacrum' (2004, pp. 303–320). Deleuze writes (2004, p. 315): 'From Lucretius to Nietzsche, the same end is pursued and attained. *Naturalism makes of thought and sensibility an affirmation*', (emphasis added). The flourishing, life-affirming nature characterized in Venus is part of a compact that the physical world has with its future. (Which is how Deleuze can say (2004, p. 308), echoing Whitehead (2004), that 'Physics is Naturalism from the speculative point of view'.) In writing which brings a discussion of nature, Nature and the imposition of totalizing powers upon creativity in the name of Being together with our earlier passages on the excessive relations between wholes and parts, Deleuze writes (2004, p. 304):

> We shall find no contradiction between the hymn to Venus-Nature and to the pluralism which was essential to this philosophy of Nature. Nature, to be precise, is power. In the name of this power things exist *one by one*, without any possibility of their being gathered *all at once*. Nor is there any possibility of their being united in a combination adequate to Nature, which would express all of it *at one time*. Lucretius reproached Epicurus' predecessors for having believed in Being, the One and the Whole. These concepts are the obsessions of the mind, speculative forms of belief in the *fatum* [destiny], and the theological forms of a false philosophy. (Original emphases.)

From identity to the One, the violence of order and a law that orders reductive obedience, to Serres and Deleuze and the creative power of diverse and heterogeneous love: we have almost everything we need from Lucretius. As always, there is something extra. Another relation noticed by Rosen and born from friendships that span and exceed given 'laws of nature'. Where 'Life', just as with 'Nature' in the passage above, is more than all its little pieces 'gathered all at once'.

From laws to life

Philosopher David Webb's (2017) 'On Causality and Law in Lucretius and Contemporary Cosmology',[23] draws attention to the cosmological urgency of friendly treaties, ontological creativity and the relations between anticipation and life. For this, Webb ranges Lucretius alongside cosmological philosopher Roberto Mangabeira Unger and physicist Lee Smolin's (2015) *The Singular Universe and the Reality of Time*. This book is not so much written cooperatively as written atomically, the parts of which are put into collisions, sometimes coalescing and at other times rebounding. Unger and Smolin's work together, spanning almost a decade, is as open with its moments of nonalignment as with its togetherness. They write in a way that stylistically manifests their theoretical concerns: notably, that there are no immutable laws that are identical across the whole range of the universe, rather evolving, local treaties that develop along with their local region, leaving a universe that is uneven.

One of the most important aspects of confluence between Lucretius, Unger and Smolin, even across two millennia, Webb shows, is a joint perspective on the relationships between the 'laws of nature' and nature itself. As currently considered by most scientists (but not the ones focused upon here) these laws seem to be developed in order to understand specific phenomena, while the initial conditions according to which such phenomena are generated 'are not themselves explained by those laws' (Webb, 2017, p. 256). Webb shows that Lucretius, Unger and Smolin offer a universe in which the laws governing its behaviour are as changeable and emergent as the universe itself and its parts. Webb (2017, p. 258) articulates the rather surprising position that they all reach, thus: 'Rather than moving from a strongly determined universe to one without meaningful laws, Unger and Smolin argue that laws and the phenomena they govern develop in tandem, their difference being more one of degree than of kind'. Whitehead (1978) calls such a process 'immanent'; and the 'transcendental empiricism' Deleuze (Sauvagnargues, 2010; Rölli, 2016; Heaney, 2018) develops from reading Whitehead and William James operates in the same manner. The conditions according to which any system exists and develops are coexistent with, and affected by, the modes of existence that they regulate;[24] and all in such a way that there is no necessity for these to be evenly, equally or identically distributed. Any system's relationship with its edges and outside possibilities – and so also that system's relationship to the future – becomes of crucial importance. This is evident in a passage from Unger and Smolin (2015, p. 27; quoted by Webb, 2017, p. 258) who write:

> Natural phenomena present themselves, according [to the conception represented in the Aristotelian tradition, through Galileo and Newton to today], within a limited range of parameters of energy and temperature. The penumbra of the adjacent possible around each phenomenon – what it can become next, given what it is then – remains restricted or thin. The laws of nature – both the effective laws operating in particular domains, and the fundamental laws or principles cutting across domains – are clearly distinct from the phenomena that they govern. It is only a short step from these conceptions to the idea that changing states of affairs are governed by unchanging laws.

Such an equally tempered universe allows for neither specific differences in behaviour nor any local, immanent, regulatory frameworks that may emerge therein; and therefore enforces a natural law as statistical mean to deliver homogeneity across all cosmological reaches. Clearly, for Unger and Smolin, the present universe appears very differently. And in such a presentation, where singular regions can exhibit relations to specific sets of natural laws, the boundaries between regions within the cooled-down universe can proliferate and become ontologically important through multiplicity and difference. The beings (living or otherwise) possible in any region will be determined differently, if they can be determined at all.

Adjacent possibles

Unger and Smolin's use of theoretical biologist Stuart Kauffman's 'adjacent possible' emphasizes further a living relationship to the future. This is an important concept for Kauffman who uses it across much of his work, from the last 30 years or so, to delineate the different areas around any living system into which it can evolve. At any moment, life includes a 'penumbra' of future states that may manifest different lines of its becoming. None can be determined by the current state and, for Kauffman, neither can they be caused. This represents a major difference between Kauffman's and Rosen's approaches to theoretical biology: even while they both remain biologists committed to relational ontologies, close to theories of complexity and critical of the reductive physics characteristic of the dominant strands of Western science as least since Newton.[25] For Rosen, causal entailment – what is entailed necessarily follows from a prior event that can be recognized as its cause – is essential in developing his position on anticipation, allowing him to posit a future cause for the present. Kauffman (2016a), on the other hand, will not even admit of entailment between a present system and an emergent future one, let alone a causal one. Rosen's position is one that he relies upon in order to advocate for a present, living system being able to *model* a future for itself as an anticipatory act. The difficulty that he finds with his contemporaries unable to accept his perspective, revolves around his assertion of an anticipated future *causing* the present (Rosen, 2012). Yet, if we follow Kauffman's recent thinking – and its alignment with the speculative metaphysics of Whitehead, as shown particularly in Chapter Four, 'A Creative Universe. No Entailing Laws' (Kauffman, 2016a, pp. 64–82)[26] – the future states afforded by actual occurrences will not necessarily be 'prestatable' (predictable) by any articulation of the present system. 'We are radically free,' Kauffman (2016a, p. 76) writes, 'and radically emergent and radically co-creative. [. . .] Indeed, we are "sucked into" the very adjacent possibilities we, largely unknowingly, create.' For Kauffman, the intensification of and movement into an adjacent possible is a necessary characteristic of creative evolution. That is, there are multiple future ontological possibilities available to any actual entity's configuration, before any particular one is actuated. It is the specific arrangement of an actual entity that articulates the possibles within its 'penumbra'; with evolution happening when this entity is 'sucked into' one of the adjacent possible states and thereby activating its own creative, becoming actual.[27]

Nietzsche (1997) asked the creative philosopher to 'clear the ground' in order to build a present alive in a hopeful future.[28] The surface prepared here, in this chapter, has shown a wide variety of creative, loving and anticipatory adjacencies: emerging 'new Actuals' (Kauffman, 2016a, p. 76); various singularities and their attendant (emerging) natural treatises; the shifting boundaries determined by the constant inter- and intra-constraining relationships that constitute all actuals (new and old). The living, creative, present requires anticipation to exist.

Rosen's work has directed our attention to biology and to characterize an anticipatory life set free from reductive homogenization; and Lucretius and Serres have allowed us to add love to our concoction of the creative complexity of becoming. Which brings us back to the beginning of this chapter, to friendship and sympathy, where a society of friends gives birth to concepts from the depths of their concerns for life. As we saw, a creator's sympathetic working of a new form is a simultaneous modulation of processes of forming and mattering distributed, at least, across the maker, materials and shapes required and, at most, across a much wider social, cultural and maybe even cosmological array of interacting contexts. In an emergent universe, the pacts and treaties needed to develop what might become are locally and immanently created; as if the mutual modulation of 'law' and mode of existence possible in any place are mutually sympathetic in love and a friendship.

Last remarks

To start to close this chapter, in the first instance, I return to Serres and the dynamic, complex approach to life he develops with Lucretius. Serres (1977, p. 139) writes: 'Turbulence is a deviation from equilibrium. And the beginning of the vortex is the minimal angle of declination. That living things trouble the order of the world means, literally, that living things are primarily turbulence'. Imagine, then, an incidental moment of deviation in the present delivered by a particle hurtling in from the future, with this deviation providing a point at which the portrayals of living things as turbulence (Serres, Lucretius) and anticipatory systems (Rosen) are mixed. A turbulent life that complexifies and creates. As Serres recounts it (1977, pp. 135–139), there are two types of physics, two natures: the martial and the loving. One leads to stagnant death and the other to fecund life.

> Immanence: the world is traversed by laws; it is, without deviation, the place of reasons. But, before developing verses, we must choose between two laws. The law of Eros, the law of death. Spring or the plague. Birds, dead bodies. The wound of love or decomposing limbs.
>
> (Serres, 1977, p. 135)

To move against strategy, then, is to move for love, for pacts and alliances, treaties and friendships. This is to be on the side of life, of life itself. Influenced by Rosen (and the other biologists mentioned above), we may regard Serres and Lucretius's Venus-infused works as 'natural philosophy' that deals in the ways in which being becomes more than itself, becomes creative. Where a lifeline to this 'more than itself' is provided by anticipation, vectoring in from the future. Here, with this 'becoming more', we have a gesture of love and sympathy for a future that also includes us. This is the anticipation where 'what-more-may-become' is able to remodel the present.

It is impossible, now, to write these words without each word preparing treaties with all the others written here. Each of this chapter's key concepts

– friendship, love, anticipation, creativity, ontology – caresses another; to finish with them, here in these final remarks, involves putting them all into motion. Again. Maybe this is the best way I can think of to characterize the recursive entailment that Rosen (1991) finds so important for life, that the words I deliver and have delivered will form the content of the works made and to come. And this done with love, anticipating ties of friendship rather than strategies for defeating enemies.

Friendship as an entity, then, can be a society of friends, an ecology of a people, to be anticipated and created from its latent potential. A friend I can become and by so being, the nature of 'friend' changes. And so on. Friendship can also be the becoming friend in the present that welcomes friendship into existence and allows it – and the elements constitutive of its society, its ecology – to develop along its, their, own lines. A friendship occurs when its possibilities for creative development – in itself and in each friend – are amplified and welcomed, encouraged and cultivated by every friendly action. A friendship anticipated and emergent, hoped for and welcomed, mutually and sympathetically created and modulated.

And love? What might it provide us in the future? Can an anticipated love provide the model according to which the actual occurrence doing the anticipating expresses its life? This must be possible, an 'adjacent possible' maybe. A penumbra that surrounds any actual life, which can coalesce around an actual life as it becomes characterized by the future, if only one has the openness to anticipate and then to welcome oneself into that space.

Acknowledgements

Many thanks are due to my family – Joanna, Ewelina, Tomek, Jacek and Bartek – for their love, support and patience: these were well needed and much appreciated. To John O'Reilly, for his friendship, criticality, editor's eye and love of philosophy allowing us to create so many, different concepts. To my colleagues and students at Central Saint Martins, University of the Arts London and Anhalt University of the Applied Sciences, Dessau, Germany who had oftentimes to meet me and my 1000-yard stare enthusing over a work-yet-to-come, thank you so much for your forbearance.

Notes

1 Aristotle (1986) presents hylomorphism in terms of the relations between matter and form. To delve into this fully, here, would be something of a digression into Aristotelian avenues that would take us to his theories of change and causality. However, there is a sense in his work that matter is inert until it is formed. Ingold (2012) gives a good overview of work that is close to the one I'm producing here. For articles addressing Aristotelian 'hylomorphism' of tangential importance to this chapter, see: Johnston (2006) – in terms of the relation between wholes and parts (mereology); Manning (2013) – on the history of the concept. Simondon's work bears some debt to Étienne Souriau's (Haumont, 2002), which is somewhat paid by Deleuze (1994, 2004). On the

relationship between Deleuze and Simondon see Bowden (2012) and Sauvagnargues (2016).

2 This is developed further in materialist work such as Barad (2007), Bennett (2010), and Connolly (2011).

3 There are so many works placing empathy at the heart of design and design thinking that it has become cliché. Influential works on design thinking (also design-driven innovation, or design for business) are: Cox (2005), Brown (2009), Martin (2009), Verganti (2009) and Johansson-Sköldberg et al. (2013). On the importance of empathy in these disciplines see, for example: Kimbell (2011), Köppen and Meinel (2015), Dalton and Kahute (2016). Empathy in these contexts is regarded as putting oneself in someone else's emotional (among others) perspective. This is different to sympathy, as the opening up of oneself to the emotions of others as an act of self-changing friendship.

4 This aligns not only with the ways that complexity biologist Stuart Kauffman discusses the catalysis of complex evolution (1993) but also with Deleuze and Guattari's own work on becoming. It is clear, in their (1994) *What is Philosophy?* that any act of becoming something should allow that something to flourish in its own becoming (see, for example, Deleuze and Guattari, 1994, p. 109). Similarly, Kauffman (1993, 2008) highlights that the catalysis of complex bio-chemical entities needs to ensure the possibilities for future creative catalysis for evolutionary change to ensue and proliferate. I discuss these issues further in Brassett (2015).

5 Donna Haraway's recent *Staying with the Trouble* (2016) develops some similar concerns. Chapter Three 'Sympoiesis' (pp. 58–98) is particularly relevant, with its disruption of the self-obsessed *autopoiesis* in favour of a more communal, sympathetic, mutual modulation afforded by *sympoiesis* into existence.

6 I discuss this in relation to design ethics and social design in Brassett (2018, 2019).

7 This enters into proximity with what John O'Reilly argues about the creative power of the 'other-wise' in this volume.

8 See Deleuze (1995, p. 87) where – in an interview on Foucault in *Libération*, collected in *Negotiations 1972–1990* under the title 'Breaking Things Open, Breaking Words Open' – he exhorts: 'Never interpret: experience, experiment'.

9 There are other philosophical traditions that deal with ethics and future populations; most notably, 'population ethics' (e.g. Blackorby et al., 1995; see also: Brassett, 2018) and 'intergenerational justice' (e.g. Sanklecha, 2017). In examining and critiquing the ethical assumptions of intergenerational justice, Sanklecha (2017, p. 230) states: 'in order to care about the future it is necessary that there *be* a future' (original emphasis); and (p. 240) 'for the present to matter, it is necessary that there is a future'. Both of Sanklecha's comments are resolutely anticipatory and a full investigation of the possible interactions between these traditions and those developed in this chapter would benefit future research.

10 Derrida (1997, p. 5) notes: 'A memory is engaged in advance, from the moments of what is called life, in this strange temporality opened by the anticipated citation of some funeral oration. I live in the present speaking of myself in the mouths of my friends, I already hear them speaking on the edge of my tomb'.

11 Which is something for another time, as it lies outside of the scope of this chapter.

12 Poli (2010, 2017) works closely on this aspect of Ernst Bloch's philosophy. Editors' note. See also Barron in this volume.

13 It is not without significance, I think, that this passage comes at the beginning of Chapter Two, 'Expression' (Whitehead, 1966, pp. 20–41). The relational ontology of Spinoza (1996) also connects with concepts of wholes and parts, bodies and elements and their affective and expressive qualities. I discuss these in terms of developing a relational ethics of design in Brassett (2018, 2019).

14 Such concepts as 'process' and 'affect' recall the writing of Souriau (2009a, b) and Whitehead (1978). For Souriau, existence – differentiated into multiple modes each on their own journeys of becoming – is not given in advance but needs to be created into constantly dynamic, endlessly varying beings. What these thinkers have in common, even

while they diverge into so many directions, is a sense of welcoming, of friendship; even of love.

15 The convention in citing passages from Lucretius's poem is to give the book number in Roman numerals, followed by line numbers.

16 In the prose translation Latham uses 'superstition' for *religio*: 'Such are the heights of wickedness to which men have been driven by superstition [*religio*]' (Lucretius, 2005, I.101). The poetry translator, Stallings, in her notes, remarks that she uses Religion and Superstition interchangeably (Lucretius, 2007, p. 240, n. 9).

17 Coming from the Greek *stratēgōs*, the head of the army (Liddell and Scott, 1889), strategy is a concept that is deeply embedded in war (von Clausewitz, 1984), with other uses outside the military context occurring only recently. Burnes (2018, p. 226) writes: 'It is commonly believed that our concept of strategy has been passed down to us from the ancient Greeks. [. . .] The concept remained a military one until the nineteenth century, when it began to be applied to the business world, though more writers believe the actual process by which this took place is untraceable [. . .].' In my presentation at the *Anticipation 3* conference in Oslo (2019), I showed a slide of the front cover of *Harvard Business Review* (2010). The image is a photograph by Jeremy Lock of a group of Australian soldiers, taken on 26 June 2007 as they train prior to deployment in either Afghanistan or Iraq. The headline reads: 'Leadership Lessons from the Military. Extreme conditions in Afghanistan and Iraq have become a testing ground for adaptive management skills that every CEO should understand' (front cover).

18 I will refer mainly to this text, providing my own translations. The English translation, Serres (2000), I will refer to when necessary.

19 Serres's description of his formative childhood and early adult years, in conversation with Bruno Latour, is extremely evocative of the violence he experienced, from the age of six (in 1936) and the Spanish Civil War, to the French colonial wars of the 1950s and 1960s, via Occupation, Liberation, Auschwitz and Nagasaki (Serres and Latour, 1995, pp. 1–42).

20 Serres's word here, '*les corps*', is also corps as in a military unit.

21 Editors' note. See Chapter 10 in this volume for more on Nietzsche and the untimely in relation to anticipation.

22 I am indebted to an audience member – in the session in which John and I test-drove some of the aspects of our work here, at the 3rd Anticipation conference, Oslo, 2019 – for remarking upon this aspect of the mythology. (I never caught their name.) In the Greek tradition, Hesiod (1988, pp. 30–31) explains: 'To Ares [Mars] the piercer of shield-hides Cytherea [Aphrodite, Venus] bore Terror and Fear [...] and Harmonia [Concordia], whom proud Cadmus made his wife.' While Roman mythology drew upon Greek, it was not a simple copying: as Budin (2004) explains with regard to the earliest syncretic relationships between Mediterranean (Cypriot and Cretan) and Near Eastern (Egyptian, Levantine and wider Semitic) deities. In relation to later Greek-Roman mythologies, it appears that the Roman versions of the shared deities were more serious, less capricious, seemingly to emphasize the greater maturity of Roman culture. Such is the difference between Mars and Ares in Roman and Greek cultures respectively: the latter almost universally despised by gods and mankind alike; the former worshipped as a serious warrior.

23 Webb's work (2006, 2017) has been of great value to my encounters with Lucretius and Serres; especially Webb's ability to encounter creatively both philosophical and scientific concepts. There is more to this, which is relevant to the terms of this chapter. You see, David Webb, John O'Reilly and I shared a house and friendships while working on our PhDs at the University of Warwick, UK in the late 1980s and early 1990s. I write this as a form of intellectual honesty and to locate an openness and sympathy to concepts in moments of collision and swerve.

24 'Regulate' is a term that Webb offers instead of 'law' (2017, p. 260) following Unger and Smolin (2015), which avoids any sense of dominance of law over the legislated. It would be interesting to examine further the territories opened up here, where issues of power, force and control are lively. Useful would be the work of Giorgio Agamben (1998, 1999).

25 See, for example, Rosen (1991, 2012) and Kauffman, (2008, 2016a, b).
26 See also Kauffman (2016b). The work of Johanna Seibt is also important, in its bringing Whiteheadian Process Theory to different scientific practices; see especially: Seibt (2010) on the relations between ontology and theories of particulars; (2018a) on ontology, process theory and biology; and (2018b) on modes of occurrence and her creation of a General Process Theory. (Indeed, see the whole collection of which Seibt, 2018a is a part: Nicholson and Dupré, 2018.)
27 The confluence of this discussion with evolutionary theory and its concepts of 'preadaptations' (see: Buss et al., 1998; Ardila, 2016) and 'exaptations' (Gould and Vrna, 1982; Stengers, 2011, p. 330) is striking, but its examination is for another time.
28 Editors' note. See Chapter 10 on anticipation and the concept of the contemporary in this volume.

References

Agamben, G. (1999) 'On Potentiality', in Agamben, G. (ed.), *Potentialities. Collected Essays in Philosophy*. Trans. D. Heller-Roazen. Stanford, CA: Stanford University Press, pp. 177–184.

Agamben, G. [1995] (1998) *Homo Sacer: Sovereign Power and Bare Life*. Trans. D. Heller-Roazen. Stanford, CA: Stanford University Press.

Ardila, A. (2016) 'The Evolutionary Concept of "Preadaptation" Applied to Cognitive Neurosciences', *Frontiers in Neuroscience*, 10, Article 103. doi:10.3389/fnins.2016.00103.

Ariew, A., Cummins, R. and Perlman, M. (eds.) (2002) *Functions*. Oxford: Oxford University Press.

Aristotle (1976) *The Nichomachean Ethics*. Trans. J.A.K. Thomson. Revised H. Tredennick. Penguin Classics. Harmondsworth, Middx.: Penguin Books Ltd.

Aristotle (1986) *De Anima (On the Soul)*. Trans. H. Lawson-Tancred. Penguin Classics. London: Penguin Books Ltd.

Asmis, E. (2008) 'Lucretius' New World Order: Making a Pact with Nature', *Classical Quarterly*, 58(1), pp. 141–157.

Asmis, E. (1982) 'Lucretius' Venus and Stoic Zeus', *Hermes*, 110, pp. 458–470.

Barad, K. (2007) *Meeting the Universe Halfway. Quantum Physics and the Entanglement of Matter and Meaning*. Durham, NC and London: Duke University Press.

Beistegui, M. de (2012) 'Science and Ontology. From Merleau-Ponty's "Reduction" to Simondon's "Transduction"', in De Boever, A., Murray, A., Roffe, J. and Woodward, A. (eds.), *Gilbert Simondon: Being and Technology*. Edinburgh: Edinburgh University Press, pp. 154–175.

Bennett, J. (2010) *Vibrant Matter. A Political Ecology of Things*. Durham, NC: Duke University Press.

Berardi, F. 'Bifo' (2017) *Futurability. The Age of Impotence and the Horizon of Possibility*. London and New York, NY: Verso.

Blackorby, C., Bossert, W. and Donaldson, D. (1995) 'Intertemporal Population Ethics: Critical-Level Utilitarian Principles', *Econometrica*, 63(6), (November), pp. 1303–1320.

Bloom, A. (1987) *The Closing of the American Mind*. New York, NY: Simon and Schuster.

Bowden, S. (2012) 'Gilles Deleuze, a Reader of Simondon', in De Boever, A., Murray, A., Roffe, J. and Woodward, A. (eds.), *Gilbert Simondon: Being and Technology*. Edinburgh: Edinburgh University Press, pp. 135–153.

Brassett, J. (2019) 'Creating Different Modes of Existence: Toward an Ontological Ethics of Design', in Muratovski, G. and Vogel, C. (eds.), *Philosophical Frameworks and Design Processes*. 'Re:Research' Series, vol. 2. Bristol: Intellect Books Ltd., pp. 17–30.

Brassett, J. (2018) 'Creating Affective Social Design. An Ethical and Ontological Discussion', *Cubic Journal*, 1, Special Issue: 'Design Social', pp. 146–159.

Brassett, J. (2015) 'Poised and Complex. The Becoming Each Other of Philosophy, Design and Innovation', in Marenko, B. and Brassett, J. (eds.), *Deleuze and Design*. 'Deleuze Connections' Series. Edinburgh: Edinburgh University Press, pp. 31–57.

Brassett, J. and O'Reilly, J. (2018a) 'Collisions, Design and The Swerve', in Vermaas, P. and Vial, S. (eds.), *Advancements in Philosophy of Design*. 'Design Research Foundations' Series. Cham, Switzerland: Springer, pp. 71–98.

Brassett, J. and O'Reilly, J. (2018b) 'Retour à 2078: Réflexions sur l'anticipation et le contemporain', *Revue de Prospective et Stratégie*, 1, Édition Spéciale: 'Voir Loin', (December), pp. 11–22.

Brassett, J. and O'Reilly, J. (2015) 'Styling the Future. A Philosophical Approach to Scenarios and Design', *Futures*, 34, Special Issue: 'Scenarios and Design', pp. 37–48.

Brown, T. (2009) *Change by Design: How Design Thinking Transforms Organizations and Inspires Innovation*. New York, NY: HarperCollins Publishers.

Budin, S.L. (2004) 'A Reconsideration of the Aphrodite-Ashtart Syncretism', *Numen*, 51(2), pp. 95–145.

Burnes, B. (2018) *Managing Change: A Strategic Approach to Organisational Dynamics*. 7th edn. London: Pearson Education.

Buss, D.M., Haselton, M.G., Shackelford, T.K., Bleske, A.L. and Wakefield, J.C. (1998) 'Adaptations, Exaptations, and Spandrels', *American Psychologist*, 52(5), pp. 533–548.

Calabratta, G., Gemsa, G. and Karpen, I. (eds.) (2016) *Strategic Design. Eight Essential Practices Every Strategic Designer Must Master*. Amsterdam: BIS Publishers.

Clausewitz, C. von [1832] (1984) *On War*. Eds. and Trans. M. Howard and P. Paret. Princeton, NJ: Princeton University Press.

Clayton, P. and Davies, P. (2006) *The Re-emergence of Emergence*. Oxford: Oxford University Press.

Clucas, S. (2005) 'Liquid History. Serres and Lucretius', in Abbas, N. (ed.), *Mapping Michel Serres*. Ann Arbor, MI: University of Michigan Press, pp. 71–83.

Colombat, A.P. (1996) 'November 4, 1995: Deleuze's Death as Event', *Man and World*, 29, pp. 235–249.

Combes, M. [1999] (2013) *Gilbert Simondon and the Philosophy of the Transindividual*. Trans. T. LaMarre. 'Technologies of Lived Abstraction' Series. Cambridge, MA and London: The MIT Press.

Connolly, W.E. (2011) *A World of Becoming*. Durham, NC and London: Duke University Press.

Corning, P.A. (2005) *Holistic Darwinism: Synergy, Cybernetics, and the Bioeconomics of Evolution*. Chicago, IL: University of Chicago Press.

Cox, G. (2005) *Cox Review of Creativity in Business: Building on the UK's Strengths*. Norwich: Her Majesty's Stationery Office.

Crawford, T.H. (2015) 'Thinking Hot: Risk, Prehension and Sympathy in Design', in Marenko, B. and Brassett, J. (eds.), *Deleuze and Design*. 'Deleuze Connections' Series. Edinburgh: Edinburgh University Press, pp. 84–106.

Dalton, J. and Kahute, T. (2016) 'Why Empathy and Customer Closeness is Crucial for Design Thinking', *Design Management Review*, 27, pp. 20–27.

Deleuze, G. [1969] (2004) *The Logic of Sense*. Trans. M. Lester with C. Stivale. Ed. C.V. Boundas. London and New York, NY: Bloomsbury.

Deleuze, G. [1995] (2001) 'Immanence: A Life', in Deleuze, G., *Pure Immanence. Essays on A Life*. Trans. A. Boyman. New York, NY: Zone Books, pp. 25–33.

Deleuze, G. [1986] (1995) 'Breaking Things Open, Breaking Words Open', in Deleuze, G., *Negotiations 1972–1990*. Trans. M. Joughin. New York, NY: Columbia University Press, pp. 83–93.

Deleuze, G. [1968] (1994) *Difference and Repetition*. Trans. P. Patton. London: Athlone Press.

Deleuze, G. [1970] (1988) *Spinoza: Practical Philosophy*. Trans. R. Hurley. San Francisco, CA: City Lights.

Deleuze, G. and Guattari, F. [1991] (1994) *What is Philosophy?* Trans. G. Burchill and H. Tomlinson. London: Verso Books.

Deleuze, G. and Guattari, F. [1980] (1988) *Capitalism and Schizophrenia 2. A Thousand Plateaus*. Trans. B. Massumi. London: Athlone Press.

Deleuze, G. and Guattari, F. [1972] (1984) *Capitalism and Schizophrenia 1. Anti-Œdipus*. Trans. R. Hurley, M. Seem and H.R. Lane. London: Athlone Press.

Dennett, D.C. (1987) *The Intentional Stance*. Cambridge, MA and London: The MIT Press.

Derrida, J. (1995) 'Il me faudra errer tout seul', *Libération*, 37(7 November). Available at: https://next.liberation.fr/culture/1995/11/07/il-me-faudra-errer-tout-seul_149753 (Accessed: 26 May 2020).

Derrida, J. [1994] (1997) *The Politics of Friendship*. Trans. G. Collins. London and New York, NY: Verso.

Doherty, R., Wrigley, C. and Matthews, J. (2014) 'From Valuing Design to Designing Value', *Proceedings of NordDesign Conference 2014*, 27–29 August, pp. 600–609.

Duch, W. (2018) 'Hylomorphism Extended: Dynamical Forms and Minds', *Philosophies*, 3(4), p. 36. doi:10.3390/philosophies3040036.

Epicurus [c. 3rd–4th C BCE] (2012) 'Leading Doctrines', in Epicurus, *The Art of Happiness*. Trans. G.K. Strodach. London: Penguin Books Ltd., pp. 173–179.

Fairbanks, A. (1907) *The Mythology of Greece and Rome*. New York, NY: D. Appleton. Available at: https://www.questia.com/library/1016708/the-mythology-of-greece-and-rome (Accessed: 29 February 2020).

Feibleman, J.K. (1954) 'Theory of Integrative Levels', *The British Journal for the Philosophy of Science*, 5, pp. 59–66.

Freud, S. [1920] (2003) *Beyond the Pleasure Principle and Other Writings*. Trans. J. Reddick. London: Penguin Books Ltd.

Gnoli, C. and Poli, R. (2004) 'Levels of Reality and Levels of Representation', *Knowledge Organization*, 31(3), pp. 151–160.

Graves, R. (2011) *The Greek Myths. The complete and definitive edition*. London: Penguin books.

Gould, S.J. and Vrba, E.S. (1982) 'Exaptation – A Missing Term in the Science of Form', *Paleobiology*, 8, pp. 4–15.

Guattari, F. (1989) *Les trois écologies*. Paris: Éditions Galilée.

Hall, J. (2013) 'Thought for the Week: What is a Quaker Concern?' *The Friend. The Quaker Magazine* (Online). Available at: https://thefriend.org/article/thought-for-the-week-what-is-a-quaker-concern (Accessed: 23 March 2020).

Haraway, D.J. (2016) *Staying with the Trouble. Making Kin in the Chthulucene*. Durham, NC and London: Duke University Press.

Haumont, A. (2002) 'L'individuation est-elle une instauration? Autour de pensées de Simondon et Souriau', in Chabot, P. (ed.), *Simondon*. Paris: Librairie Philosophique J. Vrin, pp. 69–88.

Heaney, C. (2018) 'Pursuing Joy with Deleuze: Transcendental Empiricism and Affirmative Naturalism as Worldly Practice', *Deleuze and Guattari Studies*, 12(3), (July), pp. 374–401.

Hesiod [c. late 800 BCE] (1988) *Theogony and Works and Days*. Trans. M.L. West. Oxford: Oxford University Press.

Ingold, I. (2012) 'Toward an Ecology of Materials', *Annual Review of Anthropology*, 41, pp. 427–442.

Inwood, B. and Gerson, L.P. (eds.) (1994) *The Epicurus Reader. Selected Writings and Testimonia*. Trans. B. Inwood and L.P. Gerson. Indianapolis, IN: Hackett Publishing Inc.

Johansson-Sköldberg, U., Woodilla, J. and Çetinkaya, M. (2013) 'Design Thinking: Past, Present and Possible Futures', *Creativity and Innovation Management*, 2, pp. 121–146.

Johnston, M. (2006) 'Hylomorphism', *The Journal of Philosophy*, 103(12), (December), Special Issue: 'Parts and Wholes', pp. 652–698.

Kauffman, S.A. (2016a) *Humanity in a Creative Universe*. Oxford and New York, NY: Oxford University Press.

Kauffman, S.A. (2016b) 'Res Potentia and Res Extensa Linked, Hence United, by Quantum Measurement', in Eastman, T.E., Epperson, M. and Griffin, D.R. (eds.), *Physics and Speculative Philosophy. Potentiality in Modern Science*. 'Process Thought' Series, No. 27. Berlin: De Gruyter, pp. 42–52.

Kauffman, S.A. (2008) *Reinventing the Sacred*. New York, NY: Basic Books.

Kauffman, S.A. (1993) *The Origins of Order. Self-organization and Selection in Evolution*. New York, NY and Oxford: Oxford University Press.

Kimbell, L. (2012) 'Rethinking Design Thinking: Part II', *Design and Culture*, 4(2), (July), pp. 129–148.

Kimbell, L. (2011) 'Rethinking Design Thinking: Part I', *Design and Culture*, 3(3), (November), pp. 285–306.

Köppen, E. and Meinel, C. (2015) 'Empathy via Design Thinking: Creation of Sense and Knowledge', in Plattner, H., Meinel, C. and Leifer, L. (eds.), *Design Thinking Research. Understanding Innovation*. Cham, Switzerland: Springer, pp. 15–28.

Latour, B. (2008) 'A Cautious Prometheus? A Few Steps Toward a Philosophy of Design (with Special Attention to Peter Sloterdijk)', in Hackney, F., Glynne, J. and Minton, V. (eds.), *Networks of Design. Proceedings of the International Conference of the Design History Society*. Boca Raton, FL: Universal-Publishers, pp. 2–10.

Liddell, H.G. and Scott, R. (1889) *An Intermediate Greek–English Lexicon*. Oxford: Oxford University Press.

Lockwood, T. and Walton, T. (eds.) (2008) *Building Design Strategy. Using Design to Achieve Key Business Objectives*. New York, NY: Allworth Press and the Design Management Institute.

Longinus (1972) 'On the Sublime', Trans. D.A. Russell, in Russell, D. A. and Winterbottom, M. (eds.), *Ancient Literary Criticism: The Principal Texts in New Translations*. Rev. edn. Oxford: Oxford University Press, pp. 460–503.

Louie, A.H. (2016) 'The Imminence Mapping Anticipates', in Nadin, M. (ed.), *Anticipation Across Disciplines*. 'Cognitive Systems Monographs' Series, No. 29. Cham: Springer International Publishing, pp. 163–185.

Louie, A.H. (2011) 'Essays on More than Life Itself', *Axiomathes*, 21(3), pp. 473–489.

Lucretius [55BCE] (2007) *The Nature of Things*. Trans. A.E. Stallings. London: Penguin Books.

Lucretius [55BCE] (2005) *On the Nàture of the Universe*. Trans. R.E. Latham. Revised J. Godwin. London: Penguin Books.

Manning, G. (2013) 'The History of "Hylomorphism"', *Journal of the History of Ideas*, 74(2), (April), pp. 173–187.

Martin, R. (2009) *The Design of Business: Why Design Thinking is the Next Competitive Advantage*. Boston, MA: Harvard Business Press.

Maturana, H. and Varela, F. [1972] (1980) *Autopoesis and Cognition: the Realization of the Living. Boston Studies in the Philosophy of Science*. Eds. R.S. Cohen and M.W. Wartofsky. Dordrecht: D. Reidel Publishing Co.

Miller, R., Poli, R. and Rossel, P. (2013) Working Paper No. 1. *The Discipline of Anticipation: Exploring Key Issues*. Paris: UNESCO.

Moorman, M.D. (2009) 'Lucretius' Venus and Mars Reconsidered', *Lyceum*, X(2), (Spring). Available at: http://www.lyceumphilosophy.com/?q=node/109 (Accessed: 29 February 2020).

Nadin, M. (ed.) (2016) *Anticipation Across Disciplines*. 'Cognitive Systems Monographs' Series, No. 29. Cham, Switzerland: Springer International Publishing.

Nadin, M. (2015a) 'Book Review: The Reflection of Life: Functional Entailment and Imminence in Relational Biology, by A. H. Louie', *International Journal of General Systems*, 44(1), pp. 111–122.

Nadin, M. (2015b) 'Anticipation and Creation', *Libertas Mathematica*, 35(2), pp. 1–16.

Nadin, M. (2012a) 'Prolegomena: What Speaks in Favor of an Inquiry into Anticipatory Processes?', in Rosen, R. (ed.), *Anticipatory Systems: Philosophical, Mathematical and Methodological Foundations*. 2nd edn. New York, NY: Springer, pp. xv–lx.

Nadin, M. (2012b) 'The Anticipatory Profile. An Attempt to Describe Anticipation as Process', *International Journal of General Systems*, 12(1), pp. 1–33.

Nadin, M. (2010a) 'Anticipation: Annotated Bibliography', *International Journal of General Systems*, 39(1), pp. 35–133.

Nadin, M. (2010b) 'Anticipation and Dynamics: Rosen's Anticipation in the Perspective of Time', *International Journal of General Systems*, 39(1), pp. 3–33.

Nicholson, D.J. and Dupré, J. (eds.) (2018) *Everything Flows. Towards a Processual Philosophy of Biology*. Oxford: Oxford University Press

Nietzsche, F. (1997) *Untimely Meditations*. Ed. D. Breazeale. Trans. R.J. Hollingdale. Cambridge: Cambridge University Press.

Poli, R. (2017) *Introduction to Anticipation Studies*. 'Anticipation Science' Series, No. 1. Cham, Switzerland: Springer AG.

Poli, R. (2011) 'Steps Toward an Explicit Ontology of the Future', *Journal of Futures Studies*, 16(1), pp. 67–78.

Poli, R. (2010) 'The Many Aspects of Anticipation', *Foresight*, 12(3), pp. 7–17.

Prigogine, I. and Nicolis, G. (1977) *Self-Organization in Non-Equilibrium Systems*. Hoboken, NJ: Wiley.

Prigogine, I. and Stengers, I. (1984) *Order Out of Chaos. Man's New Dialogue with Nature*. London: Flamingo.

Rölli, M. (2016) *Gilles Deleuze's Transcendental Empiricism: From Tradition to Difference*. Ed. and Trans. P. Hertz-Ohmes. 'Plateaus – New Directions in Deleuze Studies' Series. Edinburgh: Edinburgh University Press.

Rosen, R. (2012) *Anticipatory Systems: Philosophical, Mathematical and Methodological Foundations*. 2nd edn. New York, NY: Springer.

Rosen, R. (2000) *Essays on Life Itself*. New York, NY and Chichester, W. Sussex: Columbia University Press.

Rosen, R. (1991) *Life Itself. A Comprehensive Inquiry into the Nature, Origin, and Fabrication of Life*. 'Complexity in Ecological Systems' Series. New York, NY and Chichester, W. Sussex: Columbia University Press.

Rossel, P. (2010) 'Making Anticipatory Systems More Robust', *Foresight*, 12(3), pp. 73–86.

Sanklecha, P. (2017) 'Our Obligations to Future Generations: The Limits of Intergenerational Justice and the Necessity of the Ethics of Metaphysics', *Canadian Journal of Philosophy*, 47(2–3), pp. 229–245.

Sauvagnargues, A. (2016) *Artmachines. Deleuze, Guattari, Simondon*. Trans. S. Verderber with E.W. Holland. Edinburgh: Edinburgh University Press.

Sauvagnargues, A. (2010) *Deleuze: L'empirisme transcendental*. Paris: Presses Universitaires de France.

Schiller, F. [1801] (1982) *On the Aesthetic Education of Man. In a Series of Letters*. Eds. and Trans. E.M. Wilkinson and L.A. Willoughby. Oxford: The Clarendon Press.

Seibt, J. (2018a) 'Ontological Tools for the Process Turn in Biology. Some Basic Notions of General Process Theory', in Nicholson, D.J. and Dupré, J. (eds.), *Everything Flows. Towards a Processual Philosophy of Biology*. Oxford: Oxford University Press, pp. 113–136.

Seibt, J. (2018b) 'What is a Process? Modes of Occurrence and Forms of Dynamicity in General Process Theory', in Stout, R. (ed.), *Process, Action, and Experience*. Oxford: Oxford University Press, pp. 120–148.

Seibt, J. (2010) 'Particulars', in Poli, R. and Seibt, J. (eds.), *Theory and Applications of Ontology: Philosophical Perspectives*. Dordrecht: Springer, pp. 23–55.

Serres, M. (2000) *The Birth of Physics*. Trans. J. Hawkes. Ed. D. Webb. Manchester: Clinamen Press.

Serres, M. (1977) *La Naissance de la physique dans le texte de Lucrèce. Fleuves et turbulences*. Collection « Critique ». Paris: Les Éditions de Minuit.

Serres, M. and Latour, B. [1990] (1995) *Conversations on Science, Culture, and Time*. Trans. R. Lapidus. Ann Arbor, MI: University of Michigan Press.

Simondon, G. (2009) 'The Position of the Problem of Ontogenesis', Trans. G. Flinders, *Parrhesia*, 7, pp. 4–16.

Simondon, G. [1964/1989] (2013) *L'individuation à la lumière des notions de forme et d'information*. Paris: Éditions Jérôme Million.

Simondon, G. [1958] (1989) *Du modes d'existence des objets techniques*. Paris: Aubier.

Souriau, E. [1943] (2009a) *Les différents modes d'existence*. Paris: Presses Universitaires de France.

Souriau, E. [1956] (2009b) 'Du mode d'existence d'un œuvre à faire', in Souriau, E. (ed.), *Les différents modes d'existence*. Paris: Presses Universitaires de France, pp. 195–217.

Spinoza, B. de [1677] (1996) *Ethics*. Trans. E. Curley. London: Penguin Classics.

Spuybroek, L. (2011) *The Sympathy of Things: Ruskin and the Ecology of Design*. Rotterdam: V2_Publishing.

Stengers, I.(2011) *Thinking with Whitehead: A Free and Wild Creation of Concepts*. Cambridge, MA: Harvard University Press.

Stengers, I. (1997) *Cosmopolitiques. Tome 6. La Vie et l'Artifice: visages de l'émergence*. Paris: Éditions La Découverte.

Stengers, I. and Prigogine, I. (1997) 'The Reenchantment of the World', in Stengers, I. (ed.), *Power and Invention. Situating Science*. Trans. P. Bains. 'Theory Out of Bounds' Series, vol. 10. Minneapolis, MN and London: University of Minnesota Press, pp. 33–59.

Tuomi, I. (2019) 'Chronotopes of Foresight: Models of Time-Space in Probabilistic, Possibilistic and Constructivist Futures', *Futures and Foresight Science*, 1(2), e11. doi:10.1002/ffo2.11.

Unger, R.M. and Smolin, L. (2015) *The Singular Universe and the Reality of Time*. Cambridge: Cambridge University Press.

Verganti, R. (2009) *Design-Driven Innovation: Changing the Rules by Radically Innovating What Things Mean*. Cambridge, MA and London: Harvard Business Press.

Vidunas, R. (2019) 'Delegated Causality of Complex systems', *Axiomathes*, 29, pp. 81–97.

Webb, D. (2017) 'On Causality and Law in Lucretius and Contemporary Cosmology', in Greenstine, A.J. and Johnson, R.J. (eds.), *Contemporary Encounters with Ancient Metaphysics*. Edinburgh: Edinburgh University Press, pp. 254–269.

Webb, D. (2006) 'Michel Serres on Lucretius. Atomism, Science, and Ethics', *Angelaki. Journal of the Theoretical Humanities*, 11(3), (December), pp. 125–136.

Whicher, A., Swiatek, P. and Cawood, G. (2015) *Design Policy Monitor 2015. Reviewing Innovation and Design Policies across Europe*. Cardiff, UK: PDR/Cardiff Metropolitan University.

Whitehead, A.N. [1920] (2004) *The Concept of Nature*. Mineola, NY: Dover Publications Inc.

Whitehead, A.N. [1927–8] (1978) *Process and Reality. An Essay in Cosmology, Corrected Edition*. Eds. D.R. Griffin and D.W. Sherburne. New York, NY: The Free Press.

Whitehead, A.N. [1938] (1966) *Modes of Thought*. New York, NY: The Free Press.

8 Inventive devices and public issues

The *Air Pollution Toile*

Lucy Kimbell

Introduction

When my mother moved into sheltered housing, we found ghost frames on each of the walls in her otherwise carefully cleaned London home where the pictures and mirrors had been. Visiting the flat after she had moved out, these marks on the wall remained as traces of my mother's life and tastes and captured physically her memories, and those of friends, family and visitors, to remain until the next residents moved in and repainted. They were also traces of the environment in which she lived for 20 years in this ground-floor London flat. Positioned on the corner of two side roads in a relatively quiet neighbourhood, the ambient pollution in the flat resulted from passing vehicles powered by petrol or diesel, by burning leaves, wood or gas for domestic heating, as well as resulting from the use of everyday cleaning products, smoking, alongside many other factors. The walls starkly revealed the small particles accumulated there over several years. Her brain, heart, lungs and other organs probably contain similar traces. We have no visual record of the particulate matter or damage from polluting gases in my mother's body. But she lives with, what researchers say, are some of its effects: the Alzheimer's disease that meant she could no longer live alone without support, the cardio-vascular disease that slows her down and the cough that never quite seems to go away.[1]

I found these physical traces of air pollution in my mother's flat a year after I created the *Air Pollution Toile* (Kimbell, 2018), a project I was commissioned to make for a group exhibition titled *Future Knowledge* at the Modern Art Oxford gallery (Oxford, UK) in 2018 (see Figure 8.1). The *Air Pollution Toile* is a wallpaper concept that gives viewers (in the gallery) and those living with or seeing the wallpaper (installed in buildings such as homes or offices) an opportunity to engage with the issue of air pollution. Briefly, this artwork takes the form of a concept for wallpaper that changes over time, in response to pollutants in its immediate environment, and an accompanying sample book which shows audiences how this pollution-responsive wallpaper changes, using photographs of interiors.

The wallpaper's imagery operates at two levels: it combines scenes from everyday life in the UK, showing some of the activities through which air

Figure 8.1 Installation view of *Air Pollution Toile* at Modern Art Oxford, UK, September–October 2018. © Lucy Kimbell

pollution is produced and how people are exposed to it; and it includes images of three diseased organs (a lung, a heart, a brain) particularly affected by such pollution, made visible through three different medical imaging technologies. Using inks and adhesives, the proposed wallpaper would monitor specific air pollutants in its immediate environment and change in response, with imagery of organs, gradually becoming more visible over time.

At the time of writing this chapter, the wallpaper is 'just' a concept or, more specifically, using the language associated with design practice, a non-functional exploratory prototype. This is because it was not feasible to research, develop, resource and manufacture 'working' responsive wallpaper, given the few months available to me to create a project in response to an invitation from the gallery in early 2018. Indeed, at present, as I discovered, chemistry researchers do not know how to bring the concept to life.[2] Instead, for the exhibition I combined a display of the wallpaper installed on three gallery walls and the wallpaper sample book on a shelf. The discussion that follows, therefore, emphasizes the potentiality of the concept, rather than its actuality, embedded within broader discussions about the capacity of creative outputs such as this wallpaper to represent public and social issues and anticipate futures.

As with other creative projects responding to or dealing with the environment – in terms of, for example, the climate emergency, species and habitat loss, or migration resulting from disaster and displacement – the potential here is to go beyond current awareness and understanding of an issue, to open up possibilities and intervene into the issue. While avoiding assumptions that there is any linear pathway through these responses, research in psychology has shown that emotional reactions to environmental issues can produce a tendency to act (Roosen et al., 2018). A review of the ways that the visual arts can engage audiences through experiencing artworks in the topic of climate change summarized these capacities as follows:

> Artworks can increase the viewer's involvement by: making use of narrative or metaphor, prompting awareness, attention, and reflection, enhancing group identity development and encouraging or strengthening a shift in social norms, providing a way to visualize climate change and giving the audience a personal experience of the issue, provoking change and transformation and, finally, inducing a (positive) emotional response, especially when this includes inspiration.
>
> (Roosen et al., 2018, p. 98)

Psychological perspectives emphasize individual experience and choice. In this chapter, I want to explore the potential of creative outputs to connect people viewing or experiencing them with social or public issues and the anticipation of courses of action to change things. To do this, I will briefly examine literatures in design studies (Fry, 2009; Mazé, 2019), design anthropology (Gunn et al., 2013; Kjaersgaard et al., 2016) and science and technology studies (Barry, 2001; Calvillo, 2018; Marres et al., 2018), to make connections between concepts of ontological design and inventive research, and use these to discuss the particular kind of wallpaper I created. Making the wallpaper itself preceded this account. What follows might also be read as a tracing between creative practice and the work of contextualization and theorization of such a practice. In so doing I explore the limits of design to be considered as an anticipatory practice that brings futures into view.

Design's changing relation to futures

Practice within contemporary design – at least in the western European/North American versions institutionalized within organizations developing products and services, the consultancies that serve them, and the universities that educate the people who design them – has long been tied up with futures.[3] Such relations can be understood within a broader context, resulting from the ways in which design production, consumption and everyday life are constituted within neoliberal economies characterized by globalization, deregulation, financialization and ownership of intellectual property (Brassett, 2015; Julier, 2017). Understandings of design typically rely on the idea that design results in change. However, there are often unacknowledged assumptions about the scale, scope or direction of such changes and limited recognition of resulting inequalities. Neither is there extensive analysis of the nature of the relations between design practice and futures. Within discussions in research on these topics, there are several traditions which offer different perspectives on the relationship between design and futures (Hill and Candy, 2019). A brief review of these highlights key contributions and ongoing areas of debate.

The increasing intersections between the domains of practice and research known as 'design' and 'futures'[4] are evidenced by a number of publications, initiatives and events (for example: Selin et al., 2015; Brassett and O'Reilly, 2015; Candy and Potter, 2019; Mazé, 2019). Emphasizing the ways that the practices and outputs of design bring experiences into view, Stuart Candy and Jake Dunagan (2017) point to how they materialize aspects of the future in the present. Traditions of making and exhibiting artefacts in design and the arts (for example: Mazé, 2007; Dunne and Raby, 2013) and the material turn in anthropology (for example: Drazin and Küchler, 2015), emphasize creative and material practices in the exploration of futures. Such capacities in design practice position this expertise as an important resource for developing 'futures literacy', understood as 'conscious human anticipatory activities, either explicit or implicit, [for] generating the imaginary futures needed to understand and act in the present' (Miller et al., 2018, p. 5). Within the various studies of futures, there is growing interest in the creative, visual and material practices of the arts, including design (for example: Selin, 2015; Candy and Potter, 2019). Creating stories (for example: Miller, 2007), fictions (for example: Bleecker, 2009; Hales, 2013), performance and installation (for example: Kuzmanovic and Gaffney, 2017) and creative forms of fieldwork (for example: Morrison and Chisin, 2017). serve in different ways to engage people and non-human actors and open up participation in the work of anticipation. Involving participants of different kinds in making, producing, experiencing, disseminating and interpreting creative and performance-based artefacts associated with art and design can broaden anticipatory capacities for exploring futures and acting in relation to them.

Researchers connecting design (and the arts more broadly) and the social sciences have opened up the complex relations between these fields, with implications for the work of 'futuring' (for example: Gunn et al., 2013; Yelavich and Adams, 2014; Smith et al., 2016). In the social sciences, Arjun Appadurai (2013) argues for a shift in anthropology's project to make stronger contributions to human flourishing, by drawing together perspectives from planning, sustainability and design. A review of the emerging dialogues between design and anthropology proposes 'design anthropology' as a transdisciplinary field of research and practice that addresses differing temporalities, materialities and politics of future-making and their inherent relations to pasts and presents (Kjaersgaard et al., 2016). Also linking design and anthropology, Yoko Akama, Sarah Pink and Shanti Sumartojo (2018) argue that combining perspectives from anthropology and design enabled making uncertainty and possibility present, available to and productive for participants coming together to explore and address complex issues. In sociology, an account of 'inventive' social research proposed by Noortje Marres, Michael Guggenheim and Alex Wilkie (2018) reworks the long-established distinctions in social research between representation and intervention. Drawing on readings of the mathematician, physicist and philosopher Alfred North Whitehead, Marres et al. (2018) foreground the generative dialogues between sociology and other domains of practice and knowledge, such as design, the arts and architecture. Located at the intersections between anthropology and the arts, Adolfo Estallela and Tomás Criado (2018) highlight sites for fieldwork that foreground the aesthetics of collaboration and problematize the production of social science knowledge. In different ways these contributions emphasize productive relations between design and social research, which open up established distinctions between representation and action, often through combining different forms of knowledge and insight.

Several of these discussions acknowledge the work implicated in designing and anticipating futures. A critical perspective found in discussions of 'ontological' design (for example: Willis, 2007; Fry, 2009; Brassett, 2018) raises questions about the constitutive relations between design practices and how futures are imagined in designing and designs, situated as they are within globalized modern industrial ecosystems and within neoliberalism. Tony Fry (2009) points to the grounding of (industrial or product) design in capitalism which has resulted in, and continues to drive, the ongoing climate emergency resulting in what he calls 'defuturing'. Ramia Mazé (2019) shows how designing futures in the present is always political. She notes the hegemonic tripartite structure of time – modern and Western paradigms of clock time, linear progression and positivist predictability – implicated in dominant traditions of Western/Northern design practice and research (Mazé, 2019). Arguing that designers should recognize the affective character of their practice, Jamie Brassett (2018) notes that the work of designing reinforces or disavows the various agential possibilities of the types of existence that emerge. Drawing on different traditions, these discussions of design emphasize its materializing agency and multiplicity in exploring futures, along with the political and ontological work in so doing.

In short, the occasions for, modes of and means for exploring futures in the present are multiplying; resulting from fruitful connections between design and related fields, in particular futures studies, sociology and anthropology. This brief review has highlighted the constitutive role of materiality in design's anticipatory agency. However, the context of neoliberalism and its unequal and racialized consequences – such as the climate emergency and effects of COVID-19 – serve to emphasize the need to be reflexive about the direction of future-making through design and the need for awareness of the politics materialized within designs and designing. Furthermore, perspectives from science and technology studies point to the potential for the outputs of social research and of designing to be devices that offer both more and less than representation (Barry, 2001; Marres et al., 2018). Bridging representation of a social world, and intervention in it, an inventive device can be understood as opening up possibilities that enable an anticipatory form of representation. By having such a capacity, devices such as the wallpaper 'produce' and materialize data in novel ways to open up relations between publics, audiences and participants, anticipating future ways of living and being in response to public issues such as air pollution.

Against this background I now develop a case study of my creative output, the *Air Pollution Toile*, and discuss it using the concepts introduced here. While there are other ways to discuss this artefact, analysing it in this manner enables critical reflection and reflexivity about the extent to which design – in the traditions in which I am situated as privileged person working in a university in western Europe – serves to anticipate futures.

Air quality: a public health emergency

In late 2018, the World Health Organization (WHO), the international organization developing global and regional research and policy and regulation for health, declared air pollution a global public health emergency. According to them (WHO, 2018a), 91 per cent of the world's population live in areas with air pollution above WHO's recommended limits. Such pollution can cause, complicate or exacerbate many adverse health conditions: with 4.2 million people a year dying from ambient pollution and another 3.8 million a year dying from exposure to air pollution in the home and billions more harmed in one way or another (WHO, 2018a). Two years later, researchers began to find links between air pollution and experience of the COVID-19 virus. Not only do researchers suspect links between locations with levels of air pollution and increased severity of COVID-19-related illness, it is also linked to increased deaths and possibly to the spread of the disease itself (Carrington, 2020).

The WHO announcement built on decades of work by activists, researchers, clinicians, policy-makers, planners and public servants assembling growing evidence about the reductions in air quality and the negative implications that result for human and animal health and for plants (WHO, 2018a, b). Impacting on people of all ages, in all regions and across all social groups, air pollution causes more harm to some people than others (National Institute of Clinical Excellence, 2020), with children and older people disproportionately affected (Committee on the Medical Effects of Air Pollution, 2004). Air pollution impacts on every organ and every cell (Schraufnagel et al., 2019). In particular, in addition to the lungs, it damages the heart and brain; impacting and exacerbating diseases such as asthma, pneumonia, heart attacks, dementia, strokes and cancer (Schraufnagel et al., 2019). In addition to the effects on human health, air pollution damages biodiversity, animals and crops (WHO, 2018a).

The term 'air pollution' includes several distinct forms. One main form is gases, such as nitrogen dioxide or sulphur dioxide, often produced by burning fossil fuels, or resulting from industrial processes, but also chemicals that are routinely found in the home or workplaces such as volatile organic compounds. Another main form is particulate matter: tiny, invisible particles, which can make their way into human or animal bodies. This matter can have industrial origins (resulting from manufacturing, driving or burning fuels, for example) or biological origins (such as spores or mould). Particulate matter is categorized by diameter: PM10 denotes particles which are less than 10 micrometres in diameter; particles of this size can enter membranes and the upper airways, causing coughs or tearing. Finer particles, PM2.5 – smaller than 2.5 micrometres – make their way into the lungs. The tiniest, ultrafine particles, can move through the membranes separating the lungs and the blood system, enter cells and then are moved round the body via the bloodstream with the potential to reach nearly all of the cells in the body, including brain cells (Peters et al., 2006). Because these particles can carry elements or compounds which are toxic, they therefore cause damage to human and animal bodies.

In the past decade, air pollution has become increasingly visible as a contemporary international public health and social justice issue. In his discussion of air quality monitoring in a London borough in relation to a European-funded project, scientific research and broader policy debates, Andrew Barry (2001) argues that there are different kinds of air quality object, each assembled from different kinds of discursive, technological, scientific and other resources. External sources of pollution include manufacturing, agriculture, construction and transport. But dangers lurk indoors too. Indoor air quality is becoming understood as a hidden and important determinant of health. The quality of indoor air varies with sources of pollutants, air temperature, humidity and ventilation (Parliamentary Office of Science and Technology, 2010). Sources of indoor air pollution include cleaning products, household appliances and goods, and activities such as cooking, heating and smoking. Proximity to external sources of pollutants such as roads also changes interior air quality. Researchers modelling interior air pollution consider and adjust for factors such as opening windows, human activity such as vacuuming, as well as the physical fabric of buildings and spatial arrangement of dwellings (Garnett, 2020). Indoor air pollution is disproportionately higher in developing countries, particularly due to the use of solid fuels, including biomass, for cooking and heating inside homes (WHO, 2010).

In Europe, much of the policy and public health focus to date has focused upon defining air quality standards (European Environment Agency, 2019), building frameworks and capacities for data-gathering, monitoring and reporting in local and regional government: the UK's local air quality management guidelines (Defra, 2019a, 2019b), for example. Policy responses to reduce air quality in the hands of local government include measures such as encouraging or supporting car sharing, making changes to the built environment, traffic planning and creating low emission zones within which high-polluting vehicles are prohibited or disincentivized from travelling (Defra, 2019c). In terms of regulations, these do not apply equally to all forms of air pollution. According to the Alzheimer's Society (2019a), a research and advocacy charity based in the UK, there is less monitoring of (and therefore less information about) the effects of ultrafine particles than those of the larger PM2.5, PM10 and gases such as ozone. Other developments have included citizens becoming more visible in discussing air pollution and its consequences for example through citizen science (Gabrys et al., 2016) and forming publics to provoke change when government authorities act illegally or irresponsibly (Calvillo, 2018).

These policy and regulation activities have often focused on outdoor – or 'ambient' – air pollution. In the UK, for example, news media routinely report the daily or even current levels of ambient air quality based on data gathered and reported by national and local government bodies. But there is growing awareness of the need to understand and address indoor (or household) air pollution. Risks and impacts associated with interior air pollution, even in high-income countries, are less visible than for ambient pollution. For example, a report on the UK highlighted how domestic fires in 1.9 million UK households, burning wood, coal and other solid fuels, generate 40 per cent of the total of dangerous, very fine, PM2.5 particles (Emden and Murphy, 2018). This figure is more than double the emissions from UK industrial combustion (16 per cent) and more than three times as much as from road transport (12 per cent) (Emden and Murphy, 2018). Table 8.1 gives a summary of some of the key pollutants, their sources and effects on health, associated with domestic or indoor air pollution in the UK. It shows how ordinary consumer products and appliances – typically found in UK homes and associated with everyday

Table 8.1 Main indoor air pollutants in the UK

Pollutant	*Example sources*	*Example health effects*
Nitrogen oxides	Heating and cooking appliances such as gas boilers and wood-burning stoves, vehicles burning petrol or diesel	Reduced lung function, increased risk of heart and lung disease
Carbon monoxide	Heating and cooking appliances	Headache, dizziness, vomiting, and nausea, heart disease
Particulate matter	PM10 (coarse particles): cooking, break and tyre wear, road dust, cleaning PM2.5 (fine particles): diesel engines, bonfires, cooking	Reduced lung function, increased risk of heart and lung disease, dementia
Environmental tobacco smoke	Cigarettes, cigars, pipes	Chronic obstructive pulmonary disease, lung cancer, asthma, heart disease
Volatile organic compounds and ozone	Personal care products, printers, paint	Asthma, allergies, cancer
Radon	Ground	Lung cancer
Allergens	House mites, animals, moulds	Asthma

Various sources: Committee on the Medical Effects of Air Pollution (2004); Parliamentary Office of Science and Technology (2010); Schraufnagel et al. (2019); National Institute of Clinical Excellence (2020).

practices – result in pollutants which are dangerous to health. In short, indoor air pollution is prevalent as well as mundane.

Making air pollution public

As with other environmental issues (Cape Farewell, 2020), air pollution has seen interest from artists, designers and other creatives. Space does not allow a fuller analysis of the different ways that creative practice treats air pollution but, in this summary of some recent projects, I make a distinction between work that represents it, intervenes into it or problematizes the issue (Marres et al., 2018). Many creative projects do more than one of these, of course, but to clarify the argument I highlight distinctions between the three.

As an example of a creative project that represents air pollution, UK-based artist Michael Pinsky's installation *Pollution Pods* (2018) takes the form of five geodesic domes, emulating polluted environments in five cities. The air inside each dome is modified to simulate air quality in five places, based on levels of particulate matter, nitrogen dioxide, sulphur dioxide and carbon monoxide. The audience moving through the sequence of five domes experiences the differences in air quality between Tautra in Norway, London, New Delhi, Beijing and São Paulo. Commissioned in the context of a cross-disciplinary research project, this installation aims to use art to connect climate science, climate psychology and the general public (Climart.info, 2019b). According to the project's website, the aim of the artwork is to offer the audience an experience which will change people's behaviour:

> The experience of walking through the pollution pods demonstrates that these worlds are interconnected and interdependent. Our need for ever cheaper goods is reflected in the ill-health of many people in world and in the ill-health of our planet as a whole. In this installation we can feel, taste and smell the environments that are the norm for a huge swathe of the world's population. Perhaps the visceral memory of these toxic places will make us think again before we buy something else we don't really need.
> (Climart.info, 2019a)

The pollution pods use data from the cities whose current levels of five pollutants they simulate in the location of the exhibition. The artwork can be considered a form of representation that requires translation, shifting data produced in five places to the site of the installation and bringing the experience of air pollution there, here; but in so doing, this piece positions the issue of air pollution as being about air and experience. Visitors are invited into the pods to breathe simulations of the air somewhere else, without revealing anything about the infrastructures, policies and practices which result in these differences. Visitors from a European city where air pollution levels are lower than in much of urban Asia may briefly breathe air similar in quality to that of New Delhi, for example. But the varying experiences of people in that city and the long-term consequences for their health, as a result of persistent health inequalities, cannot easily be translated.

As an example of a creative practice that aims at intervention, UK-based office EcoLogicStudio (2018) created a living curtain to be installed in buildings, *Photo.Synth.Etica*, which uses photosynthesis in algae to remove carbon dioxide and produce oxygen. In this design, unfiltered urban air is introduced at the bottom of the module. Air bubbles rise through the liquid and in the course of this, carbon dioxide molecules and air pollutants are captured and stored by the algae, to grow into reusable biomass. Photosynthesized oxygen is then released from the top of each module. Similarly, Netherlands-based Studio Roosegaarde (2019) created a large-scale urban tower (2015) and bicycle (2018) that are intended to produce 'Smog Free Air'. Installed in the Netherlands, China and Poland, the 7-meter high aluminium tower designed for public spaces such as parks uses patented ionization technology to remove pollutants from the air. At a different scale to that adopted by architects, product designers are also developing responses. New products such as the Airbubbl (2019) air purifier produced by London-based start-up Air Labs are shifting the burden of improving air quality from governments towards individuals, by giving them a consumer product to improve their immediate environment. The Airbubbl enables people to remove PM2.5 and PM10 particles, nitrogen dioxide and other pollutants in the confined and heavily polluted environment of a car. All three examples are interventions into air pollution using technology to improve the air quality in the immediate vicinity. But they do not raise questions about the infrastructures, policies and practices that make such devices necessary.

A third approach is to problematize air pollution. For example, UK-based designer and sociologist Jennifer Gabrys's *CitizenSense* project (which ran from 2013–18; see, for example: CitizenSense, 2016) explores the relationships between the technologies and practices of environmental sensing and citizen engagement (Gabrys et al., 2016). *CitizenSense* set up small-scale projects in which people directly engaged with sites of contamination or pollution including noise, air, soil and water pollution. One output from the project was a toolkit for 'citizens' to analyse and download citizen-generated air-quality data points collected in Pennsylvania, which has an extensive natural gas industry and infrastructure. The tool is designed to open up the work of identifying sites of concern to local residents and to enable them to participate in making air pollution into a public issue. Along with other practical experiments sharing digital tools for people to use, *CitizenSense* contextualized, questioned and developed new understandings about the possibilities and limitations of democratized environmental action through 'citizen sensing' practices. These projects make air pollution visible; but more than simply representing air quality, they change the relations between participants and air quality data, raising questions about infrastructures, policies and practices associated with air pollution, its causes and effects.

These projects exist within different traditions and in relation to different publics. Whereas for some, data can be translated to communicate with publics, for others, the construction and articulation of the air quality issue itself, and the data this rests on, are open to question. Some projects aim to intervene directly to reduce pollution in the immediate environment, but without changing the relations between individuals experiencing the project and the air pollution issue. For some, the scale of the issue and response is individualistic, whereas other projects emphasize collective agency. For some, the infrastructures, policies and practices that result in different levels of air quality are bracketed; for others, the tracing and reconfiguring of relations between various actors is central to the creative work. It is this latter set of practices that can be described as producing an inventive device that is 'both more and less than a realistic representation of the world' (Barry, 2001, p. 155). On the one hand, a visualization can bring into view previously hidden aspects of air quality; on the other, it can also intervene, not just into local air quality, but into public understanding of and engagement with the issue. Doing so involves reassembling the relations including data, visualizations, practices, policies, infrastructures and publics associated with the issue.

Anticipating indoor air pollution

In a context in which household air pollution is a growing concern, the *Air Pollution Toile* (see Figures 8.1, 8.2, 8.3 and 8.4) is a concept for a wallpaper to represent but also problematize air quality, its causes and future consequences, for those who are exposed to it.[5] The wallpaper is intended to gradually change visually over time in response to three interior air pollutants in its immediate vicinity. In so doing, it proposes a new way for audiences who come into contact with the wallpaper in a gallery (and following further development, when installed in homes and offices), to understand, track and assess the cumulative amount of air pollution in the immediate environment where it is installed, visually and materially linking causes, conditions and effects. In response to the presence and concentration of particular pollutants, the wallpaper's inks, combined with adhesives that capture particulate matter, will change the imagery over time.

In the gallery, alongside the wallpaper is a sample book, of the kind commonly produced by wallpaper manufacturers and found in shops where you buy wallpaper. This heavy, canvas-bound, large-format book includes several samples of the *Air Pollution Toile*, with the wallpaper in various stages of change. The first page shows a version of the wallpaper with its scenes from everyday UK life. The accompanying in-situ image suggests the wallpaper is recently installed, and the wallpaper has not yet reacted to local pollutants (see Figure 8.3). Subsequent pages show pairings of a sample of the wallpaper and an in-situ image, text declaring how long the wallpaper has been installed and giving a simple summary of the levels of nitrogen oxides, particulate matter and carbon monoxide in the vicinity.

The final section of the sample book includes a sample of wallpaper for urban India, in which the line drawings of everyday life are based on photographs of

Figure 8.2 Panel of *Air Pollution Toile (Urban Europe)* wallpaper following exposure to air pollution over several years. Graphics by Suky Best. © Lucy Kimbell

Figure 8.3 Close up of section of *Air Pollution Toile (Urban Europe)*. Graphics by Suky Best. © Lucy Kimbell

Figure 8.4 Close up of section of *Air Pollution Toile (Urban Europe)* showing image of diseased heart produced by inks responding to local air pollution. Graphics by Suky Best. © Lucy Kimbell

people in India. The inclusion of one sample from India acknowledges the reality that the highest levels of air pollution are found in low- and middle-income countries. Around three billion people live in households that predominately burn fuel such as wood or coal for cooking, typically in low-income or middle-income countries, exposing them to pollutants as part of everyday life (Science Daily, 2018). In a list of the world's most polluted cities for PM2.5 – just one pollutant, often used as a stand-in to communicate the issue – seven Indian, two Pakistani cities and one Chinese city are in the top ten (Air Visual, 2018). My main focus when creating this wallpaper was indoor air pollution in UK; the images accompanying this chapter are more specifically named the *Air Pollution Toile (Urban Europe)*. But air pollution is not the same everywhere; different ways of living and working and associated technologies, forms of organization and regulation result in different levels of pollutant and different consequences. Listing future variants in the sample book – and including within it a sample from a planned *Air Pollution Toile (Urban India)* wallpaper with India-based imagery – acknowledges the specific and contingent variations in air pollution that exist around the world, with their differing and racialized consequences.

As well as adding detail to scenes of everyday activity through which air pollution is produced and how people are exposed to it, the inks and adhesives in the wallpaper gradually reveal imagery from the main organs in the body damaged by air pollution: a lung, a heart and a brain (Royal College of Physicians, 2016) (see Figure 8.4). These design decisions resulted from desk research to understand more about the effects on the human body and selected these three organs. The wallpaper gathers and materializes hyperlocal data and produces a visual account of interior air pollution over time. As with the body, these changes are irreversible.

Deceptive wallpapers, everyday toiles

As a wallpaper, this inventive device draws on historical associations between this form of interior decoration and the shaping of cultural and social behaviours. Historians of interiors have shown the ways that Chinese traditions of wallpaper intersected with developments in European furnishing and decoration, resulting in rolls of paper being used in interiors around the end of the seventeenth century (Hoskins, 2005). Some note a long association between wallpaper and deception (Stewart, 2016). By bringing into European rooms illustrations of life that are 'other' (for example, Chinoiserie in eighteenth- and nineteenth-century Europe), visualizing fake architectural details (for example, dado rails) or combining simple repeated motifs into *trompe-l'oeil* panoramas, wallpaper was understood to deceive and trick. Contemporary criticism of late-eighteenth-century wallpaper objected to the fact that it relied on imitation, appearing to be large-scale drapery in rooms where cloth previously had hung (Jacqué, 2009). Historians have also shown how dangerous wallpaper could be to the people who lived in the rooms where it was installed. Use of arsenic as a colouring additive in manufacturing in the nineteenth century was widespread, resulting in green wallpapers that made people extremely unwell, sometimes terminally so (Whorton, 2010).

The graphical language known in shorthand as 'toile' has long been associated with contemporary events. 'Toile de Jouy' is the name for a graphical style associated with furnishings and interiors since the mid-eighteenth century. With the global trade in manufactured cotton goods between Europe and the places Europeans occupied and extracted resources from, there was a growth in interior decoration using this material for furnishings and upholstery in European homes (Russell, 2012). One early innovator in printing onto cotton was a Swiss German engraver and entrepreneur called Christophe-Philippe Oberkampf, who set up a printing press in Jouy-en-Josas, south of Paris in 1760. Designs produced in this early factory included engravings of details of everyday life, printed on cotton in monochrome with wooden blocks (Gril-Mariotte, 2009).

The highly detailed designs proved popular, the factory grew fast and its style became widely adopted, copied and adapted in many other contexts. Common details include animals, people and scenes from pastoral life. Often designers used current events as material for the detail in a print (Gril-Mariotte, 2009). With closely observed details from everyday life and from current events, toile de Jouy brought images of social and technological change into people's homes (Russell, 2012). Toile became a kind of visual journalism tracking and also recording interest in current events (Gril-Mariotte, 2009). For example, Oberkampf commissioned three designs of air balloons for upholstery in 1783–4, around the time there was growing interest in early flying machines then being developed and tested in France (Gril-Mariotte, 2009). There are links, too, between political events and toile designs. As political changes happened across Europe and the Americas in the late eighteenth century, engravings about these events also became inspiration for toiles. One example is Oberkampf's design from 1780–1 demonstrating French support for American resistance to British rule in what became the War of Independence (Gril-Mariotte, 2009). In that design, France is represented as a mature woman, providing help to the emerging America in its struggle against the British, in the form of 6000 French troops sent to fight. US revolutionaries also used toile designs to highlight their politics in the fight against the British.[6] Later still, in the 1930s, designers in the US used toile designs to demonstrate patriotism (Russell, 2012): a mid-twentieth century design attributed to Elisabeth Draper[7] based on the life of President Eisenhower brought together symbols of his presidency, background and institutions he was associated with such as Columbia University, West Point and NATO (Cook, 1998). Contemporary toile designs have reinvigorated this tradition of bringing public issues into view (Hemmings, 2005). Additionally, however, some designers raise questions and disrupt expectations, as is evident in the work of UK-based textile design studio Timorous Beasties and US-based interior designer Sheila Bridges. These designers are not simply representing the world they see around them; they want to intervene into it and pose questions.

The Scottish design duo known as Timorous Beasties, Paul Simmons and Alistair MacAuley, who met as students of textile design at Glasgow School of Art, have created several toile designs. First was the *Glasgow Toile* in 2004, which updated the toile format for twenty-first-century urban contexts, realized as wallpaper and fabric. Available in blue, pink and green, this visually dense design shows grim scenes from the city's life including drug addicts, prostitutes, modern buildings and dirt. Describing their later *Edinburgh Toile*, the website notes:

> When commissioned by the Edinburgh International Festival [EIF] in 2009, Scotland's capital became a cornerstone in Timorous Beasties' urban toile collection. Featured on the cover of the EIF events magazine, projected on the Edinburgh Omni Centre, framed in Jenners department store display window, and broadcast on taxis and billboards across the city, our Edinburgh toile was part of the very fabric of the city. Controversial images of revellers urinating at the Greyfriar's Bobby fountain, homelessness, begging, and a traffic cone crowning the statue of the eminent Enlightenment philosopher, David Hume, provoked some reaction in conservative circles.
>
> (Timorous Beasties, 2019)

This description highlights the varied distribution of this graphical output, beyond domestic settings, as well as the importance of provocation for these designers. Timorous Beasties aim to use their wallpapers and textiles to translate between domains that are usually kept apart: the luxury hotels, high profile events and wealthy homes for which their work is commissioned and where it is installed; and the impoverished streets of the city; as well as making links between tourism, heritage and poverty in the urban environment. As with the Glasgow and Edinburgh variants, familiar landmarks locate their *London Toile* (2006, in the collection of the Smithsonian Cooper Hewitt National Design Museum) in space and time, but can trick the viewer who looks only at the whole and not the detail:

> If a traditional toile was replaced by the Timorous Beasties' design in an interior design magazine spread, the switch would likely go unnoticed. However, if one takes the time to examine the images presented, it is only then that the, perhaps too realistic, scenes of urban life can be recognized. This quality – from afar giving off the feel and look of traditional toile and up-close unsettling the viewer with the unexpected – is what makes *London Toile* a clever reimagining of a traditional textile that also serves as commentary on present-day culture.
>
> (Elevado, 2013)

In contrast to the tradition of architectural deception in early European wallpapers, this approach draws people in to disrupt expectations about social realities and heritage sites. The repetition of scenes of everyday life is familiar, as a toile, but the social realities they capture provoke, rather than mask, the social inequalities associated with 'heritage'.

Similarly, US-based interior designer Sheila Bridges developed a toile to challenge assumptions about how African Americans are represented. Her design, realized in wallpaper, fabrics and homewares, called *Harlem Toile de Jouy* (2005, 2006, 2020), is also in the permanent collections of the Smithsonian National Museum of African American History and Culture and the Smithsonian Cooper Hewitt National Design Museum. Bridges (2006) explains:

> As an African American living in Harlem, I have always been intrigued and inspired by the historical narrative of the decorative arts, especially traditional French toile with its pastoral motifs from the late 1700s. I'm entertained by the stories these patterns tell and the questions they sometimes raise. But after searching for many years for the perfect toile for my own home, I decided that it quite simply didn't exist. I created Harlem Toile de Jouy initially as a wallcovering then expanded the collection to include fabrics, bedding, plates, glassware, umbrellas and clothing. This design [. . .] lampoons some of the stereotypes deeply woven into the African American experience.

Bridges' design shows everyday scenes of African American people in eighteenth-century European and North American dress: women doing their hair, a man and a woman dancing, young men playing basketball with a basket hung from a tree (Herringshaw, 2014). These scenes and the rich background colours – deep purple, pistachio, Icelandic blue – on which the black and white drawings lie, insert common ways that contemporary African American lives are represented into colonizing European interior design traditions. By combining modern elements, such as a large radio cassette player, with the dress of wealthy people from the eighteenth century, Bridges invites the viewer to consider which people were able to benefit from political, social and economic changes in the eighteenth century and which were excluded. The toile includes two people dancing in front of a triumphal arch similar to the one in Washington Square Park built to celebrate the 100th anniversary of George Washington's inauguration as president, itself modelled on Paris's Arc de Triomphe. These figures, enjoying themselves dancing, contrast with stiffness and formality of the arch they decorate, associated with a War of Independence that resulted in only white, property-owning men becoming recognized as citizens of the new United States of America. Rather than proposing an 'authentic' African American experience, the *Harlem Toile de Jouy* asks the viewer to consider to what extent African Americans are understood to be equal citizens of the nation they live in. Bridges' toiles extend beyond being representations of everyday life and events updated for the contemporary era: highlighting the urban rather than the pastoral; basketball and drugs rather than woodlands, animals and birds. They are designed to be dissonant, bringing tightly curated, politically powerful visualizations of day-to-day African American life into the world of high end, small batch, luxury wallpaper production.

Inspired by developments in interior and textiles design such as the ones described above, the *Air Pollution Toile* aims to go beyond representations of air pollution and to problematize how people relate to data about air quality. A common aim of those working in the public management of air pollution is to make data visible so people can understand it. News media and government websites routinely use images of people wearing face masks as well as information graphics to communicate levels of pollutants and changes over time. However, as Bruno Latour has argued, 'making things public' goes beyond making representations of 'matters of fact' (Latour, 2008, p. 2). Negotiating the historic association between wallpapers, deception and provocation, the *Air Pollution Toile* wallpaper does not aim to reproduce scientific data about air pollution. Instead as an inventive device, it offers a different kind of truth via data-gathering and monitoring of air pollution in its immediate environment, gradually and partially bringing into view the material bodily consequences for three human organs (Figure 8.4).

The *Air Pollution Toile* includes line drawings of everyday scenes from British life through which pollutants are created or people are exposed to them, or their consequences such as ill-health: a parent taking the kids to school in the car; a child on a bicycle; people in an open plan office; cows being milked within an industrialized agricultural system; electricity pylons; scientists in a lab; people in an National Health Service waiting room; a family at home preparing a meal in the kitchen including imported fruit. The consequences of air pollution are shown through detailed imagery of three organs, produced by three medical imaging technologies: an MRI scan of a brain with Alzheimer's; a cardiogram of a diseased heart; and a pathology photograph of a lung affected by anthracosis.[8] The style of toile de Jouy distributes agency across these different settings and objects, including its very materials including inks, adhesives and pollutants, allowing the public issue of air pollution to emerge from the wallpaper (Brassett and O'Reilly, 2015).

Drawing from insights that sociologist Annemarie Mol (2002) develops in *The Body Multiple: Ontology in Medical Practice* – that technologies and clinical practices produce different kinds of illness – the choice of this imagery acknowledges that air pollution and the resulting illnesses are multiple. In this context, the *Air Pollution Toile* offers a socio-material account of air pollution which makes links between the social practices through which pollution is produced and encountered resulting from particular infrastructures, policies and practices, and the effects on the human body. The piece uses vernacular forms – toile wallpaper and a sample book – in accentuating air pollution as a domestic matter. When installed in a home or office, the resulting imagery would be situated and emergent through an unfolding process that changes over days, months and years. Rather than seeking to represent data, the wallpaper brings together different aspects of air pollution, linking data gathering and representation, production of and exposure to pollutants, public health and everyday life as well as science and interior design. Its graphical language uses traditions associated with toile de Jouy to bring into view contemporary realities and political issues in a domestic setting.

Rather like our bodies – including even placentas in pregnant women, which carry the traces of pollutants women have been exposed to (Randall, 2019) – the *Air Pollution Toile* is transformed through the accumulation of pollution to which it has been exposed, going beyond representation of data, to become materially affective. This *Toile* resists current concerns with making visible monitoring of levels of key pollutants in the public sphere, as currently managed by local authorities or national governments, or the new generation of personal sensors, wearable smart designs or representative artworks. Instead the *Air Pollution Toile* slows down or even resists the act of interpretation, avoiding producing an easy-to-read graphical association between levels of pollution in the room where the wallpaper is installed, and its visual display. A viewer looking at this wallpaper will not easily be able to answer the question: 'What level is pollution at right now, here?' Rather, the design emphasizes the accumulation of air pollution through its visual design that response to levels of three pollutants, but not through a direct translation. Further, it directly connects causes, conditions and possible effects, by combining imagery associated with how air pollution is produced and the institutions, policies and practices resulting in people being exposed to it in ordinary domestic and work settings, and the resulting effects on three organs in the body.

In this way, the *Air Pollution Toile* wallpaper makes a double move. On one hand, adopting the public recording history of toile de Jouy, it establishes new relations between domestic settings, scientific research and industrial production, resulting in a novel kind of air pollution object being brought into view. One the other, it obscures air pollution. Enacting the deceptive associations of wallpaper, it is not easy to read. It does not convert pollutants to scales that citizens can take in: with, for example, red representing high levels of pollutants and therefore very bad for health, amber medium/acceptable and green low levels/good. Indeed, the version of the wallpaper shown here would probably never exist like this. This is because the wallpaper, if realized, would be contingent on local levels of pollutants over time, resulting, in part, from the day to day practices of the people using that room, as well as any environmental pollutants that may seep in, for example, through windows or doors. The wallpaper renders air pollution multiple.

Anticipating my futures

Researchers do not yet agree on the causes of Alzheimer's disease; studies are ongoing to understand its causes and to develop treatments. However, publicly available guidance notes that there are several risk factors from the individual – such as, genetics and behaviour such as diet, consumption of alcohol and levels of exercise – to the environmental; including air pollution (Alzheimer's

Society, 2019c). I have lived in London for over 25 years, moving through this polluted city on foot, as a cyclist, runner and mother pushing a buggy and occasional driver, as well as on public transport. I have often travelled at rush hour in the mornings and evenings when there are many more vehicles on the roads and have used standard products such as gas boilers to heat my home. It is therefore reasonable to assume I have been exposed to pollutants including particulate matter and toxic gases, sometimes intensively so. There is no easy way for me, or for any researcher or clinician, to determine the degree of this exposure or its impacts. While I could avail myself of the opportunity to undergo genetic testing which might reveal markers for Alzheimer's[9] (the likely explanation for my mother's condition) or heart disease (the cause of my father's sudden death aged 54), it might well be that it's the air pollution, intersecting with genetic risks, that kills me.

Billions of people are implicated in such a future of illness and early death as a result of air pollution. But unlike me, the vast majority of them live in low or middle-income countries. The causes of air pollution, its consequences and future impacts, are now identified as a significant global environmental and health issue, linked too with COVID-19. The effects spill over into the lives of many individuals and communities, unequally. The spread of these invisible toxic pollutants harms everyday living, being, caring, doing and knowing. Anticipating my own end of life leads me to imagine my daughter reflecting on a future we might one day share:

> When my mother moved into sheltered housing, we found ghost frames made of tiny particles on each of the walls in her otherwise carefully cleaned London home where the pictures and mirrors had been . . .

Notes

1 See WHO (2018a, b) for more detail on air pollution and its health effects.

2 During the development of the Air Pollution Toile I spoke with analytical chemist Professor Nicole Pamme (University of Hull, UK), who visited the exhibition in Oxford. We have since continued our dialogue and have applied for research funding to enable researchers and PhD students to develop the wallpaper further.

3 Escobar (2018) makes an important intervention into contemporary design research and practice in the global North by linking critical design studies with traditions of thinking and activism in Latin America rooted in nonlinear, non-Eurocentric relations between community, land and time.

4 There are close relations between the terms: 'foresight' and 'futures', including journals by these names and decades of research and practice. Here I use the term 'futures' to emphasize the multiplicity of approaches, formats and practices which explicitly imagine futures.

5 I am indebted to artist Suky Best who was employed as an assistant on this project. She first suggested I consider toile as the graphical language for the wallpaper (I knew the graphical style, but I did not know its name). She developed the wallpaper's graphical language in dialogue with me in response to my brief and my early sketches and collaged mock-ups, and she produced the digital files which were manufactured.

6 Editors' note. It is interesting to remember that with the 1712 introduction into Britain of a Wallpaper Tax – 1d per yard, raised by ¼d in 1714 in order to finance the costs of

war, as design historian Martin King (2007, pp. 65–66) notes of his investigations into his own home at 53 Cross Street, Islington, London – wallpaper became a site of consumer-level political activism akin to contemporary brand or product boycotts. He continues, that at the end of the century, after the tax raised to 1¾d per yard to fund the lost war in the American colonies, its avoidance was one of many acts of civil disobedience at that time that included the Gordon Riots of 1788. Wallpaper-related activism was an everyday, domestic form of tax avoidance, alongside smuggling.

7 Cook (1998) argues there was what she calls a 'ghost' designer, textile artist Katherine Sturges Knight, who was paid to produce the design, but who is not accredited as the designer of the Eisenhower toile. There is an echo of this in my working with Suky Best, who provided substantial input into the realization of the wallpaper concept and its graphical expression and who has assigned her copyright of the Air Pollution Toile to me. While I acknowledge her contribution, what also requires acknowledgement is the power relations through which I have resources as an employed academic design researcher, while Suky relies on freelance work, art commissions and sales as an independent artist.

8 In the few months I had available to develop and produce the wallpaper for exhibition, I decided not to pursue involving participants, including possibly my mother, to access images of their organs. Instead, I licensed the photographs from a commercial image bank.

9 The Alzheimer's Society (2019b) says predictive genetic testing is only possible for inherited Alzheimer's disease, in which rare mutations in three genes are implicated, and fronto-temporal dementia, which has known mutations in at least six genes. According to the charity, the most common form of Alzheimer's disease affects about 520,000 people in the UK, mostly over the age of 65 but there are no approved predictive genetic tests for this form of the condition.

References

Airbubbl (2019) *Airbubbl. A Breath of Fresh Air.* Available at: https://airbubbl.com (Accessed: 5 October 2019).

AirVisual (2018) *World's Most Polluted Cities. PM2.5.* Available at: https://www.airvisual.com/world-most-polluted-cities (Accessed: 5 October 2019).

Alzheimer's Society (2019a) *Report: Is There a Link between Air Pollution and Dementia?* Available at: https://www.alzheimers.org.uk/for-researchers/report-on-link-between-air-pollution-and-dementia (Accessed: 1 October 2019).

Alzheimer's Society (2019b) *Alzheimer's Society's View on Genetic Testing.* Available at: https://www.alzheimers.org.uk/about-us/policy-and-influencing/what-we-think/genetic-testing (Accessed: 1 October 2019).

Alzheimer's Society (2019c) *Risk Factors and Prevention.* Available at: https://www.alzheimers.org.uk/about-dementia/risk-factors-and-prevention#riskfactors (Accessed: 1 October 2019).

Akama, Y., Pink, S. and Sumartojo, S. (2018) *Uncertainty and Possibility: new approaches to future making in design anthropology.* London and New York: Bloomsbury Publishing.

Appadurai, A. (2013) *The Future as Cultural Fact: Essays on the Global Condition.* London: Verso Books.

Barry, A. (2001) *Political Machines.* London: Athlone.

Bridges, S. (2005) *Harlem Toile de Jouy* [Ink on Paper]. Smithsonian National Museum of African American History and Culture. Available at: https://nmaahc.si.edu/object/nmaahc_2018.72.1?destination=edan-search/collection_search%3Fedan_q%3DHarlem%2520Toile%2520de%2520Jouy (Accessed: 11 January 2020).

Bridges, S. (2006) *Harlem Toile de Jouy* [Products, Various]. Available at: https://www.sheilabridges.com/product-category/harlem-toile/ (Accessed: 24 June 2020).

Bridges, S. (2020) *Sidewall, Harlem Toile de Jouy* [Screen Printed on Paper]. Cooper Hewitt, Smithsonian Design Museum. Available at: https://collection.cooperhewitt.org/objects/18711505/ (Accessed: 24 June 2020).

Bleecker, J. (2009) 'Design Fiction: A Short Essay on Design, Science, Fact and Fiction', *Near Future Laboratory*. Available at: http://blog.nearfuturelaboratory.com/2009/03/17/design-fiction-a-short-essay-on-design-science-fact-and-fiction/ (Accessed: 10 January 2020).

Brassett, J. (2018) 'Creating Affective Social Design: An Ethical and Ontological Discussion', *Cubic Journal*, 1(1), pp. 172–185.

Brassett, J. (2015) 'Poised and Complex. The Becoming Each Other of Philosophy, Design and Innovation', in Marenko, B. and Brassett, J. (eds.), *Deleuze and Design*. 'Deleuze Connections' Series. Edinburgh: Edinburgh University Press.

Brassett, J. and O'Reilly, J. (2015) 'Styling the Future. A Philosophical Approach to Design and Scenarios', *Futures*, 74, pp. 37–48.

Calvillo, N. (2018) 'Political airs: From monitoring to attuned sensing air pollution', *Social Studies of Science*, 48(3), pp. 372–388.

Candy, S. and Dunagan, J. (2017) 'Designing an experiential scenario: *The People Who Vanished*', *Futures*, 86, pp. 136–153.

Candy, S. and Potter, C. (2019) 'Design and Futures: Introduction to the Special Issue', *Journal of Futures Studies*, 23(3), (March), pp. 1–2.

Cape Farewell (2020) Home Page. Available at: https://capefarewell.com/ (Accessed: 4 May 2020).

Carrington, D. (2020) 'Is Air Pollution Making the Coronavirus Pandemic Even More Deadly?' *The Guardian*, 4 May. Available at: https://www.theguardian.com/world/2020/may/04/is-air-pollution-making-the-coronavirus-pandemic-even-more-deadly (Accessed: 10 May 2020).

CitizenSense (2019) *AirSift PM2.5 Data Analysis Toolkit*. Available at: https://citizensense.net/kits/airsift-pm2-5/ (Accessed: 20 June 2018).

Climart.info (2019a) *The Artist*. Available at: https://www.climart.info/pollutionpods (Accessed: 23 September 2019).

Climart.info (2019b) *Climart Project Description*. Available at: https://www.climart.info/about (Accessed: 24 June 2020).

Committee on the Medical Effects of Air Pollution (2004) *Guidance on the Effects on Health of Indoor Air Pollutants*. London: Department of Health. Available at: https://webarchive.nationalarchives.gov.uk/20120104230915/http:/www.dh.gov.uk/prod_consum_dh/groups/dh_digitalassets/@dh/@ab/documents/digitalasset/dh_096801.pdf (Accessed: 24 June 2020).

Cook, C. (1998) 'The Mystery of the Eisenhower Toile', *Textile Society of America Symposium Proceedings*, 156. Available at: https://digitalcommons.unl.edu/tsaconf/156 (Accessed: 20 July 2020).

Department for Environment Food and Rural Affairs (Defra) (2019a) *UK and EU Air Quality Limits*. Available at: https://uk-air.defra.gov.uk/air-pollution/uk-eu-limits (Accessed: 23 September 2019).

Department for Environment Food and Rural Affairs (Defra) (2019b) *Local Air Quality Management (LAQM) Support*. Available at: https://laqm.defra.gov.uk (Accessed: 29 September 2019).

Department for Environment Food and Rural Affairs (Defra) (2019c) *Action Planning: Measures*. Available at: https://laqm.defra.gov.uk/action-planning/measures/measures.html (Accessed: 29 September 2019).

Drazin, A. and Küchler, S. (eds.) (2015) *The Social Life of Materials: Studies in Materials and Society*. London: Bloomsbury Academic.

Dunne, A. and Raby, F. (2013) *Speculative Everything: Design, Fiction, and Social Dreaming*. Cambridge, MA and London: The MIT Press.

EcoLogicStudio (2018) *Photo.Synth.Etica*. Available at: http://www.ecologicstudio.com/v2/project.php?idcat=3&idsubcat=71&idproj=174 (Accessed: 10 September 2019).

Elevado, M. (2013) *From Romantic to Raw: Toile Transformed*, 17 July. Cooper Hewitt, Smithsonian Design Museum. Available at: https://www.cooperhewitt.org/2013/07/17/from-romantic-to-raw-toile-transformed/ (Accessed: 24 June 2020).

Emden, J. and Murphy, L. (2018) *Lethal but Legal: Air Pollution from Domestic Burning*. Available at: http://www.ippr.org/research/publications/lethal-but-legal (Accessed: 1 March 2020).

Escobar, A. (2018) *Designs for the Pluriverse: Radical Interdependence, Autonomy, and the Making of Worlds*. Durham, NC and London: Duke University Press.

Estalella, A. and Criado, T. (eds.) (2018) *Experimental Collaborations: Ethnography through Fieldwork Devices*. Oxford: Berghahn Books.

European Environment Agency (2019) *Air Quality Standards*. Available at: https://www.eea.europa.eu/themes/air/air-quality-concentrations/air-quality-standards (Accessed: 12 May 2020).

Fry, T. (2009) *Design Futuring: Sustainability, Ethics and New Practice*. Oxford: Berg.

Gabrys, J., Pritchard, H. and Barratt, B. (2016) 'Just Good Enough Data: Figuring Data Citizenships through Air Pollution Sensing and Data Stories', *Big Data and Society*. doi:10.1177/2053951716679677.

Garnett, E. (2020) 'Breathing Spaces: Modelling Exposure in Air Pollution Science', *Body and Society*, 26(2), pp. 55–78.

Gril-Mariotte, A. (2009) 'Topical Themes from the Oberkampf Textile Manufactory, Jouy-en-Josas, France, 1760–1821', Trans. P.J. Warner, *Studies in the Decorative Arts*, 17(1), pp. 162–197.

Gunn, W., Otto, T. and Smith, R.C. (eds.) (2013) *Design Anthropology: Theory and Practice*. London: Bloomsbury.

Hales, D. (2013) 'Design Fictions an Introduction and Provisional Taxonomy', *Digital Creativity*, 24(1), pp. 1–10.

Hemmings, J. (2005) 'Paradise Lost? Contemporary Toile de Jouy', *Embroidery. The Textile Art Magazine*, September/October, pp. 36–37.

Herringshaw, G. (2014) *A New Twist on Toiles*. Cooper Hewitt, Smithsonian Design Museum. Available at: https://www.cooperhewitt.org/2014/08/20/a-new-twist-on-toiles/ (Accessed: 23 September 2019).

Hill, D. and Candy, S. (2019) 'Change the Model', *Journal of Futures Studies*, 23(4), (June), pp. 123–128.

Hoskins, L. (2005) *The Papered Wall: The History, Patterns and Techniques of Wallpaper*. London: Thames and Hudson.

Jacqué, B. (2009) 'Drapery Wallpapers by Dufour and Dufour & Leroy, 1808–1830: Imitation or Creation?', trans. T. Chen, *Studies in the Decorative Arts*, 17(1), pp. 68–95.

Julier, G. (2017) *Economies of Design*. London: Sage.

Kjaersgaard, M.C., Halse, J., Smith, R.C., Vangkilde, K.T., Binder, T. and Otto, T. (2016) 'Introduction', in Smith, R.C., Vangkilde, K.T., Kjaersgaard, M.G., Otto, T., Halse, J. and Binder, T. (eds.), *Design Anthropological Futures*. London: Bloomsbury, pp. 1–16.

Kimbell, L. (2018) *Air Pollution Toile* [Installation: Wallpaper and Sample Book]. *Future Knowledge* [Exhibition]. Modern Art Oxford, 22 September–28 October 2018.

King, M. (2007) 'Essay on Stencil Wall Painting', in Cosh, M. and King, M. (eds.), *53 Cross Street. Biography of a House*. London: Islington History and Archaeology Society, pp. 51–71.

Kuzmanovic, M. and Gaffney, N. (2017) 'Enacting Futures in Postnormal Times', *Futures*, 86, pp. 107–117.

Latour, B. (2008) 'A Cautious Prometheus? A Few Steps toward a Philosophy of Design (with Special Attention to Peter Sloterdijk)', in Hackney, F., Glynne, J. and Minton, V. (eds.), *Networks of Design. Proceedings of the 2008 Annual International Conference of the Design History Society*, University College Falmouth, 3–6 September. Boca Raton, FL: Universal Publishers, pp. 2–10.

Marres, N., Guggenheim, M. and Wilkie, A. (2018) 'Introduction: From Performance to Inventing the Social', in Marres, N., Guggenheim, M. and Wilkie, A. (eds.), *Inventing the Social*. Manchester: Mattering Press, pp. 17–37.

Mazé, R. (2007) *Occupying Time: Design, technology, and the form of interaction*. Stockholm: Axl Books.

Mazé, R. (2019) 'Politics of Designing Visions of the Future', *Journal of Futures Studies*, 23(3), (March), pp. 23–38.

Miller, R. (2007) 'Futures Literacy: A Hybrid Strategic Scenario Method', *Futures*, 39(4), pp. 341–362.

Miller, R., Poli, R. and Rossel, P. (2018) 'The Discipline of Anticipation: Foundations for Futures Literacy', in Miller, R. (ed.), *Transforming the Future: Anticipation in the 21st Century*. Paris and Abingdon: UNESCO and Routledge, pp. 51–65.

Mol, A. (2002) *The Body Multiple: Ontology in Medical Practice*. Durham, NC and London: Duke University Press.

Morrison, A. and Chisin, A. (2017) 'Design Fiction, Culture and Climate Change. Weaving Together Personas, Collaboration and Fabulous Futures', *The Design Journal*, 20(Suppl. 1), pp. S146–S159.

National Institute of Clinical Excellence (2020) *Indoor Air Quality at Home NICE guideline [NG149]*, 8 January. Available at: https://www.nice.org.uk/guidance/ng149 (Accessed: 24 June 2020).

Parliamentary Office of Science and Technology (2010) 'UK Indoor Air Quality', *POST Note*, 366(November). Available at: https://post.parliament.uk/research-briefings/post-pn-366/ (Accessed: 2 August 2020).

Peters, A., Veronesi, B., Calderón-Garcidueñas, L., Gehr, P., Lung, C.C., Geiser, M., Reed, W., Rothen-Rutishauser, B., Schürch, S. and Schulz, H. (2006) 'Translocation and Potential Neurological Effects of Fine and Ultrafine Particles a Critical Update', *Particle and Fibre Toxicology*, 3(13). doi:10.1186/1743-8977-3-13.

Pinsky, M. (2018) *Pollution Pods*. Available at: http://www.michaelpinsky.com/project/pollution-pods/ (Accessed: 15 September 2019).

Randall, C. (2019) 'The Five: Airborne Pollutants in Our Bodies', *The Guardian*, 22 September. Available at: https://www.theguardian.com/technology/2019/sep/22/the-five-airborne-pollutants-in-our-bodies-placenta-heart-lungs-liver (Accessed: 10 January 2020).

Royal College of Physicians (2016) *Every Breath We Take: The Lifelong Impact of Air Pollution*. Report of a Working Party. London: RCP.

Roosen, L.J., Klöckner, C.A. and Swim, J.K. (2018) 'Visual Art as a Way to Communicate Climate Change: A Psychological Perspective on Climate Change–Related Art', *World Art*, 8(1), pp. 85–110.

Russell, J. (2012) 'James Russell on the History of Toile de Jouy', *The Design Tabloid*, 20 February. Available at: https://thedesigntabloid.com/2012/02/20/james-russell-on-the-history-of-toile-de-jouy/ (Accessed: 10 January 2020).

Science Daily (2018) 'Cooking with Wood or Coal is Linked to Increased Risk of Respiratory Illness and Death', *ScienceDaily*. Available at: https://www.sciencedaily.com/releases/2018/09/180921092447.htm (Accessed: 3 March 2020).

Schraufnagel, D., Balmes, J., Cowl, C., De Matteis, S., Jung, S., Mortimer, K., Perez-Padilla, R., Rice, M., Riojas-Rodriguez, H., Sood, A., Thurston, G., To, T., Vanker, A. and Wuebbles, W. (2019) 'Air pollution and noncommunicable diseases. A review by the forum of International Respiratory Societies' Environmental Committee, Part 1: The Damaging Effects of Air Pollution', *Chest*, 155(2), pp. 409–416.

Selin, C. (2015) 'Merging Art and Design in Foresight: Making Sense of Emerge', *Futures*, 70, pp. 24–35.

Selin, C., Kimbell, L., Ramírez, R. and Bhatti, Y. (2015) 'Scenarios and Design: Scoping the Dialogue Space', *Futures*, 74, pp. 4–17.

Smith, R.C., Vangkilde, K.T., Kjaersgaard, M.G., Otto, T., Halse, J. and Binder, T. (eds.) (2016) *Design Anthropological Futures*. London: Bloomsbury.

Stewart, J. (2016) 'A History of Wallpaper's Deception', *The Atlantic*, 1 April. Available at: https://www.theatlantic.com/technology/archive/2016/04/the-deception-of-wallpaper/476295 (Accessed: 10 August 2018).

Studio Roosegaarde (2019) *Smog Free Tower*. Available at: https://www.studioroosegaarde.net/project/smog-free-tower (Accessed: 10 September 2019).

Timorous Beasties (2019) *Edinburgh Toile* [No Details]. Available at: https://www.timorousbeasties.com/project/edinburgh_toile/ (Accessed: 29 September 2019).

Timorous Beasties (2006) *London Toile* [Screen Print on Linen]. Available at: https://collection.cooperhewitt.org/objects/18711355/ (Accessed: 29 September 2019).

Timorous Beasties (2004) *Glasgow Toile* [Product Shop, Various]. Available at: http://www.timorousbeasties.com/shop/wallcoverings/86/glasgow%20toile%20wallpaper (Accessed: 29 September 2019).

Whorton, J. (2010) *The Arsenic Century: How Victorian Britain was Poisoned at Home, Work and Play*. Oxford: Oxford University Press.

Willis, A.-M. (2007) 'Ontological Designing – Laying the Ground', in Willis, A.-M. (ed.), Design Philosophy Papers. *Collection 3*. Ravensbourne, Australia: Team D/E/S Publications, pp. 80–98.

World Health Organization (2018a) *Air Pollution*. Available at: https://www.who.int/airpollution/en/ (Accessed: 20 August, 2019).

World Health Organization (2018b) *First WHO Global Conference on Air Pollution and Health*, 30 October–1 November. Available at: https://www.who.int/airpollution/events/conference/en/ (Accessed: 20 August, 2019).

World Health Organization (Europe) (2010) *WHO Guidelines for Indoor Air Quality. Selected Pollutants*. Bonn: The WHO European Centre for Environment and Health.

Yelavich, S. and Adams, B. (eds.) (2014) *Design as Future-Making*. London: Bloomsbury.

9 The anticipatory power of the *objectile*

Derek Hales

Introduction

In the literature emerging from the field of anticipation studies there are expressions of that confluence of design and futurological practices which can be caricatured as speculative. Less present is a related encounter with the notion of 'anticipatory plagiarism'. The chapter locates this paradoxical construct in its pataphysical dimensions to compose a nomadic distribution of concepts and from which one concept, productive to thinking anticipation, might be further mobilized: the Deleuzian *objectile*. In this chapter, the modulatory power of speculative design as a creation of anticipations is demonstrated as an atemporal anticipatory plagiarism of Alfred Jarry's imaginary present, in which a clinamen is pleated into a compositional plane of architectural fiction and its techno-graphic artefacts.

'Pataphysics

The science of pataphysics[1] is described by Alfred Jarry in *Exploits and Opinions of Dr. Faustroll, Pataphysician* ([1911] 1996), as investigating 'that which is superinduced upon metaphysics, whether within or beyond the latter's limitations, extending as far beyond metaphysics as the latter extends beyond physics' (Jarry, 1996, p. 22). In Jarry's proto-surrealist fictions he described a variety of mechanical designs or contrivances and over the course of the twentieth century, Jarry's and other pataphysicians' devices have been an influence upon speculative fiction and anticipation authors.[2] In post-war Paris, a group of writers and thinkers founded *The Collège de 'Pataphysique* in 1947.[3] The *Collège* 'occulted' itself between 1975–2000.[4] Related to this period of occultation is the development of the pataphysical solutions of the *Ouvroir de Litterature Potentielle* (*Oulipo*) (Motte, 2007; Duncan, 2019) from within which the practice of Anticipatory Plagiarism is defined.

Anticipatory plagiarism

Following Lautréamont – who wrote in *Poésis II* ([1870] 1994, p. 240), that: 'Plagiarism is necessary. Progress implies it. It closely grasps an author's words, uses his expressions, deletes a false idea, replaces it with the right one'[5] – thus, I propose plagiarism as a mode of anticipatory practice, developing this notion from the group of writers and mathematicians comprizing the *Oulipo*. Founded by François Le Lionnais and Raymond Queneau in 1960, the *Oulipo* devoted itself to inventing, analysing and sometimes applying constraints for the making of literature. The key strictures of this proposed anticipatory plagiarism are not the possible, preferable, probable future but are to be found in Le Lionnais's two manifestos (both published in 1973) of the *Oulipo*, in which an act of plagiarism by anticipation is an act in which a past style merely replicates what a future style has already initiated. In this same context one can get a glimpse of the kind of temporal rupture involved from the sense in which new directions, 'totally unknown' open up an algebra (combinatoria) and topology (proximities, openings, closures).

The discovery of Pataphysics, its definition, principles and tenets

Critical and speculative designer Anthony Dunne refers to a pataphysical realm in *Hertzian Tales* (1999) without fully conveying its significance for design. Dunne's reference remains a suggestion, expressing preference for the probable futures of the science of the artificial, rather than considering designed realities more significantly as modulated by the future anticipation of possible worlds. Developing the pataphysical realm at the intersection of design with the science of imaginary solutions, has parallels with what Deleuze and Guattari (1994) term the 'planes of the chaoids'. Furthermore, this might also create the becomings of technical objects within a realm contiguous to that occupied by Klee's *Twittering Machine* (1922) and Duchamp's *Large Glass* (1915–23). That these various works had been anticipated and plagiarized, quite atemporally, by the Baroque furnishing of possible worlds with imagined technical objects, is among the discoveries of pataphysics; and its concepts, principles and tenets enable the further demonstration of the extent of this anticipatory plagiarism later in this chapter. We need still, however, to explore Jarry's worlds more fully.

Possible worlds

Whilst Jarry's interest in possible worlds is most evident in the related play, *Caesar Antichrist* (1992), it is in the tale of Faustroll's voyage that Jarry's association of possible worlds of the mathematical 'fourth dimension' and the abstract space of pataphysics is most anticipated. It is from these plays and tales, and largely via the artist Marcel Duchamp, that pataphysical ideas enter architectural discourse (Kiesler, 1937; Henderson, 2018) as well as an anticipatory plagiarism of intertextuality and the abstract spaces of Deleuzian 'numerical images' and *objectiles*. 'Colleges' of pataphysics and other quasi-institutional manifestations of the science of imaginary solutions were established across Europe after 1947. The *Collège de 'Pataphysique* (Paris 1947–1975 and again from 2000–present) attached and attracted many artists to its membership[6] and it is via these artists and their continued artistic development of pataphysical notions that pataphysical tenets become recorded in diverse registers of science fiction, continental philosophy, cybernetics and architectural theories of the machine. Foremost of these principles of pataphysics associated with the technical object, is the clinamen.[7] To the Pataphysician, Faustroll's *Clinamen* is both a concept, clinamen, and exemplifies the paradigmatic pataphysical machine and technological object. Furthermore, the concept clinamen possesses an interdisciplinary mobility[8] and it is a pataphysical *Clinamen* that serves as a prototype for science fictional, philosophical and architectural machines.

Equivalences

Artist Thieri Foulc et al. (2014, p. 20) writes: 'Equivalence lies at the very heart of 'Pataphysics, something that is made clear in the earliest pages of *Exploits and Opinions of Dr Faustroll*, those that concern the library of the Inamovable Curator'. Foulc's reference to a library here is to the naming of 27 books equivalent to those embedded in Faustroll's voyage itself[9] and of the subsequent distribution:

> Across the foliated space of the twenty-seven equivalents, Faustroll conjured up into the third dimension: From Baudelaire, E.A. Poe's Silence, taking care to retranslate Baudelaire's translation into Greek. From Bergerac, the precious tree into which the nightingale king and his subjects were metamorphosed, in the land of the sun. From Luke, the Calumniator who carried Christ on to a high place. From Bloy, the black pigs of Death, retinue of the Betrothed. From Coleridge, the ancient mariner's crossbow, and the ship's floating skeleton, which, when placed in the skiff, was sieve upon sieve.
>
> (Jarry, 1996, p. 190)

A distribution unfolding as a set of objects, one and sometimes more, from each of the 'interleaved' dimensions of these books and an anticipated communication between their pages as equivalent universes supplementary to Faustroll's.

There are certain pataphysical concepts that will present philosophical difficulties to anticipations of chronology and synthesis. The pataphysical principle of equivalence would be problematic, for example, if anticipating a repetition without difference. But the pataphysically equivalent, as the infinite aberrance of adding or subtracting supplementary worlds, seems to be compatible with the anticipation of a disjunctive synthesis, 'and . . . or' rather than 'as-if', of worlds 'superimposed upon and supported by the passive syntheses of imagination' (Deleuze, 1994, p. 71; see also: Deleuze and Guattari, 1984). To which, we can add from Foulc et al. (2014, p. 9): 'the world is a gigantic aberrance, which, additionally and universally, is based upon an infinity of other aberrances. Whatever we may say about it is a fiction of a fiction'.

The pataphysical *postulate of equivalence* assumes that realms of art and science (and one must assume philosophy) contain all infinites. For Deleuze and Guattari such infinities as the chaoids[10] are *forms of thought and creation*. Deleuze and Guattari (1994, p. 198) seem to be in accord with the pataphysical postulate of equivalence in *What is Philosophy?* when they write: 'Thinking is thought through concepts, or functions, or sensations and no one of these thoughts is better than another, or more fully, completely, or synthetically "thought"'. They also seem to concur with Jarry in the equivalence between these daughters, between their realms or the planes of the chaoids and between 'other heterogeneous elements, which are still to be created on other planes: thought as heterogenesis' (Deleuze and Guattari, 1994, p. 199). It is worth placing these 'heterogeneous elements' in their context, with the lines that precede them:

> With its concepts, philosophy brings forth events. Art erects monuments with its sensations. Science constructs states of affairs with its functions. A rich tissue of correspondences can be established between the planes. But the network has its culminating points, where sensation itself becomes sensation of concept or function where the concept becomes concept of function or of sensation, and where the function becomes function of sensation or concept. And none of these elements can appear without the other being still to come, still indeterminate or unknown. Each created element on a plane calls on other heterogeneous elements, which are still to be created on other planes: thought as heterogenesis.
>
> (Deleuze and Guattari, 1994, p. 199)

From the pataphysical thought of Hugill (2014) on such 'rich tissue' of pataphysical equivalence, is its interest in Jarry's algebraic and geometric passage beyond the metaphysical to pataphysical abstract spaces. Further equivalences can be created or anticipated between, say, Jarry's pataphysical 'sieve' and Deleuze's tabular grid, the *objectile* (clinamen) between which the pataphysician might say there exists a syzygy between *Clinamen* and Bergson's (1944, p. 367) 'nets' of arithmetic and 'meshes' of geometry. In which nets: 'the space of our geometry and the spatiality of things are mutually engendered by the reciprocal action and reaction of two terms which are essentially the same' (Bergson, 1944, pp. 213–214). To achieve pataphysical thought's equivalencies is to be necessarily entangled in archives of pataphysics. There can be no claim to the completeness of any survey of the literature on pataphysics as the material is inordinate, which is to say, its archives are too pataphysical.

Yet further influence and interest in this respect is with the then emerging scientific thought on the fourth dimension, with which Jarry is sometimes associated for his interest in the abstract mathematical spaces of Riemann (Klieger-Stillman, 1982, pp. 72–81; Henderson, 2018). Elsewhere, this intent has been discussed in relation to the assimilation of the *objectile* to the pataphysical *Clinamen* and a mobilization of the *objectile* freed from the strata of the architectural.[11] The contextualization of pataphysical objects in this chapter, however, has these spaces in mind in further consideration of the *objectile* as a part of the Deleuzian intent to overturn Plato and so further dislodge the *objectile* from the stratum of architectural idealism. For implicitly futural Ideas – if not Plato's, then Neoplatonic and hylomorphic[12] – can be said to hold sway over the architectural object and its relation to futurity.

A desired complexification of this informs the selection of exemplary works of anticipatory plagiarism evident in Italian architect and theorist Marco Frascari's architectural imaginary, in which the divinatory power of the architectural object (that he describes as 'technographic') will be détourned to offer an architectural imaginary that is *objectile* and non-hylomorphic in its modulation. First though, we will briefly reprise relations between Deleuze, Guattari and pataphysical machines.

Deleuze, Guattari and pataphysical machines

Ideas about pataphysical machines, so deeply rooted in architectural imaginaries of the twentieth century,[13] have already been extended to design imaginaries of the twenty-first (Hales, 2015). Deleuze, notably, although writing on Jarry and his fictional characters the pataphysicians Ubu and Faustroll, was never himself recorded as being a member of the *Collège de 'Pataphysique*.[14] It will be worthwhile briefly recalling before going much further, that all machines are first pataphysical: that Faustroll's pataphysical machine *Clinamen* is a kind of ur-machine of all imaginary machines.

Deleuze's (1990) essay on 'Lucretius and the Simulacra'[15] may have provided the impetus for Serres's (2000) *Birth of Physics*, his study of Lucretius (Berressem, 2005, p. 54). It is at least as likely that Deleuze's two essays on Jarry and pataphysics (1997, pp. 91–98; 2004, pp. 74–76), especially his reading of Jarry's *Faustroll* and its *Clinamen*, are a source for this influence on Serres (2000). Furthermore, when Deleuze and Guattari (1994, p. 38) write, that the 'atom will traverse space with the speed of thought' – citing it as a quotation from Epicurus – they develop an overturning of 'the classical image' in which 'error does not express what is by right the worst thing that can happen to thought, without thought being presented as "willing" truth, as oriented toward truth, as turned toward truth' (Deleuze and Guattari, 1994, p. 53). This is a question for the creation of anticipations[16] and so also a matter for the science of imaginary solutions. Deleuze's (1983) *willing truth*[17] is that of Nietzsche's science:

> So, the faith in science, which after all undeniably exists, cannot owe its origin to such a calculus of utility; rather it must have originated in spite of the fact that the disutility and dangerousness of 'the will to truth' or 'truth at any price' is proved to it constantly.
>
> (Nietzsche, 2010, p. 281)

However, it is the *willing error* of the pataphysical *Clinamen*'s fall as a generative falling, a falling of a combinatorial power back into the plane of composition, that will be most productive to this chapter. It is just such a misreading of something like a clinamen in the science of imaginary solutions as pataphysical that will lead to Deleuze's implication in what became known as the 'Sokal Affair'. Alan Sokal and Jean Bricmont in *Fashionable Nonsense* (1998, p. 1) target Deleuze for 'the repeated abuse of scientific concepts and terminology' and is particularly scathing in their review of Deleuze and Guattari's *What is Philosophy?* accusing them of 'pseudo-scientific invention'.[18] Where Deleuze will frequently obscure and combine his sources, Jarry will occasionally make deliberate and combinatorial 'errors' of attribution which provide a provisional definition of the sense of plagiarism in the anticipatory mode proposed by the *Oulipo*. There is one such error which has resonance with Deleuze, Jarry's (1989, p. 83) attribution of a 'Leibnizian definition' for 'that perception is a hallucination which is true'. A pseudonymous 'Faustroll' (Faustroll, 2006, p. 119) suggests that Leibniz's definition of error in *Brevis Demonstratio Erroris Memorabilis Cartesii et Aliorum Circa Legem Naturae* (1686) may well have been Jarry's target in bringing a sense of Leibniz's attack on Cartesian physics to a pataphysics of time-spiralling vortices. In this same chapter, Jarry writes that: 'There is nothing in the back of infinity [. . .] because movements are transmitted in rings. It is established that the stars describe narrow ellipses, or at least elliptical spirals' (Jarry, 2006, pp. 73–74).

Jill Fell (2005, p. 210) identifies Jarry's use of the concept 'foliated space' as happening somewhere between the article '*La Vérité bouffé*' (1903) and *Faustroll* in which Jarry parodies Riemann's mathematical concepts, with the anticipation of something like a black humoured[19] metafictional idea of foliation, an anticipation of the intertext or a hypertextual picturesque in the interleaved space of the pages of Faustroll's equivalent books. There also exists an atemporal anticipation, a contiguity or syzygy, between Jarry's idea of foliated space (*espace feuilleté*) and the holey space that Deleuze and Guattari propose in *A Thousand Plateaus* (1988, pp. 413–415) and which is again present in *The Fold* (Deleuze, 1993, p. 5) in its 'infinitely porous' matter. This also seems to entail a triangulation of Jarry's equivalent volumes ('folios' or books) with the non-Euclidean surface unfolding as the mathematician Riemann's concept of 'the manifold'.[20] Which is to say, a triangulation of superimposed surfaces or realities and those spaces and surfaces of Faustroll's various and clinamenic journeys. It should also be noted that Deleuze's theory of multiplicity is itself drawn from Bergson's ontological folding of Riemann's foliation of surfaces[21] which would doubtless have appealed to Jarry just as Jarry's pataphysical geometries of n-dimensions seem to have appealed to Duchamp, Borges and Lovecraft, though in different measure.[22] It has been suggested that Jarry 'found something deliciously subversive' and 'revolutionary' (Henderson, 2005, pp. 493 and 236) about the new geometries, with their challenge to so many long-standing 'truths'.[23]

If foliation, as a kind of pleating between fictional and actual worlds, expresses a transcendental-materialist, ontological, aspect of the possible pataphysical world, then, between the supplementary, plurality of worlds is a kind of difference. Which is itself, an anticipation of the modal logic of possible worlds which David Lewis (1986, p. 16) claims exists between two people, 'if one inhabits a Riemannian and the other a Lobachevskian spacetime'. This relation brings together Lewis's ideas about possible worlds and Jarry's fictional worlds[24] of possible universes 'supplementary to this one', with the Deleuzian fold and *objectile* (1993, p. 19).

Furthermore, Faustroll's machines – for example the Painting Machine or his invention of a Time Machine (Jarry, 1968, 1996) – with their clinamenic and time-spiralling gyre, render any straightforwardly chronological anticipation somewhat redundant. Through this, we will discuss the paradoxical condition of time expressed in the notion of 'anticipatory plagiarism'. The pataphysician's machines are accompanied by an anticipation of indeterminacy and a fourth dimension interleaved with ahistorical, atemporal and fictional worlds. Philosophical interest in the clinamen[25] can be described, for the purposes of this chapter – and its intersection with the clinamenic presence in other chapters of the present volume – by a definition provided, neither by Lucretius nor Deleuze, but by philosopher of technics Gilbert Simondon (2009, p. 4):

> Anything that can serve as the basis for a relation is already of the same mode of being as the individual, whether it be an atom, an external and indivisible particle, *prima materia* or form. An atom can enter into relation with other atoms via the clinamen. It constitutes thereby an individual, viable or not, through the infinite void and the becoming without end. Matter can receive a form, and within this form-matter relation lies the ontogenesis.

Certain equivalences between Simondon's 'clinamen' and Deleuze's *objectile* can be expressed through their mutual concern with form-matter relations. Such equivalences also reconfirm previous substantiations of a pataphysical assimilation of one to the other (for example, Hales, 2015), as well as demonstrating an anticipatory plagiarism by Lucretius[26] of Deleuze's use of the concept *objectile*:

> The new status of the object no longer refers its condition to a spatial mold – in other words, to a relation of form-matter – but to a temporary modulation that implies as much the beginnings of a continuous variation of matter as a continuous development of form.
>
> (Deleuze, 1993, p. 19)

Similarly, the pataphysical syzygy, which is generally taken to reference its astronomical definition as an alignment of stars (Shattuck, 1996, p. xvii), is also subject to an anticipatory plagiarism of combinatorial equivalence. There is no need to deviate from this orthodox definition greatly. Indeed, the evidence in *Faustroll* points strongly to the astronomical influence of planetary alignment, perhaps in its hermetic dimensions: 'caught unawares by the ebb tide, since the period of syzygy was nearing its end' (Jarry, 1996, p. 62). Nevertheless, Jarry's later association of the syzygy of words (Jarry, 1996, p. 98) and hence with the syzygy of fictional worlds and the syzygies of words in interleaved spaces (i.e. between pages), suggests that given Jarry's interest in the abstract spaces of the mathematics of the second half of the nineteenth century, the meaning of the syzygy is more appropriately considered in mathematical rather than astronomical terms. In mathematics, a syzygy had come by the late nineteenth century to take on the meaning *a relation of relations* (Krieger, 2003, p. 14); and it is this mathematical syzygy which is employed in the *Oulipo*. Despite this, we now turn to the *syzygy of relations* in mathematician David Hilbert's 'syzygy theorem' (from 1890), in which relations form syzygies, modules of the theorem, and hence between modules, relations of relations. Such relations anticipate the poetry of the *Oulipo* in the anticipatory plagiarism of James Joseph Sylvester prior to Hilbert's Syzygy, in the combinatorial ideas contained in his mathematical theorem *On A Theory of the Syzygetic Relations of Two Rational Integral Functions* (1853) and paralleled in his poetry.[27]

Anticipatory materialities of the *objectile*: a syzygy

Whilst acknowledging the risk in bypassing potential and significant antagonism between an anticipatory politics of pre-emption[28] and the philosophy of accelerationism, this (as diagrammed by philosopher Nick Land in *Templexity* (2014) in the form of a gyre, a spiral line in a real or complex Hilbert space, or a Lobachevskian space[29]) is in complete syzygy with a time-spiral plagiarized and anticipated by the pataphysician Pere Ubu in the spiral form of his *Gidouille*.[30] Furthermore, whilst it might appear incongruous, even controversial, to include the 'templexity' of Nick Land's writings on Capital in a discussion on speculative design, it should be recalled that Land's (1995) essay – 'Cyberspace Anarchitechture as Jungle-War' – for the *AD* special issue 'Architects in Cyberspace', edited by Neil Spiller and Martin Pearce (1995), sits alongside Dunne and Raby's (1995) article 'Fields and Thresholds'. Anticipation, for Land and cultural theorist Sadie Plant (Plant and Land, 1994), 'will have been', actualization as 'anastrophe' of the future coming together in its potentialities, coalescing unseen and unfelt in the present. (This is in opposition to catastrophe which, they say, is 'the past coming apart'.)

Speculative Design's relation to a pataphysical sense of time, chance and contingency might thus be extrapolated to an exit velocity from the constraining of the futurological impulse of the present. This, the temporality of the 'future after the future' as Berardi (2011) has it, is an uncanny acceleration of the passage from the novum of anticipatory fiction to anticipatory objects which, though more familiar extend the estrangement effect noted in the mobility of concepts between pataphysical and cybernetic cultures of control, exemplified in the thought of Deleuze and Guattari (Hales, 2015, 2019). Luciana Parisi, in *Contagious Architecture* (2013), suggests the spatiotemporal structures of anticipation be characterized as incomputable, corresponding to neither mathematical nor to physical inputs and yet from which anticipation is a productive preresponse to the incomputable. Parisi (2013, p. 90) further notes that '[c]ontrol, as Brian Massumi has brilliantly explained, is a mechanism of anticipation, whereby the apprehension of unknown variables indirectly works to determine the reality of the present'. We will reserve fuller comment on this for the moment. Suffice to say, for present purposes, that the technogenetic modulation of objectility, as the encounter held in anticipation, has a certain pataphysical equivalence with the preresponse of the anticipatory plagiarist and the becoming of a predictable uncertainty, an over-stimulated analepsis, a forward looking and affective condition in an active reorientation of the temporality of influence.

Taking up the 'anarchitechtural' framing of a nascent speculative design by the pataphysician Neil Spiller (1995) in editing *Architects in Cyberspace,* what follows provides a temporally skewed path for Anticipatory Plagiarism. In doing this we will foreground, within anticipation studies, the phrase 'architecture fiction' in order to construct a syzygy across pataphysical dimensions of design fiction, critical and speculative design. That once burgeoning set of futurological practices resuscitated after the end of the future – or if not fully resuscitated, then at least anticipated as revival – since the end of futures studies.[31]

The earliest identified reference to 'architecture fiction' is Robert Harbison's (2000); who originally used the phrase in 1977 in connection with the incunabula *Hypnerotomachia Poliphili* (1499): an influential dream text and one anticipatory of many others. As such – an imagined architecture informing Rabelais's Abbey of Thélème as built by the giant Gargantua – this 'architecture fiction' should be considered interleaved with Dr Faustroll's set of 'equivalent books' and thus to have plagiarized in anticipation architects as diverse as Le Corbusier, Carlo Scarpa, Nigel Coates and Neil Spiller.[32]

While Harbison (2000) regards fifteenth century incunabula as an *architecture fiction,* Kazys Varnelis credits science fiction writer and design theorist Bruce Sterling with coining the term in 2006. This re-invention of 'architecture fiction' creates, simultaneously, a peculiar temporality:

> instead of absorbing into itself, a Dada Capitalist architecture would look out into the world, creating architecture fiction, a term that Bruce Sterling coined after reading 'A Handful of Dust', a piece on modernism by J.G. Ballard to suggest that it is possible to write fiction with architecture.
>
> (Varnelis, 2009, online)

Further compounding the erasure of Harbison's architectural humanities approach from speculative and anticipatory histories of architecture fiction in speculative design, Mark Dery (2011) also attributes this to Sterling: 'Architecture fiction anticipates the future present [. . .] the field becomes almost infinitely more exciting when you realize that architectural projects, by definition, entail the reimagination of how humans might inhabit the earth'.[33] Harbison's thinking through the capriccio, the incunabula and Ballard's own clinamen – a swirling swerve of turbulent influences (Serres, 2000) – serve to augment Sterling's fascination with an atemporal image. This paradoxical image of futurity he notes in a serendipitous convergence between three books, one each by Freud, Baudelaire and H.P. Lovecraft; a convergence requiring 'serious postmodern semiotic analysis' (Sterling, 2010) which we will necessarily forego, favouring a schizoanalytic one.[34] There will be more to say about schizoanalysis before concluding, but it is sufficient to say now that by this we mean a productive complexification and rejection of those aspects of psychoanalysis that Deleuze and Guattari (1984) say do not work, whilst recognizing the unconscious as a machine-processing that is both productive of desire and schizophrenizing in its liberation of these desirous flows from the strictures of the State. We also mean to say that this is a practice capable, in the context of the present chapter, of creating syzygies between the designer as author, critic and theorist of possible worlds.

If the anticipatory power of an architecture fiction's futurity is atemporal as intimated by Sterling, further consideration can, via Dunne's idea of the architect's 'dreaming objects' anticipate a prior opening onto the architecture fiction of literary second or dream worlds. Dream worlds anticipated in the *Hypnerotomachia Poliphili* and the machines of Salomon de Caus (1615) an obsession with the artificial which Jarry shared with Raymond Roussel and the pataphysicians and Oulipians of the mid twentieth century; and one which also informs my ongoing schizoanalysis of the Deleuzian *objectile*.[35]

Atemporality

In the flattened thickness of the present, is that temporality of the post-futurological event of anticipatory artefacts, 'rupture' and 'anomaly', rather than the 'Future' of modernity as such, the atemporal: in the atemporality of anticipation and the indeterminate which is to say, in modern times, futurists, futurologists and those researching the future through scenarios and other instrumental means, seek to transport dreams and anticipation to a new dimension. A dimension that renders the political and aesthetic interest as playgrounds (Dunne and Raby, 2013, p. 3). Simultaneously, anticipation exceeds the bounds of Korzybski's time-binding concept which, like Freud's, belonged to the time of modernity. Korzybski writes (1958, pp. 59–60):

> And because humanity is just this magnificent natural agency by which the past lives in the present and the present for the future, I define humanity, in the universal tongue of mathematics and mechanics, to be the time-binding class of life.[36]

Time-binding for the futurist would make the future 'take time' in the present. These would be to Freud, timeless, in that the unconscious system, 'is not ordered temporally, [is] not altered by the passage of time' (Noel-Smith, 2016, p. 142). But if not so ordered, is the unconscious then negentropic, creativity itself?[37] and with the capacity to do what Bergson (1932, p. 211) describes as the 'act of placing in matter a freely creative energy'?

The speculative design process is one of transporting the 'now' of the present, into another time, into the temporalities of the future as well as of dreams and memories. It appears that speculative design deals with time in this transitive melancholic way, a downward transitivity of the temporalities of dreams 'downgraded' to the future as but a hope or wish. As if, in treating the future as an object – with Korzybski's 'time-bound' rubbing the clinamen raw[38] against Freud's description of psychoanalysis as 'time-consuming' – *the future* of speculative design had begun to obsolesce. Hope, for the future as fantasy, as anticipation, or as a preference for the suspension of time and desire, becomes its own exhaustion and excess. Psychoanalytic time is complex and varied (and time-consuming) in a context where paleofutures and the suspension of futurity coalesce as the atemporal. Sterling and Dunne diverge with respect to Freud's linear concept of time; with their design fictional positions in their respective philosophical orientations bifurcating about this point. Neither Sterling's atemporal nor Dunne and Raby's dreaming hopefulness participate in the treatment of memory or futurity as narrative in linear succession, as does Freud. Furthermore, Alfred Schütz's 'wide-awake' imagining of the *as-if,* to which he refers in his essay 'On Multiple Realities' (1945), provides a direct reference to Hans Vaihinger, whose fictionalism is of value to anticipation creators in developing argumentation over models of possible and fictional worlds. Rather than characterize makers of design fiction and speculative designers as Freud's dreaming analysands, they can be counter-posed with Schütz's 'wide-awake' actors operating to create anticipations across multiple realities.[39] The concept of time for the speculative designer, wide-awake and operating in the pataphysical imaginary present, as modified by Jarry, would be closer, then, to Deleuze's model of time and duration anticipated by Bergson.

Freud's atemporal, an anticipatory plagiarism of Jarry's imaginary present, might not be the same thing, the same object, as given temporal models in anticipation studies, but each have mutual regard for the possibility and impossibility of time and futurity, a concern which can be sensed in the pataphysical anticipation of an *imaginary present* in Franco 'Bifo' Berardi's lost future which might adequately stand in for the temporality of anticipation: that of an *infinitely expanding present*. 'In the late century,' Berardi (2011, p. 51) writes, 'SF imagination of the future vanished, became flat, narrow and dark, and finally turned into an infinitely expanding present'. What Jarry had termed an 'imaginary present' is the apparition of a clinamen in time, one in which the temporality of anticipation of a near future is mythographic and entropic.

Architecture fiction as pataphysical machine

In *Hertzian Tales* (1999), Dunne extends the lineage of design prototyping to the architectural object, the radical pedagogies of architecture and its publications exemplified in the citation of *Objects in the Landscape* (Irie, 1993). Dunne

(1999, p. 75) suggests that to explore the role of the designer as author 'it is necessary to turn first to architecture'. He seems to confirm the ongoing significance of architecture to his practice in his commentary on a book selection he provides for the *designersbooks* website (Dunne, 2013).[40] Here, he expresses an interest in books that delve into 'parallel worlds, imagination, and unreality'. Amongst these is Daniel Libeskind's *Between Zero and Infinity* (1981), of which Dunne (2013) explains:

> I was given this book when I was in my first year at the Royal College of Art for a project we were doing about drawing the invisible. It opened my mind to the possibility of designing things that didn't have to be real, didn't even have to be built, but could happily exist as a drawing or an idea.

Furthermore, whilst it may be a product of a paranoiac methodology of anticipation to align Libeskind's book title with the claim by Jarry (1996, p. 114) – 'God is the tangential point *between zero and infinity*' (emphasis added) – the similarity is inescapable. Whilst Libeskind's *three architectural machines* are perhaps not typical of the architectural projects listed in *Hertzian Tales*, they are representative of what we could call 'pataphysical anticipations of architectural machines'. The unspoken presence of Libeskind's machines in Dunne's recollection also forges relations with the reading/memory/writing machines of Roussel, their representation by pataphysicians and presence in the techniques of the *Oulipo* (Motte, 2007; Duncan, 2019). These also foreground a connectivity with the renaissance imaginaries of both architectural machines and the landscapes of memory informing Jarry's pataphysics: variously, Roussel's fictional garden from *Locus Solus* (1970; originally published 1914) and the combinatorial and anticipatory relation this has with the *Hypnerotomachia Poliphili* in Carlo Scarpa's (1969) *Brion Cemetery* and Neil Spiller's ongoing project *Communicating Vessels*. Roussel will serve as our pivot to discussion of Anticipatory Plagiarism but, prior to this, we take a brief excursion into the legal territory of innovation, patents and its creative claim over the anticipation of future properties.

Pataphysical, fictional and anticipatory patentry

Rem Koolhaas's fictional patents[41] stand in contradistinction to the anticipatory plagiaristic. These rather more straightforwardly reveal concern with intellectual property and the patenting of conceptual and abstract spatial ideas that is itself quite pataphysical.

Three related pataphysical enterprises will suffice to substantiate this observation. First, that the London Institute of Pataphysics's *Department of Reconstructive Archaeology* established an *Office of Patentry*, putting out a call 'for creators of inutilitous inventions and innocent machines' (2003). Second, that the Canadian pataphysician bpNichol – with apparent disregard to the occultation of the *Collège* from 1975–2000 – in 1992 founded a *Pataphysical Hardware Company*[42] publishing the company's catalogue of pataphysical products. Third, that Jarry, in a section titled '*Connaissance utiles et inventions nouvelles*' of Ubu's *Second Almanac* (1901), includes a 'confidential letter' from Pere Ubu, to one Monsieur Possible, of the *Bureau des Inventions et Brevets* (Dubbelboer, 2017).

Pataphysics in the architectural humanities

Regarding the architectural imaginaries of the *objectile*, Bernard Cache's (1995) numerical and computational image was a repression of his subsequent turn toward the imaginative media of the architectural treatise and the architectural fiction of Philbert de L'Orme (1567). De L'Orme's treatise has both Epicurean and Rabelaisian sources, which appealed to Jarry. Whilst the pataphysical topologies of interest are not those theatrical ones of Salazar-Sutil (2013), the rendering of space and object, time and environs of Jarry's pataphysics are topologically continuous with de L'Orme's architectural treatise. Each is depicted in a landscape garden or second world: from Robert Harbison's *Eccentric Spaces*, to gardens of architecture fiction, the *Hypnerotomachia Poliphili*, Jarry's *Messaline* (1985), the Garden of Lucullus, Roussel's *Locus Solus* (1970) and Moorcock's (1965) *Pleasure Gardens of Phillipe Sagittarius*. These architectural fictions of landscape[43] – a primary resource for architectural imagining –, are interleaved with the possible worlds of the *objectile*. Cache's renaissance influence and that of Jarry's coincide. At least, insofar as a Rabelaisian architecture fiction, the Abbey of Thelema from *Gargantua and Pantagruel* (1999) influenced both Jarry and de L'Orme. To provide a partial justification for these tangents, putting conjecture to one side, Dunne and Raby also recognize in their genealogy of speculative design (2013): architecture fiction, paper architecture and, by free association, the architectural treatise. These associations also include an equivalent text to the *Hypnerotomachia Poliphili*, the *Codex Seraphinianus* (1981), which they describe as a 'cult book surrounded by mystery. It portrays an imaginary world' (Dunne and Raby, 2013, p. 111).

Further 'imaginary worlds' of architectural fiction include Michael Webb's (1987) quizzical submersible from *Temple Island*, pictured as a 'Pepperland' regatta of the mobile point of view and the indeterminacy of memory;[44] David Greene's *rokplug, LogPlug* and *L.A.W.u.N* projects (all from 1969; see: Greene and Hardingham, 2008); and Neil Spiller's cyborgian island geographies of *Communicating Vessels* (1998); as well the works of Francois Roche and R&Sie. These pataphysical landscapes are each possessed of a peculiar pastoralism which Greene calls 'cybucolic' (Greene and Hardingham, 2008) and Spiller 'cybaroque'.[45] They are the principal architectural inheritors of an apocalyptic, visionary and phylogenetic line that the pataphysical clinamen contributes to the corpus of architectural humanities writing. A cultural incursion, a veering and pataphysical swerve recognisable as an encounter between post-Deleuzian philosophy, pataphysics and speculative *objectiles* (Hales, 2015, 2019).

This encounter shows how we can replace design thought as a 'science of the artificial' (Simon, 1988) with a pataphysical design as the 'science of imaginary solutions'. And thus pataphysics offers to anticipatory culture another of its definitions: 'Pataphysics is the science of these beings and current or future devices (*engins*) with the Power of their use (*discipulus*)' (Jarry, 2001, p. 202).[46] An interest in the becoming of 'future beings' and their pataphysical '*engins*' can be assimilated within a more general interest of the architectural humanities between architecture and the machine. It is usual for accounts of this pataphysical interest (Olshovsky, 2011, 2012) to begin with Vitruvius's (1914) treatise *On Architecture*: particularly, Book X, on Machines and Book VIII on Hydraulics.[47] And Jarry, who maintained he was not the inventor of pataphysics merely its discoverer, likewise, is a mis-reader of Lucretius, Leibniz and Bergson, introducing deliberate errors in his application of their ideas. Henceforth, Vitruvius's reading of Lucretius and the hydraulic and speculative machines of Books VIII and X (if the pataphysical logic of equivalent books established by Jarry and listed in *Faustroll* is adopted) will be taken as commencing another series of equivalents to Jarry's books, interleaving their depiction of pataphysical machines.

Jarry also invests some sense of the 'medieval marvellous' (Lightsey, 2007) in the machine, which then empowers Faustroll's time-machine. We have seen the *Clinamen* as foremost amongst Jarry's machines. Other machines, devices and '*engins*' include: the apparatus dispensing the perpetual motion food of *The Supermale* (Jarry, 2009); its flying gyroscopic train engine; the Supermale himself, a proto-cyborg and bachelor machine; as well as 'a future architecture yet to happen' (Jarry, 2009, p. 26). Faustroll, as well as inventing *Clinamen*, also provides instructions for the construction of a *time-machine* as we have also already noted. These all manifest as a seam of the 'as-if' pataphysical realm – available to the twentieth century in consequence of a 'resurrection' of the clinamen Warren Motte (1986) attributes to Jarry – and are a resource largely untapped by speculative designers.

Practitioners of what can be termed a pataphysical mode[48] of architectural writing, draw on the association of architectural and pataphysical machines – see, for example: Frascari (1991, pp. 61–76); Vidler (1992, p. 160); Teyssot (1994, pp. 8–35); Perez-Gomez and Pelletier (1997, p. 293); and Evans (2000, p. 273). In these authors' architectural histories, all written under a pataphysical charm, the term 'clinamen' is only rarely invoked. However, in *Clinamen*'s guise as the machine, monstrous and anomalous, a pataphysical clinamen makes deep, if fleeting and hidden incursions into the disciplinary apparatus of architecture.

Much of this is can be substantiated only after the influence of Marcel Duchamp and architectural analyses of the *Large Glass* are recognized (Kiesler, 1937).[49] Nonetheless, in this register of the corpus of architectural writing, the combinatorial, the monstrous or mutated and the exceptional or anomalous image of material error in architecture, parallels similar concepts as they developed in the literatures of anti-constraint. A clinamen immanent to the chance operations of the *Oulipo*[50] as to artistic and software cultures; a coincidence observable in Szeemann's (1965) *Light and Movement: Kinetic Art and New Tendencies in Architecture* (Phillips et al., 2016).

Thus, pataphysical fiction is informed by a tradition of architectural machine fantasies from Colonna to the 'pataphysical conspiracies' of Jean-Jaques Lequeu (Birkstead, 2009, p. 267) and, in turn, informs architectures of the 'cybaroque' and 'cybucolic' already encountered. These influences extend to radical design and architecture with sources in cold-war communications theories and cybernetics – as shown in the pataphysics issue of the *Evergreen Review* (1960). Further, a concern to escape control, embrace error and the indeterminate manifests in the concern to re-mobilize the *objectile* pataphysically within the field of architecture. An *objectile* prepared as the creative and liberatory power of the aleatory and developed through a temporal complexification of the anticipatory. In pursuit of this and in addition to the specifically architectural implications in the works of Spiller, Greene and R&Sie(n), we can add a marginal reference by architect Peter Cook (1983) to a 'pataphysical promenade' in *Strange Pavilions of the Mind: The work of Diploma Unit 10 1973–1983*: an exhibition of Nigel Coates and Bernard Tschumi's (1983) teaching Unit at the Architectural Association. This supports the supposition of a deep incursion of this fugitive pataphysical figure into radical architectural and pedagogical discourse.

The spectre of Jarry's pataphysics, then, has continued to influence pataphysical fictions after 1947 and beyond the *Collège de 'Pataphysique* and an international network of such institutions. Where Shattuck (1988) shows the pataphysical involvement of proto-modernisms, this vortex of fictional influences coalesces to confirm the architect Francois Roche's wider-ranging observation that: 'we might consider the fiction of architecture as a kind of pataphysics, as in the writing of Alfred Jarry' (Roche, 2013, p. 203).[51]

Another convergence of pataphysical machines happens with the exhibition *The Bachelor Machines* (Szeemann, 1975) and the theorist and historian Michel Carrouges's essay 'Bachelor Machines' (1954), without so far providing a fuller account in a discussion of their presence in Deleuze and Guattari's (1984) *Anti-Oedipus*.[52] Across his philosophy Deleuze had recourse to pataphysical contrivances and contraptions including, with Guattari, the notable example of paranoiac electrical machines designed by the 'talented' Robert Gie (Deleuze and Guattari, 1984, p. 17). The inclusion of these devices in *Anti-Oedipus* is most pertinent when Deleuze and Guattari cite the survey of fantastic machines conducted by Carrouges, in which he had discerned similarities between Kafka's 'Harrow' from *In the Penal Colony* (1948) and Duchamp's *The Bride Stripped Bare by her Bachelors, Even* (1915–23) and other imagined machines from art and literature. They write:

> Michel Carrouges has identified a certain number of fantastic machines – 'celibate machines' – that he has discovered in works of literature. The examples he points to are of many very different sorts, and at first glance do not seem to belong to a single category: Marcel Duchamp's painting

> *La mariée mise à nu par ses célibataires, même* ('The Bride Stripped Bare by Her Bachelors, Even'), the machine in Kafka's 'In the Penal Colony', Raymond Roussel's machines, those of Jarry's *Surmale* (*Supermale*), certain of Edgar Allan Poe's machines, Villiers's *Eve future* (*The Future Eve*), etc.
>
> (Deleuze and Guattari, 1984, p. 18)

Whilst these might be better known as Duchampian 'bachelor machines', what Deleuze and Guattari do not make explicit is that these machines – together, it might be supposed, with all Deleuze's and Guattari's (individually) and Deleuze and Guattari's (collectively) machines; indeed 'all machines' as Carrouges (1975, p. 44) would have it – are first pataphysical. These bachelor machines, Deleuze and Guattari argue, form alliances with other machines of desiring-production and the field of productivity. Similarly, the pataphysical machines of speculative designers and those others that modulate the designed artefacts of abstract culture, make fresh alliances between bachelor machines and, what Deleuze and Guattari term, the 'machinic phylum' through what we know as interfaces and the software and algorithmic processes behind these, as well as the numerical images we see and feel in the foreground.[53] Or, pataphysical *objectiles*, device and operator, participate in the probing of a space of possibilities: the *machinic* as the synthesizing of heterogeneities; and the *phylum* of self-organizing and combinatorial human and nonhuman potentialities.

In the second of Carrouges's articles on pataphysical machines, '*Les Machines Pataphysiques De Maldoror Et Leurs Groupes De Transformation*' [*Maldoror*'s Pataphysical Machines and their Transformational Groups] (1986),[54] he follows-up on his comments in *Bachelor Machines* that Lautréamont's (1994) *Maldoror* contains pataphysical 'transformation' machines. Notable here is that *Maldoror* informs the surrealist doctrine of objective chance attributable to Andre Breton and that it is Lautréamont's sense of plagiarism which informs the surrealist object. An ethnographic surrealism inheres in the cultural probes of socially motivated surrealist design fiction: 'as beautiful as a chance meeting of a sewing machine and an umbrella on an operating table' (Lautréamont, 1994, p. 193). This happens too in the objects of pataphysician Max Ernst, who explains this beautiful meeting of machines in *Maldoror* as 'a coupling of two realities irreconcilable in appearance, upon a plane which apparently does not suit them'[55] (Ernst, 1948, p. 13).

Jarry's foliated spaces and *Maldoror*'s superpositional[56] coupling of realities, in schizoanalytic terms, replace the 'as-if' perception of pataphysical science fiction with a productive surface, a field of matter-energy, in which 'both', 'as', 'and' and 'if' are operative and full of creative potential, each at the same time. To repeat an earlier observation, Deleuze's (1994, p. 71) methodological construction involves the superimposition of layers of thought: '[i]n other words, the active synthesis of memory and understanding are superimposed upon and supported by the passive syntheses of imagination'.

Architecture, technographies and auguries

Jean Baudrillard's (1993) apocalyptic augury is his oracular text *Pataphysics of the Year 2000*. Jarry's is published in 1901, the *Second Ubu Almanac*, in which real and imaginary dates are combined as potent augury of change, in parody of Nostradamus. Allegory 'passes to augury', Steven Connor (2009, p. 3) explains and – in what amounts to a pataphysical swerve – Baudrillard's *Simulacra and Simulation* (1994, p. 2) asserts that we have crossed into a space 'whose *curvature* is no longer that of the real' (emphasis added), in which experience is replaced by a spatialized and combinatorial system of objects and equivalences. For Baudrillard, reality merely imitates the model; models precede and determine the real world in what he terms the precession of simulacra. There has been a continuous movement from allegories of simulacra to the auguries of simulation in post-futurological practices and architectural fiction. William Bogard, offers an alternative account of simulation suggesting a 'pataphysics of simulation, a science of virtualisation as an imaginary solution' (Bogard, 1996, p. 23) and, further, that 'pataphysics is the description of a technology in the mode of its future past' (Bogard, 1996, p. 185, n. 10).

In his analysis of Carlo Scarpa's *Brion Cemetery* (1969[57]), the Italian architect and architectural theorist Marco Frascari (1985) excavates some of its architectural devices deeming these equivalent to the augurial staff or 'lituus'. These, Frascari suggests, function as instruments of architectural divination, which we might now extend to those Carrouges identifies in *Bachelor Machines*, with the 'medium' as three-dimensional intermediary or operator between the second and fourth dimensions. Bachelor and pataphysical machines, therefore, project a future architecture. They also bear a similarity, an equivalence even, to Ubu's Physic Stick and its alchemical presentation.[58] Frascari intends that architects, by using their instruments as augurs and auspexes, read (draw) a future architecture from the messiness of the constructed world and proposes that augury 'is a kind of corporeal time machine where the past, the present and the future are related architecturally through memory' (Frascari, 1991, p. 61). Furthermore, he describes in some detail how such architectural time machines function like the pataphysical one devised by Dr Faustroll where the past lies 'beyond the future in a theatre of memory'. Frascari continues:

> The technological figure in the Jarry machine is based on a contrasting and ironic use of materials, and metonymic use of shapes. Cheap materials are mated with expensive ones, the traditional with the unusual; ebony and ivory are coupled with nickel, quartz, and copper. A bicycle frame is the support for gyroscopes. This is a machine whose function is to nudge the visitor's thought processes to locate the past beyond the future in a theatre of memory.
>
> (Frascari, 1991, p. 61)

David Greene's architectural work *Bottery* (1969; see Green and Hardingham, 2008), along with other objects in the landscape are pataphysical anticipations charting the trajectory of what would become an 'internet of things' and would begin to investigate the flight of such technical objects from the technical virtual into, what William Gibson (2010) termed, the 'eversion of cyberspace'.[59] Greene's architectural and electronic objects are part of a general movement toward a fully speculative culture: part of the mechanism of technographic acceleration in which prototypes and images of technical objects make themselves real through their powers of modulation and anticipatory affect. Architect Robin Evans's work displays its pataphysical credentials similarly, through its sense of multidimensional realities coming into being through a range of media and, often, undergoing ontologically uncertain transitions.

> [The] procedure is therefore more like augury than writing: first form the signs, knowing only how, never what, and then look to see if they signify anything, sometimes they do, sometimes they don't, sometimes good news, sometimes bad, sometimes nothing. Such a procedure shifts the weight of meaning from behind to in front, from before to after, from the verifiable to the unverifiable.
>
> (Evans, 1983, p. 92)

We have seen the pataphysical dimension of architectural machines extend from the literary to media studies and the computational influence of pataphysical fictions (Cramer, 2005, pp. 88–92). Frascari's pataphysical 'technographies' have influence on the plane of architectural fiction and, before reaching concluding remarks for this chapter, Frascari's (1983, 1991) notion of 'technography' is proposed as a model for speculative or conjectural illustration and the diegetic prototype as a drawing.[60] Furthermore, such an 'illustration is an area in which the aesthetics of unreality in varying degrees of abstraction can be enjoyed' (Dunne and Raby, 2013, p. 102). The following will displace Frascari's 'technographic' concept colliding it with the pataphysical *Clinamen*. *Clinamen* – Faustroll's painting machine remember – and his Time Machine (Jarry, 1968, 1996) will serve to destabilize the necessary passage between realms. That is, we will complete this anticipatory plagiarism by interleaving the 'and . . . or' of worlds 'superimposed upon and supported by the passive syntheses of imagination' (Deleuze, 1994, p. 71) to disclose an alternative to Frascari's phenomenological orientation to imaginary 'being' and to oppose this hylomorphic framework for architecture fiction with one of becoming and objectility.

The Objectility of Technography

We have seen Frascari introduce, with 'technography', what he terms the 'auguries' of architecture's means of representation. Oracular and speculative, such tools have a relation to the future/past deserving of attention. Deleuze scholar Joshua Ramey (2016) draws on two of Frascari's sources – the semiotics of Charles S. Peirce (1955) and the political philosophy of Giambattista

Vico (2001) – to articulate a speculative position different to one that is loosed within, outside, beyond and upon the absolute unreason of speculative realism. Ramey's (2014, p. 31) is a Deleuzian formulation of speculation, 'a realism of contingency without an axiom of unreason'. Such practices of divination parallel both the auguries of numerical images and the various 'cargo-cults' of Ballard's speculative fiction, which tend to be based around fetishized objectiles[61] such as the numerological charts of the flight path of satellites. It is with such images in mind that we suggest Frascari's notion of architectural auguries be revisited as *objectile*. The architect as augur is expressed in *Marco Frascari's Dream House, A Theory of Imagination* (Frascari, 2017) and we will construct from this a syzygy of associations to suggest that the relation between architect and artefact, between augur and technography is *objectile*.

> However, the interpretation of divination, as an imaginative form of reading and writing, overturns this negative rendition. Following this non-rational method, architects become augurs and trace spaces to mirror the needs of future inhabitants.
>
> (Frascari, 2017, p. 98)

This argument does not seek to defend any actual divinatory practices of architects, appealing instead to imagination and the creativity of the drawings and models of the analogic architectural design process. Thus far it resembles Ramey's position. Where we depart from Frascari's representational positioning is to suggest that the architect-augur relation is one of displacement. That is, that these practices be reconsidered as *objectile* and, only then, can there be the possibility of a divinatory form of speculation adequate to architectures of contingency and indeterminacy. Furthermore, it only when this happens that architects-augurs can escape the hylomorphic determinism of an authorial construct that imposes form on matter that is still implicit in the phenomenological hermeneutics offered by Evans, Frascari, Perez-Gomez et al. On the stratum we are developing of the architectural as implicitly futural and imaginary, Ideas (if not Plato's then certainly Heidegger's hylomorphic metaphysics) fail to overturn the primacy of form over matter in the privileging of the architect's (inner) vision.

In *Dream Machines* (2017), Steven Connor applies the term 'technographies' to imagined and imaginary machines, paralleling Harbison's categories of the unbuilt and the unbuildable. These 'technographies' capture Mallarmé's disposition toward typography, Jarry's interleaving of texts and Apollinaire's calligrams. They become constellations of the problems of relation between the senses. That is to say, Connor's technographies anticipate analogous architectures through the drawing and designing of architectural images, interleaved relationships with the drawn object. This shifts the concern from anticipating the artificial or the accelerated, to the anticipatory plagiarism of a science of imaginary solutions. We would also shift the locus of technographies from the psychoanalytic to the schizoanalytic, moving from Connor's psychotechnographies to a schizoanalysis of fictional technical objects and speculative drawings. Schizoanalysis, as noted earlier in this chapter, practices, conceptualizes and models, multiplies connected flows of desire such that they can be called machines. Schizoanalysis was developed by Félix Guattari (1984) in his own psychoanalytic practice (Dosse, 2010) and used across his own writings, along with his collaborations with Deleuze (especially Deleuze and Guattari, 1984). In his last book *Chaosmosis*, Guattari (1995, pp. 58–59) writes:

> Beneath the diversity of beings, no univocal ontological plinth is given, rather there is a plane of machinic interfaces. Being crystallises through an infinity of enunciative assemblages associating actualised, discursive components (material and indicative Fluxes, machinic Phylums) with non-discursive, virtual components (incorporeal Universes and existential Territories).

In following this, we are recognizing the unconscious as schizophrenic at its core but machinic in its processes: nonunivocal and infinitely enunciative, undertaking fluid, energy-matter transcription. These *schizzes* (Deleuze and Guattari, 1984, p. 39)[62] also liberate Frascari's (1983, 1991) use of 'technography' in ways compatible with his heterodox and 'cyclical practice', but that brings together a range of representational and non-representational modes (and means) employed in the making of anticipatory artefacts. The sense in which schizoanalysis is contextualized as pataphysical, is in relation to an open-system of bachelor machines in which, what Deleuze and Guattari (1984) term 'desiring-production', creates desiring-machines and through which 'combinatoria', the production of subjectivities, are assembled.

Connor, David Trotter and James Purdhon provide useful and expansive etymological treatments of the use of technography in preparation for their Open Humanities book series, 'Technographies'. Missing from their preliminary positioning accounts, though, is Frascari's earlier tracing of the origins of the term 'technography' to the 'technometric' (Gibbs, 1972) and 'technologia' of the seventeenth century. Connor, Trotter and Prudhon, hence, miss the anticipatory plagiarism of their work by Ames's technometrics:

> Technography can be explained within William Ames's technometric framework. Technography, like calligraphy, would belong to the group of the less dignified, but eminently productive faculties, which are not in themselves unworthy if they are practiced with natural talent rather than doing them 'artificially'.
>
> (Frascari, 1991, p. 128, n. 5)

Where attention is attuned to the architectural object in this way, concern for potentialities between past, future and present can be abstracted from architecture to the artefact as a more general term and extrapolated to the making of those of an anticipatory nature in contemporary terms: the making of an anticipatory artefact is an exercise of *fabricating* a fiction (Steil, 2013). As much as it is a prototype, it is a numerological, diagrammatic and notational invention; that is, it is the union of fabrication and a possible logic in what Vaihinger (2009, p. 42) terms a 'practical fiction'. The technographic results are non-trivial, imaginary architectural artifacts for Frascari. However, the anticipated and actualized artefacts from fictional worlds, along with 'retrospective construing', are the potential from possible futures for speculative or anticipatory culture. Attention to technographic detail is significant to each, for imagining involves retrospection and prospection. Frascari (1989, p. 3) writes:

> Retrospective construing is a monstrous procedure, which, through selection and manipulation of elements of the real, produces in future users/readers a reconstruction of evident and nonevident connections with the physical and cultural context of the everyday environment.

Retrospective construing and prospective conjuring unify the made and the real, making the artificial through a process of anticipatory expression and the representation of architectural fantasy in which the imagination is stimulated. It is in this stimulus that the possibility of architectural knowledge is created and Frascari's science of the artificial (the made) becomes a science of pataphysical exchange. Where an equivalence between fabrication and fabulation and of anticipation between the retrospective and the prospective occurs. Frascari's technography conceptualizes architectural representation as a process of semiosis which, seemingly, draws upon the semiotics of the *Oulipo*, via Umberto Eco, by invoking the calligram of Mallarmé.

> They are wonderful calligrams of technological thought, the analogical expression of the processes of construction [. . .] It is evident that this manner of graphic expression is not based on a mimesis of form, but rather it is a procedure for a simulation of an architectural reality.
>
> (Frascari, 1991, p. 33)

After evoking the 'calligram', Frascari develops his inscriptive notion of the palimpsest, which he uses to describe as technographic a demonstrative expression of a hypothetical basis in 'three overlapping semiotic relationships'. In *Monsters of Architecture* (1991) Frascari uses the term 'demonstration' to describe the work of the technical architectural object. He relates this first to Vico's *monstrare* and second to Jarry's sense of the monster as an inexhaustible beauty, a progeny of monstrous doubles and paradoxes. Dr Faustroll is of this nature of seeming opposites. Similarly, Pere Ubu, bound together only by a pataphysical energy, the spiral or gyre of his Gidoulle, his belly or viscera.

The empty set of Bruce Sterling's (2013) hierarchy of disbelief has at its highest level or plane, the holy relic. If this vacant slot is populated by a monstrance, which ritually displays the sacred host – as monsters such as these demonstrate the anticipation of the unknown – then the pataphysical machines of the beyond that Carrougues finds in his further research on pataphysical devices, along with the function of the medium in *The Bachelor Machines*, demonstrate the spatial or dimensional boundaries of speculation. Or, to use another Jarryism, divulge its latent obscurities. 'It is common usage to call "monsters" an unfamiliar concord of dissonant elements: the centaur, the chimera, are thus defined for those without understanding. I call "monster" all original inexhaustible beauty' (Jarry, 1895; cited in Frascari, 1991, p. 33).

In this chapter we have been interested in Frascari's concept of the technographic and how this can be put into new syzygetical relation to the pataphysical imagination of Jarry's inventions. Which is to say, put into new syzygetical relations with the time-machine and with *Clinamen*, This is also to offer an equivalence for these in what Deleuze describes, using the neologism – or monstrously hybrid portmanteau word – *objectile*. This is, of course, a retrospective distortion of Frascari's technographic terms, one which is to see objectility as anticipating the technological object in a possible world shared by all those involved as designers and fabricators of the real. That is, this construing of Frascari's concern with the imaginary solution – with its anticipatory plagiarism from a similar source as Deleuze's – seems to anticipate a modulatory *objectile* power conferred upon the technographic artefact. Frascari writes (1991, p. 94), 'between a real artifact and the instrumental icon in the mind of someone involved in a building trade related to its construction'.[63] A list of such technographies of pataphysical machines for a future media study of anticipation, such as that empty set of categories by Sterling in his 'slider-bar of disbelief' (2013), might then include: technographies of architectural, theatrical and product models of the essay film; the actualization of events in constructed photography; diegetic prototypes for unfilmed scripts and scenarios of lost films; numerical, optical and notational speculation; reconstructed and ekphrastic archaeologies; and the anticipatory plagiarism of pataphysical prototypes.

Concluding Remarks

Whilst admitting of significant divergences between such modes of the making of anticipatory artefacts as the docu-fictions of tube-tracts[64] and the printed renaissance architectural treatise the *vedute* paintings of 'made up' architectural scenes of partly or completely imaginary, or at last imagined, elements. What are known as *capricci* and *vedute ideate*, or *veduta di fantasia*, while distanced from the field of anticipation studies are related nonetheless to demonstrations of the pataphysical *Clinamen*, the painting machine invented by Faustroll. Demonstrations of both the vistas it paints and those vistas onto future worlds to which travel is afforded by Faustroll's time machine, when one is constructed. At the closing stage of this discussion, such architecture fictions and their pataphysical 'promenades' seem absent from the confluence of creativity and anticipation. They do exemplify a phylogenetic relation though and are of a kind, as demonstrations of fiction as method – familiar to makers of anticipations – and in the ways in which these technographic fictions become real.

It is with this sense of objectility, in which an architectural fiction becomes real, that we can conclude with a final syzygetical relation: that of Le Lionnais's notion of Anticipatory Plagiarism and Luciana Parisi's mechanism of anticipation on which comment was earlier reserved. Parisi noted control as a mechanism of anticipation, whereby the 'apprehension of unknown variables indirectly works to determine the reality of the present' (Parisi, 2013, p. 90).[65] In leaving the resolution of the syzygy of relations open, the technogenetic modulation of its objectility, as the encounter of a control held in anticipation, has a certain pataphysical equivalence with the preresponse of the anticipatory plagiarist and the becoming of a predictable uncertainty. Such an openness leaves us in a state of over-stimulated analepsis, a forward looking and affective condition, an active reorientation of the temporality of influence (retrospective construing). We are fully in the swerve of causality or the entailment of its influence. Le Lionnais proposes that in Anticipatory Plagiarism there is a transformation by projection (*stéganographie*) with reference both to the protagonists moving forward and backward through time – in both analepsis and prolepsis. And in which the anticipatory plagiaristic relation between its anoulipistic discovery and where, once again, the modulatory powers of anticipation create a synthoulipistic invention. That is to say, by the act of some clinamen, an 'anoulipism' (the analysis of a past creative constraint) inspires a 'synthoulipism'. There is in this making of the anticipatory artefact an intensive synthesis of interjective scenes and the becomings of the anticipatory power of the *objectile*.

Notes

1 As is customary, I do not use the apostrophe in 'Pataphysics except where I refer specifically to the *Collège de 'Pataphysique*. Jarry insisted that the apostrophe was used to avoid a simple pun.
2 Notably the British new wave of Science Fiction of the late 1950s and 1960s.
3 For research on the *Collège*, the histories of pataphysics and details of Jarry's biography, the chapter draws upon the published works of Klieger-Stillman (1983); Fell (2005, 2010); Hugill (2012) and Brotchie (2015).
4 The reasons for occultation must remain somewhat esoteric but the extension of critique afforded to Pataphysics by Jorn (1961) as a 'religion in the making' might provide one justification. Another would be the prospect, the anticipation, of a future impact of de-occultation in the lunar context of millennial tension. Regardless, the intended occultation singularly failed to prevent the ongoing transcontinental adoption and development of pataphysical ideas. The *Oulipo* will be considered in this chapter as extending from this source.
5 The encounter with Lautréamont here follows Debord's (1956) argument, in *Methods of Détournement*, on the radicalism of Lautréamont's thought. The reprint of *Poésies II* in Aragon et al. (1919) provides the complete text and serves to contextualize its reception.
6 Voluntarily, honorarily and posthumously.
7 *Clinamen*, in title case and italicized, will be used throughout the chapter to refer to Jarry's *Clinamen*, to avoid confusion with the philosophical discussion on the concept.
8 'Mobile' in the sense of artist Mieke Bal's (2002) notion of the mobility of concepts.

9 For a survey of the 27 equivalent books see Fisher (2000).
10 Likewise, when it comes to consideration of a pataphysical realm, the realms of art, science and philosophy are referred, in *What is Philosophy?* as three 'chaoids'. These are three sisters – three daughters of chaos – each one holding dominion over one of three realms, but one realm each and no more, one each from art, science and philosophy (Deleuze and Guattari, 1994, pp. 204–208).
11 See Shattuck (1988) and Shattuck and Watson Taylor (1960, 1965). This current chapter relies on the principle of clinamen to contain both error, combinatorial potential, time-anomaly, indeterminacy and chance. I provide a revised clinamen redux, updating Motte's (1986), in Hales (2015). A detailed analysis of the clinamen in its pataphysical contexts, as well as in the context of Deleuze (1990, pp. 303–320), is given in Hales (2019). For provisional context, useful to here, the *Clinamen* of pataphysicians is: first, the concept 'clinamen' associated with Epicureanism after Lucretius's poem *De Rerum Natura*, which requires a swerve (clinamen) to disrupt the fall of atoms. Second, *Clinamen* is the painting machine invented by the pataphysician Dr Faustroll. Brotchie (2015), amongst others, attributes the pataphysical *Clinamen* to an influence on Jarry of his philosophy teacher, Henri Bergson. As Bergson (1884, 1959) had written on Lucretius's poem, this line of influence seems to be at least likely.
12 Editors' note. For another reference to hylomorphism in Aristotle and Simondon, see Brassett in this volume.
13 For an excellent account of this within a discourse on architectural machines traced back to antiquity, see Olshavsky (2012).
14 Indeed, Andrew Hugill suggests that Deleuze is not considered pataphysical enough by the *Collège* (Personal Communication, 'Speculative Hardware' Seminar, *Algomech Festival*, 2017).
15 Deleuze (1961) further reinforces the significance.
16 In the sense of anticipatory speculative fiction and anticipation studies interested in the design of technical objects.
17 See also Deleuze and Guattari (1988, p. 98) for a repetition of this and in which a 'willing truth' and a 'willing error' are coincident.
18 Sokal and Bricmont (1998) target 'postmodern' continental philosophy, or what they suggest should rather be called 'philosophico-literary criticism'. More than one philosopher of mathematics has refuted Sokal and Bricmont's claims, however – especially those against Deleuze's use of infinitesimals. See, for example, Johnson (2017, p. 45) on Deleuze's intentional and correct mathematical implementation of classical calculus.
19 For points of entry into Jarry's black humour see Greenfeld (1988) and Breton (2001).
20 Scholz (1999) notes that Riemann's concept of 'the manifold' was inaugurated in Riemann's *habilitation* presentation (1854).
21 On Bergson's relation to Riemann, see Durie (2004).
22 Lovecraft seems not to acknowledge such a source. However, Borges (1977) notes the possibility and provides a kind of missing link between Lovecraft and Jarry.
23 Henderson here, in fact, suggests this of Duchamp, but notes that Duchamp inherits this enjoyment from Jarry.
24 For a complex, non-pataphysical, discussion on relations between possible worlds theories and those of fictional worlds in literary theory, see Ronen (1994).
25 For the remainder of this chapter, *Clinamen* (italicized and capitalized) will refer to Jarry's Painting Machine (Jarry, 1996, pp. 88–94); and 'clinamen' will be Lucretius's (1977).
26 Strictly speaking, this anticipatory plagiarism is foreshadowed by that earlier one of Epicurus.
27 See Brown (2013, pp. 207–233).
28 Such an anticipatory politics might be gathered from Massumi (2007) and accelerationism in Mackay and Avanessian (2014).

29 See – and read with opposition in mind – Land (2020) in which he develops the notion of the blockchain as a philosophy of time, and from which this, in syzygy with Hilbert, Lobachevsky and Jarry, is expropriated.
30 For definitions of the *Gidouille*, see Lowe (2006). The spiral form of modernity is studied in Israel (2017); Ireland (2017) develops Land's 'templexity' in relation to the modernist poetics of *Blast* and Vorticism.
31 Berardi suggests 1977, but a less precise end of the future is signalled for example in the shift in the EU foresight programme from futures methods to its discussion of the discipline of 'Anticipations'. The notion of anticipations here is suggestive of the management of risk, whilst at the same time contains a sense of suspension and an echo of the short-lived Belgian SF magazine *Anticipations* (1945–46): reprinting American and British content from *Tales of Wonder* for a francophone audience, and instantly conveying both 'hope and fear' in post-war Europe.
32 The literature on the *Hypnerotomachia Poliphili* is extensive. Principal amongst the sources for this chapter is Lefaivre (1997).
33 A claim repeated by Rajagopal (2013).
34 The 'key premise of schizoanalysis is that desire is productive, that the world as it exists is literally a product of desire, that desire composes' (Buchanan et al., 2015, p. 4).
35 The Rabelaisian and Shakespearean aspects of this have received much scholarly attention; see, for example, Shattuck and Watson Taylor (1965, pp. 82–85). A medievalism can also be found in Jarry (1884–85) and (1886); see also Berger (1990).
36 This quotation has further significance to what follows for its reference to Korzybski's General Semantics. This same quotation is used as an epigram to Winetrout's (1970) review of *The Future of the Future* by John McHale. McHale is interesting too as a former member of the Independent Group and collaborator with the visionary engineer Buckminster Fuller; his book provides a survey of the birth of futures studies.
37 With thanks to Jamie Brassett for this observation.
38 An anticipatory plagiarism here of Bratton's (2016) work on the confluence of computational and political power.
39 See Wagner (1977) who argues that Schütz's phenomenology is informed by Bergson's philosophy of *durée*.
40 See also Raby (2013). Dunne's instrumental book selections contrast with Raby's science fiction inflected choices: *The Book of Dave* (Self, 2006), *Codex Seraphinianus* (Serafini, 1981), *Oryx and Crake* (Atwood, 2003), *Positron* (Atwood, 2012), *Super-Cannes* (Ballard, 2000), *The Wind-up Bird Chronicle* (Murakami, 2003).
41 Listed by Koolhaas et al. (2004, pp. 73–83 and 510–513) under the heading 'Universal Modernization Patents' the fifteen fictional patents are: *Social Condenser* (Parc de la Villette), *Strategy of the Void I* (Melun-Sénart), *Timed Erasures* (La Defense), *Loop Trick* (Kunsthal), *Strategy of the Void II* (TGB), *Stacked Freedoms* (Karlsruhe), *Inside-Out City* (Jussieu), *Disconnect* (Cardiff Bay Opera), *Everywhere and Nowhere* (Bordeaux), *Variable-Speed Museum* (Tate), *Inertness Modified* (Universal HQ), *Tall and Slender* (Hyperbuilding), *Skyscraper Loop* (CCTV), *Cake-Tin Architecture* (TVCC), *The End of the Road* (CBD Beijing).
42 A parafictional company 'established' in 1944.
43 For examples of how these actualize in design fictional artefacts, see Manaugh (2013).
44 See Simone's (2018) monograph on Webb for an account of this project. The 'Pepperland' reference is made by Sorkin (1991, pp. 203, 340), the association of this to The Beatles' Pepperland as 'quizzical and pataphysical' is not confirmed or denied by Webb (personal communication), but it seems to be the case that a Duchampian optics is involved.
45 A term also used for a programme of events held in 1997 addressing the interface between physical and 'virtual worlds' for the RCA by Coates.
46 Jarry's (1894) original French runs: '*La Pataphysique est la science de ces êtres et engins actuels ou futurs avec le Pouvoir de leur Usage (*discipulus*)*'. *Engins* as cunning devices or machines

might be taken, pataphysically speaking, as equivalents; and Jarry's beings and current or future devices may be deliberately misread as the becomings of actual or future beings and machines.

47 Serres (2000, p. 84) sees 'numerous intersections with Lucretius' in the book on hydraulics; and Weiner (2016, pp. 133–161) regards Vitruvius as a selective reader of Lucretius.

48 One which might be compared to arguments about genres or modes of science fiction writing.

49 Such as that by Richard Hamilton significant for its circulation in the wake of the Independent Group and their influence on British interdisciplinary cultures.

50 In an interview, Perec (1983, p. 70) says it is necessary 'to introduce an error into the system because when one establishes a system of constraints, there must also be anti-constraint in it. The system of constraints – and this is important – must be destroyed'. Similarly, for Calvino, another member of the *Oulipo*, there are combinatorics and the anti-combinatory (Duncan, 2019, p. 121).

51 Roche's observation, from the position of architectural practice, has its own anxieties of influence in Spiller's *Communicating Vessels* (1998). For Spiller's account of his debt to pataphysics and the *Hypnerotomachia Poliphili*, see Hales (2013).

52 I have discussed this convergence previously (Hales 2015) and am currently developing research on pataphysical machines with reference to Jarry's Time-Machine and Carrouges (1985).

53 The field of software and platform studies is interesting in his respect. See, for example, the relation Galloway (2012, pp. 25–53) builds between a painting as doorway and window with the computer screen and not analogically but as interface and Deleuze's 'tabular grid' (1993, p. 27).

54 This special issue of *L'Esprit Créateur*, titled 'Machines in Textes [sic], Texts as Machines', contains essays on *Oulipo*, Perec, a review of Stillman's book on Jarry, as well as Guattari's essay 'Schizoanalysis'.

55 Clifford's 'Ethnographic Surrealism' (1981) mentions Ernst. This essay is an underutilized resource for ethnographic designers and design ethnographers.

56 The reference here is to Eisenman's (2000, pp. 582–585) *architectural* sense of superposition, as much as a quantum sense.

57 This is the date of Scarpa's commission. Scarpa continued drawing the project until his death and his own internment in the cemetery at Brion; construction works were completed in 1978. The architect and collaborator with Scarpa, Pietropoli (2008), is a primary source for future research of the Scarpa archive, which, no doubt, is pataphysically inordinate.

58 This is given the following discussion in the *Baudrillard Dictionary* under its 'pataphysics' entry by Genosko (2010, p. 151): 'The idea that Baudrillard's wily and active object that takes its revenge on the subject is a pure and enigmatic crystal that functions in Baudrillard's thought as a Jarryesque principle: the magical "physics-stick" [sic] explicated by Faustroll and whose spinning flips back and forth from plus to minus, revealing the equality of opposites, infecting all registers with undecidability and randomness'.

59 'Cyberspace, not so long ago, was a specific elsewhere, one we visited periodically, peering into it from the familiar physical world. Now cyberspace has everted. Turned itself inside out. Colonised the physical' (Gibson, 2010).

60 The sense of 'drawing' used here, departs from Frascari's (2017) and, instead, employs an inflection given by Deleuze and Guattari across their work together: a drawing forth, a conjectural channelling of matter-flow. Thus, it regards designing as actualizating the virtual.

61 The phrase 'cargo cult objectiles' was deployed in the writing of curator Dane Sutherland for the exhibition and audio zine *Exploit.zzxjoanw.Gen* (2014) and in relation to O'Sullivan's *Plastique Fantastique* art group. I borrow from Ballard's sense of the science

fiction cargo cult in Frost (2013). For more on divination, design and technology see Marenko (2015).

62 This coinage of Deleuze and Guattari's draws upon the Greek *schizo* (σχίζω): 'to cleave', 'to split', 'to divide' (Liddell and Scott, 1889, p. 787).

63 See Ridgeway (2015) for a presentation of Frascari's sense of technography in relation to the imagination and outside of the Deleuze and Guattarian design philosophy proposed by this chapter.

64 Thomson and Maglioni and The Otolith Group are sources for further engagement with these terms.

65 Parisi's citation is from Massumi (2007).

References

Aragon, L., Breton, A. and Soupault, P. (eds.) (1919) *Littérature*, 3(Mai). Available at: http://sdrc.lib.uiowa.edu/dada/litterature/3/index.htm. (Accessed: 31 December 2020).

Atwood, M. (2012) *I'm Starved for You*. San Francisco, CA: Byliner Inc.

Atwood, M. (2003) *Oryx and Crake*. London: Bloomsbury.

Bal, M. (2002) *Travelling Concepts in the Humanities: A Rough Guide*. Toronto: Toronto University Press.

Ballard, J.G. (2000) *Super-Cannes*. London: Flamingo.

Baudrillard, J. (1993) 'Pataphysics of the Year 2000', *CTheory*, a011. Trans. C. Dudas. Available at: http://ctheory.net/ctheory_wp/pataphysics-of-year-2000/ (Accessed: 01 May 2020).

Baudrillard, J. [1981] (1994) *Simulacra and Simulation*. Trans. S.F. Glaser. Ann Arbor, MI: University of Michigan Press.

Berardi, F. (2011) *After the Future*. Trans. A. Bove, M. Cooper, E. Empson Enrico, G. Mecchia, and T. Terranova. Eds. G. Genosko and N. Thoburn. Chico, CA: AK Press.

Berger, H. Jr. (1990) *Second World and Green World: Studies in Renaissance Fiction-Making*. Berkeley, CA: University of California Press.

Bergson, H. (1959) *Philosophy of Poetry: The Genius of Lucretius*. Ed. and Trans. W. Baskin. New York, NY: Philosophical Library.

Bergson, H. (1944) *Creative Evolution*. Trans. A. Mitchell. New York, NY: Henry Holt.

Bergson, H. (1935) *The Two Sources of Morality and Religion*. Trans. R.A. Andra and C. Brereton, with W. Horsfall Carter. London: Macmillan and Co. Ltd.

Bergson, H. (1884) *Extraits de Lucrèce: avec un commentaire, des notes et une étudie sur la poésie, la philosophie, la physique, le texte et la langue de Lucrèce*. 11th edn. Paris: Librairie Delagrave.

Berressem, H. (2005) 'Incerto *Tempore* Incertisque Locis: The Logic of the Clinamen and the Birth of Physics', in N. Abbas (ed.), *Mapping Michel Serres*. Ann Arbor, MI: University of Michigan Press, pp. 51–50.

Birksted, J.K. (2009) *Le Corbusier and the Occult*. Cambridge, MA and London: The MIT Press.

Bogard, W. (1996) *The Simulation of Surveillance*. Cambridge: Cambridge University Press.

Borges, J.L. (1977) *The Book of Sand*. Trans. N.T. di Giovanni. New York, NY: Dutton.

Bratton, B.H. (2016) *The Stack: On Software and Sovereignty*. Cambridge, MA and London: The MIT Press.

Breton, A. (ed.) (2001) *Anthology of Black Humour*. Trans. M. Polizzotti. San Francisco, CA: City Lights Publishers.

Brotchie, A. (2015) *Alfred Jarry: A Pataphysical Life*. Cambridge, MA and London: The MIT Press.

Brotchie, A. (2003) *Call for Patents*. Office of Patentry, Department of Reconstructive Archaeology. London: London Institute of 'Pataphysics.

Brown, D. (2013) *The Poetry of Victorian Scientists: Style, Science and Nonsense*. Cambridge Studies in Nineteenth-Century Literature and Culture, No. 83. Cambridge: Cambridge University Press.

Buchanan, I., Matts, T. and Tynman, A. (eds.) (2015) *Deleuze and the Schizoanalysis of Literature*. London: Bloomsbury.

Cache, B. (2003) 'Philibert De L'Orme Pavilion: Towards an Associative Architecture', *Architectural Design*, 73(2), Special Issue: 'Surface Consciousness', (March/April), pp. 20–25.

Cache, B. (1995) *Earth Moves: The Furnishing of Territories*. Trans. A. Boyman. Ed. M. Speaks. Cambridge, MA and London: The MIT Press.

Carrouges, M. (1986) 'Les Machines pataphysiques de Maldoror et leurs groupes de transformation', *L'Esprit Créateur*, 26(4), Special Edition: 'Machines in Textes, Texts as Machines', (Winter), pp. 16–25.

Carrouges, M. (1985) 'Machines pataphysiques pour l'au-delà', *Les Etudes Philosophiques*, 1, pp. 77–89.

Carrouges, M. (1975) 'Les Machines Célibataires', in Szeemann, H. (ed.), *Le Macchine Celibi/The Bachelor Machines*. New York, NY: Rizzoli.

Clifford, J. (1981) 'On Ethnographic Surrealism'. *Comparative Studies in Society and History*, 23(4), pp. 539–564.

Coates, N. and Tschumi, B. (1983) *Discourse of Events*. London: Architectural Association.

Connor, S. (2017) *Dream Machines*. London: Open Humanities Press.

Connor, S. (2009) *Michel Serres: The Hard and the Soft* [Transcript]. Talk given at the Centre for Modern Studies, University of York, 26 November. Available at: http://stevenconnor.com/hardsoft.html (Accessed: 31 May 2020).

Cook, P. (1983) 'Strange Pavilions of the Mind. The Work of Diploma Unit 10, 1973–1983', *AA Files*, 4, (July), pp. 102–107.

Cramer, F. (2005) *Words Made Flesh: Code, Culture, Imagination*. Rotterdam: Piet Zwart Institute.

Debord, G. (1956) *Methods of Détournement*. Available at: http://library.nothingness.org/articles/SI/en/display/3 (Accessed: 26 May 2020).

Deleuze, G. (2004) *Desert Islands and Other Texts, 1953–1974*. New York, NY: Semiotext(e).

Deleuze, G. (1997) *Essays Critical and Clinical*. Trans. D.W. Smith and M.A. Greco. Minneapolis, MN: University of Minnesota Press.

Deleuze, G. (1994) *Difference and Repetition*. Trans. P. Patton. London: The Athlone Press.

Deleuze, G. (1993) *The Fold: Leibniz and the Baroque*. Trans. T. Conley. London: The Athlone Press.

Deleuze, G. (1990) *The Logic of Sense*. Trans. M. Lester and C. Stivale. Ed. C.V. Boundas. London: The Athlone Press.

Deleuze, G. (1983) *Nietzsche and Philosophy*. Trans. H. Tomlinson. London: The Athlone Press.

Deleuze, G. (1961) 'Lucrece et le naturalisme', *Les Etudes philosophiques, Nouvelle Série, 16e Année*, 1, (January–March), pp. 19–29.

Deleuze, G. and Guattari, F. (1994) *What is Philosophy?* Trans. H. Tomlinson and G. Burchell. London and New York, NY: Verso.

Deleuze, G. and Guattari, F. (1988) *A Thousand Plateaus. Capitalism and Schizophrenia 2*. Trans. B. Massumi. London: The Athlone Press.

Deleuze, G. and Guattari, F. (1984) *Anti-Oedipus. Capitalism and Schizophrenia 1*. Trans. R. Hurley, M. Seem and H.R. Lane. London: The Athlone Press.

Dery, M. (2011) *Architecture Fiction: Premonitions of the Present*. Available at: http://thoughtcatalog.com/2011/architecture-fiction-premonitions-of-the-present (Accessed: 31 May 2020).

Dosse, F. (2010) *Gilles Deleuze and Félix Guattari. Intersecting Lives*. Trans. D. Glassman. New York, NY: Columbia University Press.

Dubbelboer, M. (2017) *The Subversive Poetics of Alfred Jarry: Ubusing Culture in the Almanachs Du Pere Ubu*. London and New York, NY: Routledge.

Duchamp, M. (1915–23) *La mariée mise à nu par ses célibataires, même* (The Bride Stripped Bare by Her Bachelors, Even) [Oil, Varnish, Lead Foil, Lead Wire and Dust on Two Glass Panels]. Philadelphia, PA: Philadelphia Museum of Art.

Duncan, D. (2019) *The Oulipo and Modern Thought*. Oxford: Oxford University Press.

Dunne, A. (2013) *Booklist*. Available at: http://www.designersandbooks.com/designer/booklist/anthony-dunne (Accessed: 26 May 2020).

Dunne, A. (1999) *Hertzian Tales: Electronic Products, Aesthetic Experience and Critical Design*. Cambridge, MA and London: The MIT Press.

Dunne, A. and Raby, F. (2013) *Speculative Everything Design, Fiction, and Social Dreaming*. Cambridge, MA and London: The MIT Press.

Dunne, A. and Raby, F. (1995) 'Fields and Thresholds', *Architectural Design: Architects in Cyberspace*, Profile No. 118, pp. 60–65.

Durie, R. (2004) 'The Mathematical Basis of Bergson's Philosophy', *Journal of the British Society for Phenomenology*, 35(1), pp. 54–67.

Eisenman, P. (2000) 'Moving Arrows, Eros', in Hays, M. (ed.), *Architecture Theory Since 1968*. Cambridge, MA and London: The MIT Press, pp. 582–585.

Evans, R. (2000) *The Projective Cast: Architecture and Its Three Geometries*. Cambridge, MA and London: The MIT Press.

Evans, R. (1997) *Translations from Drawing to Building and Other Essays*. London: Architectural Association.

Evans, R. (1983) 'In Front of Lines That Leave Nothing Behind', *AA Files*, 6, pp. 88–96.

Ernst, M. (1948) '*What is the Mechanism of Collage?*', in Ernst, M., *Beyond Painting*. Trans. D. Tanning. New York, NY: Wittenborn and Schultz.

Faustroll, Dr. (ed.) (2006) *Pataphysica 4: Pataphysica e Alchimia 2*. New York, NY: iUniverse.

Fell, J. (2010) *Alfred Jarry*. London: Reaktion Books.

Fell, J. (2005) *Alfred Jarry, an Imagination in Revolt*. Cranbury, NJ: Fairleigh Dickinson University Press.

Fisher, B. (2000) *The Pataphysician's Library: An Exploration of Alfred Jarry's 'Livres Pairs'*. Oxford: Oxford University Press.

Foulc, T., Lacaze, D., and Brotchie, A. (eds.) (2014) 'Pataphysical Equivalence', *The Journal of the London Institute of 'Pataphysics*, 10, (December), p. 142.

Frascari, M. (2017) *Marco Frascari's Dream House: A Theory of Imagination*. Ed. M. Goffi. New York, NY and London: Routledge.

Frascari, M. (1991) *Monsters of Architecture: Anthropomorphism in Architectural Theory*. Savage, MD: Rowman and Littlefield.

Frascari, M. (1989) 'The *Particolareggiamento* in the Narration of Architecture', *Journal of Architectural Education*, 43(1), (Autumn), pp. 3–11.

Frascari, M. (1985) 'Carlo Scarpa in Magna Graecia: The Abatellis Palace in Palermo', *AA Files*, 9, pp. 3–9.

Frascari, M. (1983) 'The Tell-the-Tale Detail', in Deely, J.N. and Lenhart, M.D. (eds.), *Semiotics 1981*. New York, NY: Plenum Press/Springer US, pp. 325–336.

Frost, A. (2013) 'Cosmic Sentinels and Spiral Jetties: J.G. Ballard, Robert Smithson and Tacita Dean', *Ballardian*. Available at: http://ballardian.com/cosmic-sentinels-spiral-jetties-ballard-smithson-dean (Accessed: 31 May 2020).

Galloway, A.R. (2012) *The Interface Effect*. London: Polity Press.

Genosko, G. (2010) ''Pataphysics', in Smith, R.G. (ed.), *Baudrillard Dictionary*. Edinburgh: Edinburgh University Press, pp. 150–153.

Gibbs, L. (1972) 'William Ames's Technometry', *Journal of the History of Ideas*, 33(4), pp. 615–624.

Gibson, W. (2010) 'Google's Earth', *New York Times*, 31 August. Available at: https://www.nytimes.com/2010/09/01/opinion/01gibson.html (Accessed: 31 May 2020).

Greene, D. and Hardingham, S. (2008) *L.A.W.U.N Project #19: Situated Technologies and the Picturesque*. London: AA Publications.

Greenfeld, A. (1988) 'Jarry, Ubu and "*Humour Noir*"', *Romance Notes*, 28(3), (Spring), pp. 227–234.

Guattari, F. (1995) *Chaosmosis. An Ethico-Aesthetic Paradigm*. Trans. P. Bains and J. Pefanis. Sydney: Power Publications.

Guattari, F. (1984) *Molecular Revolution. Psychiatry and Politics*. Trans. R. Sheed. Harmondsworth, Middx.: Penguin Books Ltd.

Hales, D. (2019) *Assimilating the Deleuzian Objectile to a Pataphysical Clinamen: A Pataphysical Objectile for Design Research*. Unpublished PhD Thesis, Royal College of Art.

Hales, D. (2015) 'Re-designing the Objectile', in Marenko, B. and Brassett, J. (eds.), *Deleuze and Design*. 'Deleuze Connections' Series. Edinburgh: Edinburgh University Press, pp. 139–172.

Hales, D. (2013) 'Design Fictions an Introduction and Provisional Taxonomy', *Digital Creativity*, 24(1), pp. 1–10.

Harbison, R. (2000) *Eccentric Spaces*. Cambridge, MA and London: The MIT Press.

Henderson, L.D. (2018) *The Fourth Dimension and Non-Euclidean Geometry in Modern Art*. Rev. edn. Cambridge, MA and London: The MIT Press.

Henderson, L.D. (2005) *Duchamp in Context: Science and Technology in the Large Glass and Related Works*. Princeton, NJ: Princeton University Press.

Hugill, A. (2014) 'Pataphysics and Computation', *Philadelphia à la Pataphysique Conference*. Kislak Center for Special Collections, Rare Books, and Manuscripts at Van Pelt Library University of Pennsylvania, Philadelphia, PA, 21–23 March.

Hugill, A. (2012) *'Pataphysics: A Useless Guide*. Cambridge, MA and London: The MIT Press.

Irie, K. (1993) 'AA Intermediate Unit 3: Objects in the Landscape', *AA Files*, 25, pp. 80–83.

Ireland, A. (2017) 'The Poememenon: Form as Occult Technology', *Urbanomic Document* 027. Available at: https://www.urbanomic.com/document/poememenon/ (Accessed: 31 May 2020).

Israel, N. (2017) *Spirals, The Whirled Image in Twentieth-Century Literature and Art*. Modernist Latitudes Series. New York, NY: Columbia University Press.

Jarry, A. (2006) *Three Early Novels*. Eds. A. Brotchie and P. Edwards. Trans. A. Lykiard, S.W. Taylor, and P. Edwards. London: Atlas Press.

Jarry, A. (2001) 'Present and Future Visions', in Jarry, A., Brotchie, A. and Edwards, P. (eds.), *Adventures in 'Pataphysics*. Trans. A. Melville. London: Atlas Press.

Jarry, A. (1996) *Exploits and Opinions of Dr. Faustroll, Pataphysician*. Trans. S.W. Taylor. Cambridge, MA: Exact Change.

Jarry, A. (1992) *Caesar Antichrist*. Trans. A. Melville. London: Atlas Press.

Jarry, A. (1989) *Days and Nights: Novel of a Deserter*. Trans. A. Lykiard. London: Atlas Press.

Jarry, A. (1985) *Messalina: A Novel of Imperial Rome*. Trans. J. Harman. London: Atlas Press.

Jarry, A. (2009) *The Supermale*. Trans. R. Gladstone and B. Wright. Cambridge, MA: Exact Change.

Jarry, A. (1968) 'How to Construct a Time Machine', in Moorcock, M. (ed.), *The Traps of Time*. London: Rapp and Whiting.

Jarry, A. (1901) *Almanach Illustré de Pere Ubu [XXe Siecle]*, No. 2. Paris: Le Castor Astral.

Jarry, A. (1903) 'La Vérité bouffe', *La Plume*, 339, (June), pp. 646–648.

Jarry, A. (1894) 'Visions actuelles et futures'. Available at: http://claude.ognois.pagesperso-orange.fr/ubu02.htm#Visions%20actuelles%20et%20futures (Accessed: 3 January 2021).

Johnson, R.J. (2017) *The Deleuze-Lucretius Encounter*. Edinburgh: Edinburgh University Press.

Jorn, A. (1961) 'Pataphysics: A Religion in the Making', *Internationale Situationiste*, 6. Available at: https://www.cddc.vt.edu/sionline/si/pataphysics.html (Accessed: 31 May 2020).

Kafka, F. (1948) *In the Penal Colony*. Trans. W. and E. Muir. New York, NY: Schocken Books.

Kiesler, F. (1937) 'Design-Correlation [Marcel Duchamp's Large Glass]', *Architectural Record*, 81(5), (May), pp. 53–60.

Klieger-Stillman, L. (1983) *Alfred Jarry*. Boston, MA: Twayne Publishers.

Klieger-Stillman, L. (1982) 'Modern Narrative Techniques: Jarry, the Pre-Text', *SubStance*, 11(3), (Issue 36), pp. 72–81.

Klee, P. (1922) *Twittering Machine* [Oil Transfer Drawing, Watercolour and Ink on Paper with Gouache and Ink Borders on Board]. New York, NY: Museum of Modern Art.

Krieger, M.H. (2015) *Doing Mathematics: Convention, Subject, Calculation, Analogy*. 2nd edn. Singapore: World Scientific.

Koolhaas, R., McGetrick, B. and Office for Metropolitan Architecture (2004) *Content: Triumph of Realization*. Taschen: Köln.

Korzybski, A. (1958) *Science and Sanity: An Introduction to Non-Aristotelian Systems and General Semantics*. 4th edn. Lakefield, CT: Institute of General Semantics.

Land, N. (2020) 'Crypto-Current, An Introduction to Bitcoin and Philosophy', *Sum*, (10.2). Available at: http://sumrevija.si/en/sum10-2-nick-land-crypto-current-an-introduction-to-bitcoin-and-philosophy/ (Accessed: 31 May 2020).

Land, N. (2014) *Templexity: Disordered Loops through Shanghai Time*. Shanghai: Urbanatomy Electronic.

Land, N. (1995) 'Cyberspace Anarchitechture as Jungle-War', *AD Architectural Design*, Profile No.118, Special Issue: 'Architects in Cyberspace', pp. 58–59.

Lautreamont, Comte de (1994) *Maldoror*. Trans. A. Lykiard. Cambridge, MA: Exact Change.

Lautreamont, Comte de (1919) 'Poésies II', *Littérature*, 3, pp. 8–24.

Lefaivre, L. (1997) *Leon Battista Alberti's Hypnerotomachia Poliphili. Re-Cognizing the Architectural Body in the Early Italian Renaissance*. Cambridge, MA and London: The MIT Press.

Lewis, D. (1986) *On the Plurality of Worlds*. Oxford: Blackwell.

Libeskind, D. (1981) *Between Zero and Infinity: Selected Projects in Architecture*. New York, NY: Rizzoli.

Liddell, H.H. and Scott, R. (1889) *An Intermediate Greek–English Lexicon*. Oxford: Oxford University Press.

Lightsey, S. (2007) *Manmade Marvels in Medieval Culture and Literature*. New York, NY: Palgrave Macmillan US.

L'Orme, P. de (1567) *Premier tome de l'architecture. Paris: chez Federic Morel.*
Lowe, N. (2006) 'Corruption Contraption. The Gidouille as transubstantiation device', in Clements, C. (ed.), *Some Machines of Pataphysics. Pataphysica*, vol. 3. Athens, GA: mNemonic iNk, pp. 131–149.
Lucretius (1977) *The Nature of Things*. Trans. F.O. Copley. New York, NY: Norton.
Mackay, R. and Avanessian, A. (eds.) (2014) *Accelerate*. Falmouth, UK: Urbanomic.
Manaugh, G. (ed.) (2013) *Landscape Futures: Instruments, Devices and Architectural Inventions*. New York, NY: Actar.
Marenko, B. (2015) 'When making becomes divination: Uncertainty and contingency in computational glitch-events', *Design Studies*, 41, pp. 110–125.
Massumi, B. (2007) 'Potential Politics and the Primacy of Preemption', *Theory and Event*, 10(2). doi:10.1353/tae.2007.0066.
Moorcock, M. (1965) 'Pleasure Garden of Felipe Sagittarius', *New Worlds SF*, 49(154), (September), pp. 100–114.
Motte, W. (ed.) (2007) *Oulipo: A Primer of Potential Literature*. Trans. W. Motte. French Literature Series. McLean, IL, London, and Dublin: Dalkey Archive Press.
Motte, W.F. (1986) 'Clinamen Redux', *Comparative Literature Studies*, 23(4), pp. 263–281.
Murakami, H. (2003) *The Wind-up Bird Chronicle*. Trans. J. Rubin, with participation of H. Murakami. London: Vintage.
Nietzsche, F. (2010) *The Gay Science: With a Prelude in Rhymes and an Appendix of Songs*. Trans. W. Kaufmann. New York, NY: Knopf Doubleday Publishing Group.
Noel-Smith, K. (2016) *Freud on Time and Timelessness*. Cham, CH: Springer Nature.
Olshavsky, P. (2012) *Questions Concerning Architectural Machines: or 'Pataphysics in Early Modern Architecture*. Unpublished PhD Thesis, McGill University. Available at: http://digitool.library.mcgill.ca/thesisfile110410.pdf (Accessed: 06 April 2020).
Olshavsky, P. (2011) 'Situating Pataphysical Machines: A History of Architectural Machinations', in Perez-Gomes, A. and Parcell, S. (eds.), *Chora Six: Intervals in the Philosophy of Architecture*. Montreal: McGill-Queen's University Press, pp. 181–210.
Parisi, L. (2013) *Contagious Architecture: Computation, Aesthetics, and Space*. Cambridge, MA and London: The MIT Press.
Peirce, C. S. (1955) *Philosophical Writings of Peirce*. Ed. J. Buchler, New York: Dover Publications.
Perec, G. (1983) 'Entretien Georges Perec/Ewa Pawlikowska', *Littératures*, 7, pp. 69–77.
Perez-Gomez, A. and Pelletier, L. (1997) *Architectural Representation and the Perspective Hinge*. Cambridge, MA and London: The MIT Press.
Phillips, G., Kaiser, P., Chon, D., and Rigolo, P. (eds.) (2016) *Harald Szeemann*, Museum of Obsessions. Los Angeles, CA: Getty Publications.
Plant, S. and Land, N. (1994) 'Cyberpositive', in Fuller, M. (ed.), *Unnatural: Techno-Theory for a Contaminated Culture*. London: Underground, pp. 3–10.
Rabelais, F. (1999) *The Complete Works of François Rabelais*. Trans. D. Frame. Berkeley, CA: University of California Press.
Raby, Fiona (2013) *Designers and Books: Fiona Raby*. Available at: http://www.designersandbooks.com/designer/bio/fiona-raby (Accessed: 31 May 2020).
Rajagopal, A. (2013) 'Five Compelling Works of Architecture Fiction', *Metropolis* (3 December). Available at: https://www.metropolismag.com/cities/five-compelling-works-of-architecture-fiction/ (Accessed: 3 January 2021).
Ramey, J. (2016) *Politics of Divination: Neoliberal Endgame and the Religion of Contingency*. London: Rowman and Littlefield International.
Ramey, J. (2014) 'Contingency without Unreason', *Angelaki*, 19(1), pp. 31–46.

Ridgeway, S. (2015) *Architectural Projects of Marco Frascari: The Pleasure of a Demonstration*. Ashgate Studies in Architecture Series. Farnham, Surrey, and Burlington, VT: Ashgate.

Roche, F. (2013) 'Matters of Fabulation: On the Construction of Realities in the Anthropocene', in Turpin, E. (ed.), *Architecture in the Anthropocene: Encounters Among Design, Deep Time, Science and Philosophy*. Ann Arbor, MI: Open Humanities Press/ Michigan Publishing, pp. 197–208.

Ronen, R. (1994) *Possible Worlds in Literary Theory*. Cambridge: Cambridge University Press.

Roussel, R. (1970) *Locus Solus: A Novel*. Trans. R.C. Cunningham. London: Calder and Boyer.

Salazar-Sutil, N. (2013) '"Set in Poland, that is to say Nowhere": Alfred Jarry and the Politics of Topological Space', in Fischer-Lichte, E. and Wihstutz, B. (eds.), *Performance and the Politics of Space: Theatre and Topology*. London and New York, NY: Routledge, pp. 114–126.

Scarpa, C. (1969) *Brion Cemetery* [Architecture]. Trivoli, Italy: San Vito d'Altivoli.

Scholz, E. (1999) 'The Concept of Manifold, 1850–1950', in James, I.M. (ed.), *History of Topology*. Amsterdam: Elsevier, pp. 29–64.

Schütz, A. (1945) *'On Multiple Realities'*, *Philosophy and Phenomenological Research*, 5(4), (June), pp. 533–576.

Self, W. (2006) *The Book of Dave*. London: Penguin.

Serafini, L. (1981) *Codex Seraphinius*. New York, NY: Rizzoli.

Serres, M. (2000) *The Birth of Physics*. Trans. J. Hawkes. Ed. D. Webb. Manchester: Clinamen Press.

Shattuck, R. (1996) 'Introduction', in Jarry, A. (ed.), *Exploits and Opinions of Dr. Faustroll, Pataphysician*. Trans. S.W. Taylor. Cambridge, MA: Exact Change, pp. vii–xviii.

Shattuck, R. [1955] (1988) *The Banquet Years: The Origins of the Avant-Garde in France, 1885 to World War I*. Rev. edn. New York, NY: Random House USA.

Shattuck, R. and Taylor, S.W. (eds.) (1965) 'Ubu Roi', Trans. B. Wright, in Shattuck, R. and Taylor, S.W. (eds.), *Selected Works of Alfred Jarry*. New York, NY: Grove Press.

Shattuck, R. and Taylor, S.W. (eds.) (1960) 'What is Pataphysics?', *Evergreen Review*, 4(13), p. 26.

Sylvester, J.J. (1853) 'On A Theory of the Syzygetic Relations of Two Rational Integral Functions', *Philosophical Transactions*, 143, pp. 407–543.

Simon, H. (1988) 'The Science of Design: Creating the Artificial', *Design Issues*, 4(1/2), Special Issue: 'Designing the Immaterial Society', pp. 67–82.

Simondon, G. (2009) 'The Position of the Problem of Ontogenesis', Trans. G. Flanders, *Parrhesia*, 7, pp. 4–16.

Simone, A. (ed.) (2018) *Michael Webb: Two Journeys*. Zurich: Lars Müller Publishers GmbH.

Sokal, A. and Bricmont, J. (1998) *Fashionable Nonsense*. London: Profile Books.

Sorkin, M. (1991) *Exquisite Corpses. Writing on Buildings*. London: Verso.

Spiller, N. (1998) *Communicating Vessels* [Architecture]. Private Collection.

Spiller, N. and Pearce, M. (eds.) (1995) *AD. Architectural Design*, Profile No. 118, 65, Special Issue: 'Architects in Cyberspace'.

Steil, L. (2013) *The Architectural Capriccio, Memory, Fantasy and Invention*. London: Ashgate.

Sterling, B. (2013) 'Design Fiction: the Design Fiction Slider Bar of Disbelief', *Wired*, 04 November. Available at: https://www.wired.com/2013/04/design-fiction-the-design-fiction-slider-bar-of-disbelief/ (Accessed: 31 May 2020).

Sterling, B. (2010) *Atemporality and The Passage of Time* [Lectures], European Graduate School. Available at: http://egs.edu/faculty/bruce-sterling/lectures (Accessed: 31 May 2020).

Sutherland, D. (2014) *Exploit.zzxjoanw*.Gen [Audiozine]. Punctum Records.
Szeemann, H. (1975) *The Bachelor Machines*. Exhibition held at Venice Biennale [Exhibition Catalogue].
Szeemann, H. (1965) *Light and Movement: Kinetic Art and New Tendencies in Architecture*. Exhibition held at Kunsthalle, Bern [Exhibition Catalogue].
Teyssot, G. (1994) 'The Mutant Body of Architecture,' in Diller, E. and Scofidio, R. (eds.), *Flesh: Architectural Probes*. Princeton, NJ: Princeton Architectural Press, pp. 8–35.
Vaihinger, H. (2009) *The Philosophy of 'As if': A System of the Theoretical, Practical and Religious Fictions of Mankind*. Trans. C. K. Ogden. Eastford, CT: Martino.
Varnelis, K. (2009) *Defence of Architecture (Fiction)*, 2 March. Available at: http://index.varnelis.net/blog/in_defense_of_architecture_fiction (Accessed: 06 May 2020).
Vico, G. (2001) *The New Science*. Trans. D. Marsh. Harmondsworth, Middx.: Penguin.
Vidler, A. (1992) *The Architectural Uncanny: Essays in the Modern Unhomely*. Cambridge, MA and London: The MIT Press.
Vitruvius's [c.30–15 BCE] (1914) *Ten Books on Architecture*. Trans. M.H. Morgan. Cambridge, MA: Harvard University Press.
Wagner, H. (1977) 'The Bergsonian Period of Alfred Schutz', *Philosophy and Phenomenological Research*, 38, pp. 187–199.
Webb, M. (1987) *Temple Island*. London: AA Publications.
Weiner, J. (2016) 'Transcending Lucretius: Vitruvius, Atomism, and the Rhetoric of Monumental Permanence', *Helios*, 43(2), pp. 133–161.
Winetrout, K. (1970) 'Reviewed Work: *The Future of the Future* by John McHale', *ETC: A Review of General Semantics*, 27(3), pp. 369–376.

10 2078/1978. Anticipation and the contemporary

Jamie Brassett and John O'Reilly

What a future!

Friedrich Nietzsche, letter to Rohde, 15 November 1874

Introduction

In 2018 we were asked to anticipate 2078 for a piece that provided some form of foresight into that world-to-come (Brassett and O'Reilly, 2018b). We, fairly obviously to us but maybe a little perversely, chose to look first to 1978, to a moment of our teenage/proto-teenage years and punk/new wave. This was not so much a choice driven by some theoretical framework or other, but one that was stylistic: a trope that allowed us into the world of 2078. The 78 as a node joining different narratives, trends and affects. These 78s (and all the others) will always be connected, we thought, even if different flows spill through the tracks. As we are familiar with the work of philosopher-psychoanalyst couple Gilles Deleuze and Félix Guattari, we found a similar energetics informing their joint writings: especially the second volume of their 'Capitalism and Schizophrenia' series, *A Thousand Plateaus* (1988); and, in that text, particularly their opening 'Introduction: Rhizome' (pp. 3–25). Here, Deleuze and Guattari are resolutely more than individuals collaborating. Their opening words are: 'The two of us wrote *Anti-Oedipus* together. Since each of us was several, there was already quite a crowd' (Deleuze and Guattari, 1988, p. 3). They crowd and spread through their text – though, to say 'their text' undermines their desire to create an open-ended, multiplicity of a book – rambling through works, disciplines and loves, creating philosophical concepts as they go. Most importantly for us, now, is their choice to attach dates to each of the chapters[1] (apart from their introduction and conclusion). So, for example, Chapter 7 is 'Year Zero: Faciality' and Chapter 11 is '1837: Of the Refrain' (Deleuze and Guattari, 1988, pp. 167–191, 310–350).[2] For them, the date gives a specific moment at which the concepts they put into play intensify strongly and are able to communicate as pure a sense of themselves as possible. Their dates are barbed, sticky, seed-like forms that serve to attach concepts to those of us passing by. Similarly, then, do we offer 2078 and 1978 to locate our sense of contemporary in relation to anticipation, in a way that is also anticipatory:

1978 opens a way into the future and 2078 provides an anticipation for the present.

Our main concern then, in this chapter, is to present a conceptualization of the contemporary as conducted by philosopher Giorgio Agamben (2008) in such a way that it can serve as a model for acts of anticipation; and to do so in a way that emphasizes their ontologies and creativity. Here both contemporary and anticipatory positions characterize attitudes, ways of being that encounter and enhance both. That is, the ontological state of being contemporary can provide a model for the anticipator to use to adopt their stance, as Riel Miller (2011) argues should be done with the scenario planner. This is our central argument. Our chapter will develop this by first reimagining 1978 in relation to a song by The Boomtown Rats (1978) called *She's So Modern*. This is not only to find a way into the future that 2078 announces – and we will explain how this heading to the past allows us to do this – but also to encounter a historical moment we can use to 'clear the ground' upon which to anticipate. Our next move will be to look at an essay on the concept of contemporary, as an ontological category, in an essay by Agamben (2008). This is valuable because of its proximity to Nietzsche's concept of the 'untimely' (1997), particularly as he uses it to locate a position in relation to history. For Nietzsche, someone who is untimely has a creative use of history in order to build a new future in the present. We will conclude by bringing these thoughts back to anticipation.

1978 and so on

Teenagehood is the moment at which the structures and givens of foresight already embedded in the world first clash with emergent ontologies of anticipation. How is this so? For teenagers – who occupy an emergent state of becoming characteristic of the transition from childhood to adulthood – anticipation is their mode of engagement with the world. Moreover, it is into that moment of anticipation that adults and grown-up culture give the teenager lessons in foresight, knowledge and the future: what they need to study in order to give the examiner what is expected in exams; what kinds of subjects they should be studying in university in order to be successful; what they should expect as they head out into the adult world of organizations and work; what not to take, do, be; and with whom all these states should, and should not, be done. A teenager's own urge is to escape the present in order to achieve a future adulthood that is largely fantastical. And so, the well-intentioned foresight of adults tries to capture the emerging futures of anticipatory teenagehood but rushes headlong into a shared space of shared, stubborn antagonisms. Such that a 'we-know-best' based upon experience borne of habit – an anaesthetized form of living, philosopher Vilém Flusser reminds us (2002) – comes up against an anticipation borne of a deeply felt urge for creative becoming.

Our example from 1978 – the song *She's So Modern* by the Irish punk/new wave band The Boomtown Rats – constructs an embodiment of the modern, presents the modern itself as new human productive capacity of embodied, materialized interpretation: an interpretation of signs that becomes its own anticipation of a world.[3] Being modern is different from being contemporary, however. We will see these differences in stance develop via Agamben (2008), who is himself developing the contemporary from Friedrich Nietzsche's concept of the 'untimely' (1997). Briefly, in Nietzsche's *Untimely Meditations* (1997), 'On the Uses and Disadvantages of History for Life', untimeliness is an ontological attitude necessary to adopt in order to account for history in such a way as to develop a future-focused present. There is a passage in which he links explicitly anticipation, creativity and life:

> If the historical drive does not also contain a drive to construct, if the purpose of destroying and clearing is not to allow a future already alive in anticipation to raise its house on the ground thus liberated, if justice alone prevails, then the instinct for creation will be enfeebled and discouraged.
>
> (Nietzsche, 1997, p. 95)[4]

For Nietzsche here, anticipation of possible futures remains in the present as potential and needs access to a creative power in order for the encrustations of the present to be cleaned away. Not only this, but creativity itself is enabled by such a ground-clearing; or should be for these futures to emerge living, vital, vivid, even exuberant [*lebendige*]. Nietzsche is almost breathless in this passage. This is important: anticipation and hope, life and joy, clearing and creating; better futures and presents as the normative outcome of an untimely history. Agamben works with this approach to develop an ontology of the contemporary; he takes this untimeliness into the present, as an attitude to the present. We, in turn, take this contemporary back to the future.

So, now, back to 1978. The Boomtown Rats' *She's So Modern* from that year, a punk/new wave[5] vision of modernity, disruption and semiotic limits. The modern girl knows much: what to wear, what to say, what is important art (Dada, by the way) and the difference between sexual fetishism and right-wing politics.[6] This captures a sense of its time wrapped in the garb of the modern. 'She' is the pinnacle of the twentieth century: symbolized by the 1970s, captured in the rhythm and styling of 1978. 'She' is the spirit of her time; 'she' is 'a modern girl'. The 'modern' is a girl, with language. There is an important moment when the narrative is disrupted by lang-ga-ga-ga-guage in an utterly Artaudian manner (Deleuze and Guattari, 1984, 1994) – the singer/storyteller escapes from the constraints of the story to interject this anti-linguistic refrain. This, alongside the opening 'la-la-la-la-la', serves to punctuate the characterization; and which, along with the snarled tone of the delivery by singer Bob Geldof, takes us listening (and so implicated in the storytelling) to a place outside the modern and the modern girl. This is critique (and John O'Reilly in his chapter in this volume names this critical space and time, 'Other-wise'). A critical declamation that is delivered in the opening of the song in lines that also form the chorus: stated up-front and repeated throughout.

But 'she' has limits to her becoming – in both the archaic judgemental ('becoming' as 'charming') and ontological senses. These limits are the very fact of her modernity. 'She' remains, stuck in the 1970s, in 1978; 'she' is fixed and, now because of this, already out-of-date. From a certain perspective, 'she' is too right, 'she' fits too well, 'she' accommodates the given of a moment rather than shapes it. Any anticipation that remains is regularized as the active configuring of signs, materials and strategy. It merely registers and makes something different, gets ready; it is standardized according to the 'right' clothes and the 'right' codes. This future of the being-modern is identical and predictable; it is the same as before. What every teenager secretly knows, and deeply fears, are the mechanisms whereby the expressions of their own anticipation are captured, rendered semiotically meaningful, made modern and socially stable, situated and tamed. 'Her' being modern is the mark of her archaism; 'she' is no longer our now, because 'she' is only 'her' 'now'; 'she' is forever 'now' in 1978. That is where she belongs.[7]

The other option to being *so* modern and therefore *so* out-of-date, is to belong to a time but to no longer 'coincide with it' or 'adjust [. . .] to its demands', Agamben (2008, p. 41). That is, it is to be *contemporary*.

2008. The contemporary

In his essay 'What is the contemporary?' (2008) Agamben initially examines the question via Nietzsche's concept of the untimely – though other philosophers, poets and fashion designers (among others) are encountered – in order to position his own evaluation of the models of time, untimeliness and, of course, contemporariness. Agamben's characteristic erudition allows him to gather these concepts together with particular attention to the relationship between past and present, between historical and contemporary times. This is done not simply to scrape Nietzsche's bones for his best concepts[8] but to take them and put them back into play in another time, to do related but different types of work, to experience them anew and experiment with them to create novel concepts. Similarly, our intention is to replay Agamben's conception of the contemporary in such a way that it can be mobilized within anticipatory futures thinking (Rosen and Kineman, 2005; Poli, 2010, 2017; Rossel, 2010; Rosen, 2012; Roubelat, 2016). We will argue that Agamben's and Nietzsche's attitudes of contemporary untimeliness can be as equally directed towards the future in its relation to the present, as they can to the past and its relation to the present.

We have noted that Poli (2010) engages with some philosophical work: he focuses upon Edmund Husserl (1991) and Ernst Bloch (2000). While Husserl's phenomenology and some of its subsequent adherents are discussed in terms of the relationship between consciousness and the future, Poli (2010, p. 8) attends more to Bloch's work on hope as an ontology of the future in terms of open categories (that is, categories that are only partially determined). In Chapter 5 of his *Introduction to Anticipation Studies*, 'Anticipation in Philosophy'

(2017, pp. 77–100), Poli extends his philosophical focal points much further: fixing again upon Ernst Bloch and, a philosopher on whom Poli is expert, Nicolai Hartmann. The other twentieth-century philosophers Poli exemplifies in this chapter are: Bergson, Husserl, George Mead, Whitehead, Peirce, Hans Jonas, Heidegger and Deleuze. With regard to those philosophers from this list with whom we are most familiar – Whitehead and Deleuze – Poli gives Deleuze short shrift: confessing that he 'may have grossly misunderstood Deleuze – his language is notoriously obscure' (Poli, 2017, p. 88). Whitehead's process philosophy and its speculative metaphysics, however, is much closer to Poli's position, he admits (2017, pp. 82–85). We will need to investigate Poli, Whitehead and anticipation another time, but will note here that there is more to offer from Deleuze's philosophy to anticipation, especially with regard to his (Deleuze, 1993) reworking of Whitehead's concept of the event. (We note, also, that Nathaniel Barron writes on Bloch in this volume.) For us, a turn to Agamben (and Nietzsche) offers another direction to the future, one that steers close to the philosophy of Deleuze insofar as it focuses upon life, ontology and creativity (Holland, 2009).

In Agamben, the issue of ontology is no less important than in Husserl, for example. But his is characterized differently. Agamben's focus is not haunted by epistemological concerns, paring away experience to arrive at a pure, clear subjective truth, but is primed by an ontology of creation that aims to accentuate the creative potentialities available to us all. For us, such a creative approach to considering the changes that define being in all its modes – often referred to as becoming or ontogenesis – is also one that incorporates an attitude to the future.[9] Agamben outlines four key modes of being contemporary: to stand outside one's time; thus, to be able to perceive the darkness of one's time; to have the courage to act upon one's time; and so, to recognize the modern as archaic. We will take them in order.

The contemporary stands outside

First, to be contemporary requires the ability to stand outside of one's age – we have noted this already in relation to the modern and The Boomtown Rats. Agamben (2008, p. 40) writes:

> Those who are truly contemporary, who belong to their time, are those who neither perfectly coincide with it nor adjust themselves to its demands. They are thus in this sense irrelevant.[10] But precisely because of this condition, precisely through this disconnection and this anachronism, they are more capable than others of perceiving and grasping their own time.

The modern *legislates* the difference between the relevant and irrelevant, the light and dark, by configuring the relationships between power and value: determining what counts as significant; identifying what an organization needs to attend to and the ways it must perform to reach those goals. The contemporary, as Agamben shows here, can occupy a position that enables a more creative and critical response to their time: they are more 'capable', they can 'perceive' and 'grasp' on the basis of their perception. It becomes important, then, to foster anachronism and irrelevance as attitudes whereby new perceptions can be created. They are difficult skills, skills of difficulty perhaps; because to be outside the given time means engagement and entanglement with change and the new, rather than fitting in, being relevant, being modern. In his breakdown of anticipation into 'explicit' and 'implicit', Roberto Poli argues that the awareness of explicit anticipation that guides decisions and choices, avoiding harm and threat, is efficient both in directing and keeping focus on a clear goal. However, there is a blind spot to this apparently rational approach, a blind spot whose territory is located by the term 'relevance'. Relevance, here, Poli defines as cognitive:

> focused goal-oriented behavior usually gives rise to inattentional [sic] blindness, i.e. the incapacity to perceive things that are in plain sight [. . .]. Patterns constrain attention, govern the boundary of relevance, and they direct attention to pre-established foci. The more efficient the patterns, the more likely is the outcome of an over-restricted focus of attention. The more efficient the behavioral patterns are, the more rigid they become.
>
> (Poli, 2010, pp. 12–13)

For us, exceeding this 'boundary of relevance' is an affective act of ontological derangement, expressed and experienced in an 'untimely' fashion as 'contemporary'. Riel Miller argues similarly for scenario work as an escape route from the probabilistic approach of futures work built around a vision of epistemological efficacy, rather than the ontological awareness of uncertainty that may deliver valuable insight into the future. We would go even further to argue that an 'ontological awareness of uncertainty' is a criterion of valuable insight. Miller writes:

> the 'scenaric stance' achieves something that so far has largely escaped the futures studies community – the combination of a focus on the 'capacity to be free' [. . .] and a decisive break with the 'probabilistic stance' on the basis of an ontological rather than epistemological point of departure.
>
> (Miller, 2011, p. 25)

Just as with Miller's 'scenaric stance', we propose creative nondeterminism as a material act of fabricating a contemporary experiential space. This is an ontological space of immanent generative criticality, which enables, and is enabled by, the transformation of value through a two-step practice of critical-analytic decoding and speculative-creative anticipation.

Agamben's first move, then, appears to give temporal form to the classic Outsider (Wilson, 1956; and re-appraised, Jay, 2007). But the contemporary does not simply stand outside of their own time, either through their own will or dismissed by others as an irrelevance: their very difference allows them to have a particular hold on their epoch. The contemporary does not give up on their time, but stands outside in order to make a difference, a difference that The Boomtown Rats' 'modern girl' is unable to make because she coincides, to use Agamben's words, 'too well with the epoch'. He continues 'those who are perfectly tied to it in every respect, are not contemporaries, precisely because they do not manage to see it; they are not able to firmly hold their gaze upon it' (Agamben, 2008, p. 41). Thus, for Agamben, we have the contemporary regarding its epoch from the status of an anachronism.[11] For, just as they do not belong to their time, so their time does not belong to them; their 'now' is pushed away from them as they are dissociated from it in the double movement of anachronism and irrelevance. But they *care*, as we will see.

The contemporary perceives the darkness of its time

Agamben's second characterization of the contemporary takes this dissociation further. This disconnection from the age in which the contemporary is still interested provides a perspective on particular aspects of the times that those of us rooted in them do not easily see. Agamben (2008, pp. 44–45) writes:

> But what does he who sees his time actually see? What is this demented grin on the face of his age? I would like at this point to propose a second definition of contemporariness: The contemporary is he who firmly holds his gaze on his own time so as to preserve not its light, but its darkness. [. . .] The ones who call themselves contemporary are only those who do not allow themselves to be blinded by the lights of the century, and so manage to get a glimpse of the shadows in those lights, of their intimate obscurity.

The dark, the obscured and excluded become the focus of the contemporary. The Boomtown Rats' poor 'modern girl' is so blinded by the light of the age that 'she knows the right things to say' and 'the right clothes to wear'. But the contemporary is not blinded by the given codes, by what the modern makes relevant for seeing. The contemporary is beyond the modern simplicity of right and wrong, or the repeated pattern recognition of the habituated and anaesthetized. The contemporary is able to engage with the darkness of the times,[12] with the complexity that is otherwise obscured by correct vision, noticing the light only for the shadows it casts. The critical position constructed by the contemporary is not one for its own sake only. It is not epistemologically focused, for it has nothing to do with meaning or interpretation; neither is it instrumental, with tools that grant better access to the relevance of the modern. In these ways, too, the contemporary is anticipatory. The sheer ontological weight expressed through this dissociation and attraction to the darkness is palpable.

Being critical, then, both constructs the contemporary and enables a point from which to act with future orientation. The darkness 'concerns' the contemporary and 'never ceases to engage' him (Agamben, 2008, p. 45). The concept of criticality we will return to below when discussing Nietzsche's untimely, in his context of 'critical history', but we can note that for Agamben being critical – and its cognate terms 'critique' and 'crisis' – play an important role in taking political action.[13]

Similarly, the concept of 'concern' is one that is playing a role in current thinking and practice of philosophy of design – most notably, through the work of Bruno Latour (2008). Here a concern is not simply a worry, but a call to action, as Whitehead (1967, pp. 176 and 180) explains. For us, these concepts come together in an intriguing way. Contemporaneity allows for a distancing from the present in order to uncover the critical and crisis-inducing moments of its darkness and thence to act. This is an important consideration and one that brings us close to anticipation, insofar as a future-caused concern leads to (creative) action in the present. To which we will return later.

The contemporary has the courage to act

Thirdly, for Agamben, the contemporary is characterized by a certain sense of 'courage'. He explains:

> to be contemporary is, first and foremost, a question of courage, because it means being able not only to firmly fix your gaze on the darkness of the epoch, but also to perceive in this darkness a light that, while directed towards us, infinitely distances itself from us [as happens with the light from distant galaxies, moving too fast away from us ever to reach us].
>
> (Agamben, 2008, p. 46)

The courage to act creatively comes through strongly in this passage. It is not enough simply to recognize darkness in order to lead to action. Acting as a contemporary is not a simple decision, it requires bravery and resolution both to engage with darkness and find the moments upon and within which to act. These moments emit as lightness, photonic beams that can take us, if we are brave enough, on a voyage into and out of the now. For Agamben, such slippage across time in relation to the present, is exemplified well in fashion, which actively provides for time a 'peculiar discontinuity that divides it according to its relevance or irrelevance, its being-in-fashion or no-longer-being-in-fashion' (Agamben, 2008, p. 47). This is different to 'knowing the right clothes to wear', of course.

This 'peculiar discontinuity' is interesting, especially to those of us working in innovation. We are well aware of the term 'disruptive innovation', coined by Clayton Christiansen (1997) to describe a technology that is dragged from the past to produce a disjunction of present processes, allowing companies to take a technology out of time to enter, or even define, new areas for business, or new markets to ply. Indeed, the notion of disruption has become something of a cliché, something required, legislated.[14] Nevertheless, there is a way in which we can consider disruption, or any radical discontinuity of innovation, as the (strategic) creation of the becoming contemporary of technology (or any product, service or process). One would imagine, then, that any originators of a particular technology might require a degree of courage, as they face an uncertain future however well they might have mapped it strategically. This is to introduce a future-focus a little early in our piece, however. So, we might consider any organization finding an old technology (or process, product or service) to rework to a different present (different from the perspective of the technology, and so on, that is) as courageous, critical historians of the type articulated by Nietzsche; who writes:

> Men and ages which serve life by judging and destroying a past are always dangerous and endangered men and ages. For since we are the outcome of earlier generations, we are also the outcome of their aberrations, passions and errors, and indeed their crimes; it is not possible wholly to free oneself from this chain.
>
> (Nietzsche, 1997, p. 76)

Nietzsche's untimely man recognizes the darkness in the present, in himself: as dangerous and endangered and generated from the abnormalities, mistakes and even crimes of the past. That crisis and criticality for which Agamben's contemporary requires courage is for Nietzsche's untimely a service for life. It may be a step too far to imagine that organizations panning for the next disruptive innovation, as for specks of gold in the mud, are courageous contemporaries. (Maybe we could give them the benefit of the doubt? Or, maybe, innovators and the novelty they strive for are never quite irrelevant enough.)

The contemporary, then, serves the now by being outside it, recognizing its faults, its deviations, its dangerous crises and is called to act by them, by the flashes of light in the midst of the darkness. A fashionable slippage from time proves to be an immensely creative response to it. Agamben (2008, p. 49) explains further:

> the 'now,' the *kairos*[15] of fashion is ungraspable: the phrase, 'I am in this instant in fashion' is contradictory, because the moment in which the subject pronounces it, he is already out of fashion. So, being in fashion, like contemporariness, entails a certain 'ease,' a certain quality of being out-of-phase or out-of-date, in which one's relevance includes within itself a small part of what lies outside of itself, a shade of *démodé*, of being out of fashion.

Here we see, added to courage, 'a certain ease' in encompassing (but not exhaustively, or as an act of homogenization or totalization) relevance and irrelevance, being in and out of fashion, of grasping the ungraspable, to disrupt as a service to life, being concerned (as a call to action) about crises. We can now consider Agamben's fourth, and final, characteristic of the contemporary: in the production of the archaic.

The contemporary sees the present as archaic

Finally, we have the fourth and last characteristic of being contemporary. Agamben tells us that it involves the creation of the present as archaic, the recognition of the 'modern and recent' as archaic. At the same time as the contemporary dislocates from the space and the time of the now and acts both against and for it, it is located as archaic. This is not a simple distancing however; as we have seen, the contemporary also is expressed in terms of being-out-of-date, outside of now, irrelevant. In the production of the present as archaic, therefore, we have the simultaneous disruption of present time and location of untimely beings as critical (that is, in relation to crises) and creative acts. This is dangerous, for the judgement upon the now destroys it by repositioning it as archaic. The doubling back of this danger upon the contemporary person, Agamben highlights, is because the archaic situates the *arkhē* (ἀρχή), the origin, not only of our present, but of our contemporary too. Criticality becomes self-oriented and recognizes that little bit of irrelevance inside the contemporary. Dangerous and endangered the untimely contemporary person may be, but its approach to crisis is creative and is one that is angled resolutely to, or as an anticipatory 'from', the future. At this moment, it is worth noting Agamben's final sentence of this essay, for it brings us to a point where he not only reflects upon the value of his words, but in so doing, anticipates their possible future significance; it runs as follows:

> It is on our ability to respond to this exigency and this shadow, to be contemporaries not only of our century and the 'now,' but also of its figures in the texts and documents of the past, that the success or failure of our seminar depends.
>
> (Agamben, 2008, p. 54)

'Success or failure' depend upon a number of varying plausible, possible or, maybe, even probable futures that remain contingent upon the differing contexts of their reception of Agamben's work. There is a sense in which Agamben's essay, 'our seminar', is itself contemporary: standing outside of its moment of creation and delivery, with the power to produce its own devastating critique, for it may well fail. 'Success' may come from Agamben's work engendering multiple future contemporaries; that it becomes both relevant and irrelevant, in and out of time, even fashionable and not, all at the same un-time.

The contemporary, as elucidated by Agamben, as the figure, emblem and embodiment of anticipation is deeply problematic for businesses because it incorporates the figure of uncertainty. Not simply does the contemporary not look right (unlike the 'modern girl'), it creates a space of critical distance that is organizationally uncomfortable. Perhaps most uncomfortable of all, because the very visibility of the contemporary – as an alternative vision and radically different version of success – provides for any business a troublesome apparition. And doubly troublesome because the contemporary as uncertainty is an organizational signal, a commercial itch demanding a creative, disjointed and different response. The irony, of course, is that because such an organization is bound by the comforting certainty of relevance, they are incapable of becoming contemporary and the courage needed to anticipate.

2020. Anticipation and the contemporary

In a way, the concept of the contemporary appears to point in two directions: to the present and the past. We have some work to do to draw this to the future. Luckily, for us, Agamben has provided a way into this, when he writes (2008, p. 48): 'time of fashion, therefore, constitutively *anticipates* itself and consequently is always too late. It always takes the form of an ungraspable threshold between a "not yet" and a "no more"' (emphasis added). We mentioned, above, how the fashionable contemporary is able to enter into and remove itself from present time and able to make critical comments about its relevance and irrelevance. For Agamben, this exemplified its 'courage' in the face of any darkness perceived in the current time. In this quotation, however, he introduces another temporal attitude: anticipation. Such that not only does the fashionable contemporary react to the present, it does so through a 'stance' (Miller, 2011) that relates to the future. Being 'in advance of itself' places the contemporary in-between future, past and present, in and out of time. Which characterises the modalities of its ontological present, too and, we argue, is a necessary attitude of anticipation. For to describe someone (or something) anticipating is to characterise them incorporating, within themselves, a model of the future in order to reorient themselves in the present differently (Miller, 2011; Rosen, 2012; Poli, 2017; and the many others discussed in this book).

In fact, this happens in Nietzsche's work on critical history too, from which Agamben draws his own enquiry into the contemporary. In the seventh section of Nietzsche's 'On the uses and disadvantages of history for life', the second of his untimely meditations, he accentuates the future-focused, anticipatory nature of his concept of a critical history. He writes:

> If the historical drive does not also contain a drive to construct, if the purpose of destroying and clearing is not to allow a future already alive in anticipation to raise its house on the ground thus liberated, if justice alone prevails, then the instinct for creation will be enfeebled and discouraged.
> (Nietzsche, 1997, p. 95)

Criticality and creativity are key characteristics of the untimely and contemporary alike, in whatever way they are they are oriented. A critical creativity will take an empty historicism on a new journey, a future journey. But one that situates neither the historical nor the future in self-referential limits, spinning like cold, dead stars. Rather, they elucidate and provoke a present full of possibilities, of potentials, of openness, where what may become can set the terms of any present actuality.

If we are to see anticipation as being concerned with the future in order to create a different present, being engaged with the darkness of the future even, then this is in order to make an intervention today. This concern is a call to present action. Those who anticipate must simultaneously occupy a position outside their own time and see it in all its darkness – thus becoming contemporary – and do so in a manner that stands outside future possible times too. They need the courage to anticipate their own contemporariness, their own irrelevance, in order for it to be created and involved in an on-going creation. Not to be relevant is therefore an exceeding or surpassing the boundaries of any self, as philosopher Georges Bataille notes in the first of our epigrams to this chapter. As any creative anticipation challenges the accepted 'boundary of relevance' (Poli, 2010, pp. 12–13) – the given boundary of relevance in any foresight or planning process; the inevitability and necessity that pre-determines relevant outcomes; the drive for certainty in the face of uncertainty – courage is needed in adopting a critical position in relation to the darkness within the organizational, cultural, social and political paradigms within which one is situated. We cannot forget, in all of this, that Agamben's contemporary needs to be able to see the tiniest of lights amid the darkness. This is the spark to anticipate, to take up Whitehead's call to speculate, to go on an adventure. To live and create.

Imagine, then, the contemporary of 2078 outside of their time, aware of the darkness of their epoch as something unable to be lived, is experiencing the archaic attraction of a previous time, 1978 say, 2018, or even 2021 – all important moments in the course of this chapter – how might they be utterly irrelevant? Or, imagine that *our contemporary* of 2021 is not only experiencing the forces of uncountable prior, archaic moments on the present, but pressure from even more futures on the present felt as unliveable. Our untimely men, women and all in 2021, live in a doubly defined darkness of life that can never be lived but can be anticipated: as archaeologists of prior time and anticipators of future times.

Closing the introductory section of the second of his untimely meditations, 'On the uses and disadvantages of history for life', Nietzsche (1997, p. 60) ponders:

> for I do not know what meaning classical studies could have for our time if they were not untimely – that is to say, acting counter to our time and thereby acting on our time and, let us hope, for the benefit of a time to come.

From Agamben we have seen how this thought pertains also to the contemporary, who acts upon and against and for a now. Nietzsche here emphasizes the value of such a position 'for the benefit of a time to come'. Would it be over-presumptuous of us to rework Nietzsche's words, slightly, to stress this future-focus even further? And so: *we do not know what meaning futures studies could have for our time if they were not untimely; acting counter to and on our time for the benefit of our present.* Or, in other words, aligning with the work of Agamben we have been discussing here: *any courageous, critical anticipation activated to deal with a dark present, needs to model itself along the lines of the contemporary.* For to be anticipatory is also to be untimely. Now, what a future.

Acknowledgements

We would like to thank Fabrice Roubelat, General Editor of the French futures journal *Prospective et Stratégie*, for giving us permission to use the English version of the article originally published in that journal in French: Brassett and O'Reilly (2018b). This chapter is an expanded elaboration of that article.

Notes

1 They explain: 'A book has neither object nor subject; it is made of variously formed matters, and very different dates and speeds. To attribute this book to a subject is to overlook these workings of matters, and the exteriority of their relations' (Deleuze and Guattari, 1988, p. 3).

2 We are aware that an aspect of Deleuze and Guattari's creative approach to writing *A Thousand Plateaus* is to move away from the standards of book production and to regard their chapters as 'plateaus', with many different strands passing through each of them rather than the rigid, linear connections between their conclusions and introductions. We use 'chapters' not to undo their argument about disrupting the writing of totalizing books, but to keep attention on our narrative.

3 We are trying not to encounter the concept of 'modernity' and its vast sociological and cultural studies literature. This seems to approach a different set of concerns, issues and agencies than those of this chapter. For work on modernity in a number of registers, see: Appadurai (1996) and Giddens (1991).

4 A more literal translation would be: 'When, behind the historical drive, there is no drive to build, when destroying and clearing does not liberate a ground upon which to build a future already alive in hope, when justice alone is done, then the creative instinct will be invalidated and discouraged [*Wenn hinter dem historischen Triebe kein Bautrieb wirkt, wenn nicht zerstört und aufgeräumt wird, damit eine bereits in der Hoffnung lebendige Zukunft auf dem befreiten Boden ihr Haus baue, wenn die Gerechtigkeit allein waltet, dann wird der schaffende Instinkt entkräftet und entmutigt*]' (Nietzsche, 1954, §7). Thanks to Joanna Brassett for help with this translation.

5 We mention punk in terms of subcultural style in our article on design and scenario planning, Brassett and O'Reilly (2015), with reference to the seminal work on this topic Hebdige (1979). We note that Muggleton (2000) provides a more recent engagement with similar themes. See also: Sklar (2013), for a US perspective on punk style expressed through fashion; and Savage (1991) for a UK focus covering many of its cultural forms.

6 Interestingly, one of the modern girls mentioned in the song is described as a performer, comedian, who wants to be a photograph. Pop songs about celebrity and being watched seem to have been popular in 1978: along with *She's So Modern*, there was another New Wave song, by Scottish band The Rezillos *Top of the Pops* (the title references a popular music television show in the UK at the time) and Kraftwerk's *The Model*.

7 Later in the song, 'she' is described as joyfully painting moustaches on herself, with the explanation that 'she' loves the Mona Lisa. With moustaches, 'she' identifies with an earlier twentieth-century act of Punk Rock/Dada. 'She' becomes 'he' – hence our use of inverted commas.
8 As Deleuze and Guattari explain (1994, p. 83; interestingly, in relation to Nietzsche) of a certain sort of lazy philosophizing: 'Nothing positive is done, nothing at all, in the domains of either criticism or history, when we are content to brandish ready-made old concepts, like skeletons destined to intimidate any creation, without seeing that the ancient philosophers from whom they are borrowed were already doing what we would like to prevent modern philosophers from doing: they were creating their concepts and were not content to clean and scrape bones like the critic and historian of our time' (translation modified).
9 We have written about this with emphasis on the possible relations between design and scenario planning: Brassett and O'Reilly (2015). See also Zamenopoulos and Alexiou (2007), which gives a good overview of Rosen's work on anticipation in relation to design.
10 Agamben's original Italian is *inattuale*. We are intrigued with the possibilities that this word entails, and the conceptual journey it takes us through, especially in regard to terms we are discussing here. We can see some other way of dealing with the actual in Agamben's word, which would align – through Deleuze's (1991) work on Bergson – to notions of virtuality but, also, the concept of potentiality. 'Potentiality' is a term that occurs elsewhere in Agamben's work – see Agamben (1999). Further work is necessary to investigate these relationships fully.
11 As Nietzsche (1997) also argues for the untimely man. For brevity, we have not mapped in detail Agamben's dealings with Nietzsche's concept of untimely, though we have and will make some references to Nietzsche, an encounter with his relevant work will need to come another time.
12 As may be possible by a philosopher like Hannah Arendt. Brassett (2021) writes an entry titled 'Creativity' for the book *Designing in Dark Times. An Arendtian Lexicon*. Agamben, we should note, is influenced by Arendt, particularly in his political philosophical writings (De Boever, 2016, pp. 237–238); see: Agamben (1998).
13 We have noted this in our essay on Lucretius, the clinamen and design: Brassett and O'Reilly (2018a). We should also note that we recognise the way in which Whitehead characterises 'critical philosophy', favouring rather 'speculative philosophy'. This will need proper investigation in relation to anticipation and the contemporary another time but, for now, we include this: 'The divergence between the two schools [critical and speculative] is the quarrel between safety and adventure' (Whitehead, 1966, p. 173). The adventurous Whitehead is always contemporary (Stengers, 2011, p. 18).
14 Editors' note. See O'Reilly's chapter in this volume for more on disruption and its critique.
15 Liddell and Scott (1889, p. 392) have four entries for *kairós* (καιρός). The third, of time, defines it as: 'the right point of time, the proper time or season of action, the exact or critical time'.

References

Agamben, G. (2008) 'What is the Contemporary?', in Agamben, G., *What Is an Apparatus? and other Essays*. Trans. D. Kishik and S. Pedatella. Stanford, CA: Stanford University Press, pp. 1–24.

Agamben, G. (1999) 'On Potentiality', in Agamben, G., *Potentialities. Collected Essays in Philosophy*. Trans. D. Heller-Roazen. Stanford, CA: Stanford University Press, pp. 177–184.

Agamben, G. (1998) *Homo Sacer: Sovereign Power and Bare Life*. Trans. D. Heller-Roazen. Stanford, CA: Stanford University Press.

Appadurai, A. (1996) *Modernity at Large: Cultural Dimensions of Globalization*. Minneapolis, MN: University of Minnesota Press.

Bloch, E. (2000) *The Spirit of Utopia*. Trans. A.A. Nassar. Stanford, CA: Stanford University Press.

Brassett, J. (2021) 'Creativity', in Staszowski, E. and Tassinari, V. (eds.), *Designing in Dark Times. An Arendtian Lexicon*. Series: 'Designing in Dark Times'. New York and London: Bloomsbury, pp. 99–103.

Brassett, J. and O'Reilly, J. (2018a) 'Collisions, Design and the Swerve', in Vermaas, P. and Vial, S. (eds.), *Advancements in Philosophy of Design*. Series: Design Research Foundations. Cham, Switzerland: Springer, pp. 71–98.

Brassett, J. and O'Reilly, J. (2018b) 'Retour à 2078: Réflexions sur l'anticipation et le contemporain', *Revue de Prospective et Stratégie*, 9, Édition Spéciale: 'Voir Loin', (December), pp. 11–22.

Brassett, J. and O'Reilly, J. (2015) 'Styling the Future. A Philosophical Approach to Design and Scenarios', *Futures*, 74, pp. 37–48.

Christensen, C. (1997) *The Innovator's Dilemma. When New Technologies Cause Great Firms to Fail*. Cambridge, MA: Harvard Business Review Press.

De Boever, A. (2016) *Plastic Sovereignties. Agamben and the Politics of Aesthetics*. 'Incitements' Series. Edinburgh: Edinburgh University Press.

Deleuze, G. (1993) *The Fold. Leibniz and the Baroque*. Trans. T. Conley. London: Athlone Press.

Deleuze, G. (1991) *Bergsonism*. New York, NY: Zone Books.

Deleuze, G. and Guattari, F. (1994) *What is Philosophy?* Trans. G. Burchill and H. Tomlinson. London: Verso Books.

Deleuze, G. and Guattari, F. (1988) *Capitalism and Schizophrenia 2. A Thousand Plateaus*. Trans. B. Massumi. London: Athlone Press.

Deleuze, G. and Guattari, F. (1984) *Capitalism and Schizophrenia 1. Anti-Œdipus*. Trans. R. Hurley, M. Seem and H.R. Lane. London: Athlone Press.

Flusser, V. (2002) 'Habit: the true aesthetic criterion', in Flusser, V., *Writings*. Ed. A. Ströhl, trans. E. Eisel. Minneapolis, MN: University of Minnesota Press, pp. 51–57.

Giddens, A. (1991) *Modernity and Self-Identity. Self and Society in the Late Modern Age*. London: Polity Press.

Hebdige, D. (1979) *Subculture. The Meaning of Style*. New Accents Series. London: Methuen.

Holland, E. (2009) 'Introduction. Image, Text, Thought', in Holland, E.W., Smith, D.W. and Stivale, C.J. (eds.), *Gilles Deleuze: Image and Text*. London and New York, NY: Continuum Books, pp. 1–6.

Husserl, E. (1991) *On the Phenomenology of the Consciousness of Internal Time (1893–1917)*. Trans. J.B. Brough. Heidelberg: Springer Netherlands.

Jay, M. (2007) 'Still Sleeping Rough: Colin Wilson's *The Outsider* at Fifty', *Salmagundi*, 155/156, (Summer–Fall), pp. 3–12.

Kraftwerk (1978) 'Das Model' (The Model), *Die Mensch Maschine* [vinyl, 1 C 058-32 843]. Köln: EMI Electrola Gmbh.

Latour, B. (2008) 'A Cautious Prometheus? A Few Steps Toward a Philosophy of Design (with Special Attention to Peter Sloterdijk)', in Hackney, F., Glynne, J. and Minton, V. (eds.), *Networks of Design. Proceedings of the International Conference of the Design History Society*. Boca Raton, FL: Universal-Publishers, pp. 2–10.

Liddell, H.G. and Scott, R. (1889) *An Intermediate Greek–English Lexicon*. Oxford: Oxford University Press.
Miller, R. (2011) 'Being without Existing: The Futures Community at a Turning Point? A Comment on Jay Ogilvy's "Facing the Fold"', *Foresight*, 13(4), pp. 22–34.
Muggleton, D. (2000) *Inside Subculture. The Postmodern Meaning of Style*. Dress, Body, Culture Series. London and New York, NY: Berg.
Nietzsche, F. (1997) *Untimely Meditations*. Ed. D. Breazeale. Trans. R.J. Hollingdale. Cambridge: Cambridge University Press.
Nietzsche, F. (1954) 'Vom Nutzen und Nachteil der Historie für das Leben', *Unzeitgemäße Betrachtungen*. Available at: http://www.zeno.org/nid/20009229833 (Accessed: 8 May 2020).
Poli, R. (2017) *Introduction to Anticipation Studies*. Series: 'Anticipation Science', vol. 1. Berlin: Springer.
Poli, R. (2010) 'The Many Aspects of Anticipation', *Foresight*, 12(3), pp. 7–17.
Rosen, J. and Kineman, J.J. (2005) 'Anticipatory Systems and Time: A New Look at Rosenean Complexity', *Systems Research and Behavioral Science*, 22, pp. 399–412.
Rosen, R. (2012) *Anticipatory Systems: Philosophical, Mathematical and Methodological Foundations*. 2nd edn. London: Pergamon Press.
Rossel, P. (2010) 'Making Anticipatory Systems More Robust', *Foresight*, 12(3), pp. 72–85.
Roubelat, F. (2016) 'Anticipation et scenarisation de l'innovation. Enjeux, mise en œuvre, perspectives', *Prospective et Stratégie*, 7, pp. 125–139.
Savage, J. (1991) *England's Dreaming. Sex Pistols and Punk Rock*. London: Faber and Faber Ltd.
Sklar, M. (2013) *Punk Style*. 'Subcultural Style' Series. London and New York, NY: Bloomsbury Publishing Ltd.
Stengers, I. [2002] (2011) *Thinking with Whitehead. A Free and Wild Creation of Concepts*. Trans. M. Chase. Cambridge, MA and London: Harvard University Press.
The Boomtown Rats (1978) *She's So Modern* [vinyl, LUN715]. Dublin: Mulligan Records.
The Rezillos (1978) *Top of the Pops* [vinyl, SIR4001]. London: WEA Records Ltd.
Whitehead, A.N. [1933 (1967)] *Adventures of Ideas*. New York, NY: The Free Press.
Whitehead, A.N. [1938] (1966) *Modes of Thought*. New York, NY: The Free Press.
Wilson, C. (1956) *The Outsider*. London: Victor Gollancz Ltd.
Zamenopoulos, T. and Alexiou, K. (2007) 'Towards an Anticipatory View of Design', *Design Studies*, 28, pp. 411–436.

Index

Page numbers in **bold** denote tables, those in *italic* denote figures.

For Product Safety Concerns and Information please contact our EU
representative GPSR@taylorandfrancis.com
Taylor & Francis Verlag GmbH, Kaufingerstraße 24, 80331 München, Germany

www.ingramcontent.com/pod-product-compliance
Ingram Content Group UK Ltd.
Pitfield, Milton Keynes, MK11 3LW, UK
UKHW020150200926
475702UK00016B/2639

* 9 7 8 0 3 6 7 7 5 7 3 5 9 *